PLANT SCIENCE RESEARCH AND PRACTICES

MICROALGAE AND OTHER PHOTOTROPHIC BACTERIA

CULTURE, PROCESSING, RECOVERY AND NEW PRODUCTS

Plant Science Research and Practices

Additional books in this series can be found on Nova's website
under the Series tab.

Additional e-books in this series can be found on Nova's website
under the e-books tab.

PLANT SCIENCE RESEARCH AND PRACTICES

MICROALGAE AND OTHER PHOTOTROPHIC BACTERIA

CULTURE, PROCESSING, RECOVERY AND NEW PRODUCTS

LUIS G. TORRES BUSTILLOS
EDITOR

nova publishers
New York

Library of Congress Cataloging-in-Publication Data

Microalgae and other phototrophic bacteria : culture, processing, recovery and new products / editor: Luis G. Torres Bustillos, Bioprocess Department, UPIBI-Instituto Politecnico Nacional, Calle Acueducto, Col. Barrio La Laguna-Ticoman, Del Gustavo Madero, Mexico.
pages cm. -- (Plant science research and practices)
ISBN 978-1-63482-078-3 (hardcover)
1. Microalgae--Biotechnology. 2. Microalgae--Industrial applications. 3. Photosynthetic bacteria. I. Torres Bustillos, Luis G., editor. II. Series: Plant science research and practices.
TP248.27.A46M52 2015
579.8--dc23

2015001642

Published by Nova Science Publishers, Inc. † New York

CONTENTS

INTRODUCTION

Microalgae are primary producers of oxygen and exhibit an enormous potential for biotechnological applications. Microalgae cultivation is also an interesting option for wastewater treatment. These microorganisms are efficient in recovering high amounts of nitrogen, inorganic phosphorus, and heavy metals from effluents. Furthermore, microalgae are responsible for the reduction of CO_2 from gaseous effluent and from the atmosphere, reducing the green house gases effect. On the other hand, microalgae biomass can be used for the production of pigments, lipids, foods, and renewable energy.

The culture of microalgae is not a new issue, since for many decades they have been employed with aquaculture purposes. In fact, some species of *Nostoc, Arthrospira* (*Spirulina*) and *Aphanizomenon* species have been used for food for thousands of years for food applications. The treatment of wastewaters employing microalgae has been reported from decades (specifically for the elimination of phosphorus and nitrogen compounds).

Nowadays, there is an increasing interest in the production of biofuels. Bioethanol, biohydrogen, biomethane, biodiesel and other novel products can be obtained using microalgae as biomass or metabolic products (sugars, lipids, etc.). In the next future, microalgae can be oriented (via molecular engineering) to the production of the hole biofuels, such as biodiesel or bioturbosins.

Hence, commercial applications of microalgae and other phototropic bacteria are the following (but not limited to):

Applications in human and animal nutrition
Microalgal metabolites in cosmetic products
Biofuels production
Wastewater treatment
Bioplastics production

To mention a few products generated by microalgae, let us mention pigments and carotenoides (betacarotene, astaxantin, leutin, zeaxanthin, cantaxanthine, chlorophyll, phycocianin, phycoerithrin and flucoxanthin), polyunsaturated fatty acids (DHA; 22:6, EPA; C20:5; ARA; C20:4 and GAL; C18:3), vitamins (A, B1, B6, B12, biotin, rivoflavin, nicotinic acid, pantothenate, folic acid), antioxidants (catalases, polyphenols, superoxide dismutase, tocopherols), and other valuable products (antimicrobial, antifungal and antiviral agents, toxins, aminoacids, proteins, sterols, MAAs for light protection.

Production of novel and known products is a challenge from the environmental and the economical point of view. The production is limited by the generated products added value or by the massive production of themselves.

A whole process for the production of this materials comprehend the use of photobioreactors, which are more or less similar to other bioreactors employed in biotechnology, but with the particularity that they need to be designed to provide enough amounts of light (i.e., certain amount of photons per area of reactor)

Microalgae and other phototrophic bacteria is a book which will provide information for academics students, policy makers, and general public regarding the state of the art in the field, as well as detailed description of the methodologies employed for culture, processing, recovery and new products.

Aspects covered by this book are the microalgae and other phototrophic bacteria regarding to the culture, processing, recovery and new products.

Aspects such as the improvement of photobioreactors, the biofuel potential of *Botryococcus braunii*, measuring environmental impact of microalgae production and biofuel production through LCA, dewatering and dehydration of microalgae, recovery of microalgae by coagulation-flocculation-sedimentation and the characterization of the produced pastes, production of bioplastics employing residues and phototropic bacteria, review on instrumentation and automatic control for microalgae culture and microalgae in livestock nutrition, will be covered by this proposed book.

Every chapter has been written by an expert on the field, covering both the state of the art and the specific contributions of each group of authors in the worldwide literature.

In: Microalgae and Other Phototrophic Bacteria
Editor: Luis G. Torres Bustillos
ISBN: 978-1-63482-078-3

Chapter 1

MOLECULAR PHARMING IN THE CHLOROPLAST OF *Chlamydomonas reinhardtii* FOR THE PRODUCTION OF RECOMBINANT THERAPEUTIC PROTEINS

Alma Lorena Almaraz-Delgado, José Luis Castrejón Flores and Jesús Agustín Badillo Corona*
UPIBI-Instituto Politécnico Nacional, Mexico

ABSTRACT

Chlamydomonas reinhardtii is a eukaryotic green unicellular alga that contains a single chloroplast. The genome in the chloroplast contains 99 genes and can be easily transformed by particle bombardment. With the help of chloroplast biotechnology, foreign genes can be introduced into the chloroplast genome for different purposes, but primarily it has been used to produce recombinant proteins with therapeutic properties. Proteins ranging from antibodies, antigens, immunotoxins and therapeutic proteins have been successfully produced in the chloroplast, demonstrating that the technology could compete with current production platforms at competitive yields and prices. Even though almost any gene can be introduced into the genome with this technology, there are still problems that need to be solved before it can be commercially successful. In this chapter we present the current status of chloroplast biotechnology to produce therapeutic proteins in the chloroplast of *C. reinhardtii*.

1. INTRODUCTION

Chlamydomonas reinhardtii is a green unicellular alga. Cells are ovate in shape and have a polar structure with two anterior flagella and a single basal chloroplast that may partially surround the nucleus. The genus *Chlamydomonas* was first described by C.G Ehrenberg in

* Corrrespondence: E-mail: jbadilloc@ipn.mx.

1833, but it was not until 1888 that Dangeard described *Chlamydomonas reinhardtii*, named after Ludwig Reinhard. By the year 1960, *C. reinhardtii* had become the most widely used strain in laboratories around the world [1]. Nowadays it is considered an ancestor of both plants and animals while retaining many features from both, including the flagella and the photosynthetic ability [2]. The microalga can be found in environments ranging from fresh water and damp soil to the open sea. In addition, some species of this genus survive even in snow. *C. reinhardtii* is an iconic model organism from which many biological sciences have obtained important insights.

Early studies in *Chlamydomonas* have elucidated the process of photosynthesis by the discovery and isolation of chlorophyll-binding polypeptides, and the purification of protein complexes from thylakoid membranes [3,4]. This microalgae can be grown either heterotrophically in the dark using acetate as a carbon source, mixotrophically using light and acetate or phototrophically in the light with CO_2 as a sole carbon source. These features have been crucial for the study of photosynthetic processes. In 1995 Jean-David Rochaix called it the photosynthetic yeast because besides being used for the study of photosynthesis, it has also been extensively used for the study of genetics and biogenesis of organelles, flagella assembly and function, mating reactions and gametogenesis, cell wall synthesis, phototaxis, circadian rhythms, as well as carbon, nitrogen, and sulphur metabolism [5].

C. reinhardtii cells contain three genomes, in the nucleus, the chloroplast, and the mitochondria. All three genomes are amenable for transformation. Genetic modification of the chloroplast of *C. reinhardtii* was first achieved by Boynton et al. in 1988 while Kindle et al. transformed the nucleus in 1990; Remacle et al. transformed the mitochondria in 2006. Genome transformation in all three compartments is nowadays easily accomplished for different purposes. For example, mitochondrial transformation has been used mainly for site-directed mutagenesis [6-8]. Nuclear transformation has primarily been used to shed light on gene function of the following genes: *arg7* [9,10], *nit1* [11,12], *OEE1* [13], *RSP3* [14], *ATPC* [15]. Improvements to promoters that enhance gene expression [16-18] has also been carried out.

When genetic modification of the nucleus was first accomplished, it was seen as an alternative for the production of recombinant proteins that were difficult to produce in other systems or that were already produced in other systems, but that somehow represented a risk to human health. Several attempts to use *C. reinhardtii* as an expression platform for the production of economically important recombinant proteins have been made with varying success [19,20]. Unfortunately, the low level of protein accumulation along with the problems of gene silencing, inherent to nuclear transformation, have hampered the use of nuclear transformed *C. reinhardtii* at industrial scale.

In contrast, the introduction of foreign genes into the chloroplast genome of *C. reinhardtii* has demonstrated that the organelle can overcome some of the drawbacks of nuclear transformation. Some of the advantages of chloroplast transformation are that a high level of protein accumulation can be achieved and that genes are inserted into the genome by homologous recombination, which yields lines that are more homogeneous in terms of productivity. In addition, large complex proteins can be produced, and there is little or no risk of spreading transgenes to the environment.

Chloroplast transformation has been used for expressing selectable marker genes, mainly to develop genetic tools, such as the *aadA* [21], and *aphA6* genes [22] as well as reporter genes such as the *Aequorea victoria* GFP protein [23,24], and the bacterial *uidA* [25-27].

Additionally, chloroplast transformation has been employed to express genes that encode proteins with therapeutic use. Chloroplast transformation with the purpose of generating, a new product, be it a commercially important macromolecule or metabolite, or a new strain with improved traits, has become known as chloroplast biotechnology. On this we focus in this chapter, and it will be presented in detail in subsequent sections.

In what follows we review the current status of chloroplast biotechnology for the production of recombinant proteins in *C. reinhardtii*. In particular, in section 2 we describe the current and most widely used technology to generate strains with a transformed chloroplast genome. Also in section 2 we describe genetic tools and elements, mainly focusing our discussion on promoters and terminators, used to modulate gene expression in the chloroplast of *C. reinhardtii*. In section 3, we present an overview of some of the cases in which recombinant proteins have successfully been produced. In the final section of this chapter we will discuss some of the prospects in the field.

2. Chloroplast Genetic Engineerig Technology

The *C. reinhardtii* chloroplast genome consists of a 203-kb double stranded DNA molecule comprising 2 single copy regions (~ 80 kb each) and two inverted repeats (21.2 kb each). The genome, also called the plastome, contains 99 genes encoding mostly proteins involved in photosynthesis, ribosome structural proteins and subunits of the plastid RNA polymerase. A complete set of genes coding for all tRNAs and rRNA is also present [28]. The chloroplast contains the necessary apparatus for replication, transcription and translation of the genes in its genome. However, not all the components of this machinery are encoded in the chloroplast genome, and it has been estimated that a fully functional chloroplast contains some 3000 proteins, most of which are encoded in the nucleus and imported into the chloroplast after translation.

As mentioned above, chloroplast genetic engineering was first accomplished by Boynton (1988) using the particle bombardment device and has, since then, essentially remained unchanged. Cells are placed in the vacuum chamber of the particle bombardment device and bombarded at high speed with DNA coated particles. The selection process is then carried out in media with an appropriate antibiotic. Subsequently, transformed lines are transferred to fresh selection media and cultured until the state of homoplasmy is achieved, i.e., when all copies of the wild-type genome have been eliminated by selective pressure on the antibiotic, and only the copies of the genome carrying the gene of interest and the dominant marker remain. Once the genes of interest have integrated into the chloroplast genome and the state of homoplasmy has been reached, the acquired trait is inherited to the progeny [29]. Organisms in which the chloroplast genome has been modified are termed transplastomic.

Typically, vectors used for transformation are adsorbed onto the surface of gold or tungsten particles, which are then introduced into the cells and organelles using a particle bombardment device. Vectors carry what is called an expression cassette, that is, an element consisting of a transcription promoter, a protein-coding sequence and a transcription terminator for the gene of interest (Figure 1). Two regions that are homologous to endogenous chloroplast genome sequences flank the expression cassette. After bombardment, the integration of the expression cassette occurs through recombination of the homologous sequences [30], which can be intergenic and allow for insertion of transgenes without disruption of endogenous genes, or through recombination of gene coding regions, in which case disruption of endogenous genes occurs. Site-directed integration is one of the many advantages of chloroplast transformation. The complete transformation element, i.e., the expression cassette and the flanking regions, is contained in a bacterial plasmid for the ease of propagation.

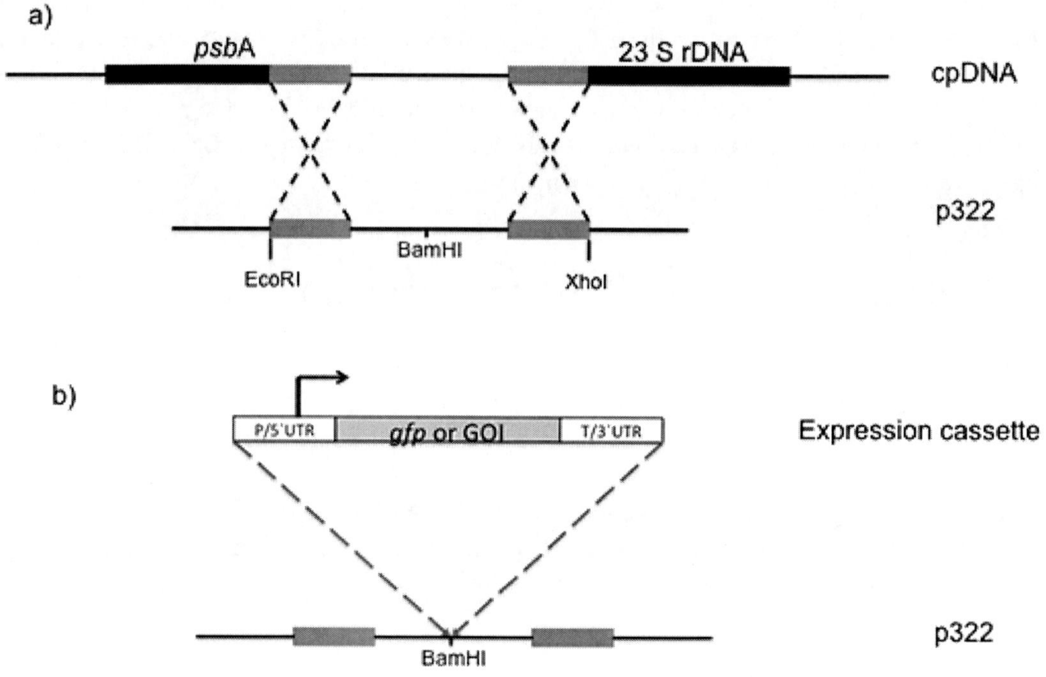

Figure 1. Schematic representation of chloroplast transformation vector p322. Chloroplast transformation vector p322 contains regions that are homologous to regions in the chloroplast genome (cpDNA). After bombardment this regions recombine and genes inserted in the *Bam*HI site integrate into the chloroplast genome (a). Expression cassettes used typically contain a promoter and 5´UTR, the coding sequence for the gene of interest (GOI), in this example *gfp,* and a terminator or 3´UTR. Expression cassettes can be inserted into transformation vector p322 to generate a final transformation vector.

In theory, any plastome sequence can be used for site-directed integration of the transgenes, but to date only a handful of sites have been used (Table 1). One particular site in the inverted repeats has been the most commonly used site for the insertion of transgenes [31,32]. To achieve this, transformation vector p322, derived from a chloroplast genome library, has been used (Chlamydomonas Center, Duke University, Durham, NC, USA). This vector carries a 5.5 kbp *Eco*RI-*Xho*I fragment from the *C. reinhardtii* chloroplast genome comprising the region *psbA-5S rRNA* inserted in pBluescript KS+ (Figure 1). Transformation vector p322 carries a convenient unique *Bam*HI site in the intergenic region, in which an expression cassette for the gene of interest can be inserted, but has the drawback of still containing remnants of the MCS from pBluescript KS+ making it thus necessary to assemble the expression cassettes in another plasmid [24,33,34].

Table 1. Sites used for insertion of transgenes in chloroplast genome of *C. reinhardtii*

Site of insertion	Gene inserted	Reference
clpP-trnL-petB-chlL-rpl23-rpl2	hTRAIL	[81]
psbH-psbN	83K7C-LC	[42]
tscA	CTB-D2	[38]
chlL	CTB-VP1	[36]
psbA-5S23S	HSV8-lsc	[33]
psbD	psbA	[47]
psbA	psbA	[47]
psbD-psbA	M-SAA	[47]
psbA5' – psbA3'	Pfs25	[45]

Unlike chloroplast transformation in plants, where transformation vectors carry a selectable marker gene, until recently, most reports of chloroplast transformation in *C. reinhardtii* had relied on the use of two vectors to achieve transformation. One of the vectors carries the transgene to be inserted, usually p322 as mentioned above, and a second plasmid, namely p228, which needs to be co-bombarded, carrying a fragment of the 16S *rDNA* gene with a point mutation in nucleotide 1123. Plasmid p228 is also in itself a transformation vector, conferring resistance to spectinomycin after recombination with the endogenous 16S *rDNA* gene. Selection of transgene-carrying transformants consists of two steps: first, selection of spectinomycin-resistant lines and subsequently selection of PCR-positives lines for the gene of interest [24,33,34].

The process of selection, although efficient, is labour-intensive and restrictive. Some of the recovered lines are sometimes resistant to spectinomycin but lack the gene of interest. To avoid such limitation, alternate chloroplast transformation vectors containing an expression cassette for either *aadA* (aminoglycoside-3″-adenylyltransferase) or *aphA*-6 (3'-aminoglycoside phosphotransferase type VI) have been developed [22,35,36]. When *aadA* and *aphA-6* are used, selection can be carried out on spectinomycin and kanamycin, respectively, and transformed lines are mostly positive for the selectable marker and the gene of interest [37,38]. The use of vectors carrying a selectable marker gene has the advantage of requiring only one vector for transformation thus making it easier to identify truly transformed lines. It is also possible to eliminate the selectable marker after transformation. Michelet (2010) has shown that two direct repeats of bacterial DNA+3'UTRof *rbcL* flanking the expression cassette for *aadA* can efficiently be used to excise the marker spontaneously when the selective pressure is released.

When it was envisioned that transformation of the chloroplast could overcome the low level of recombinant protein accumulation and the problem associated with random gene integration present in nuclear transformation, research in the field began to seek for elements that would allow for the expression of foreign genes in the chloroplast. Chloroplasts are responsible for the transcription and translation of the genes encoded in their genome. However, chloroplast gene expression is a complex process that can be influenced by transcription, mRNA processing, mRNA splicing, mRNA stability, initiation of translation and protein turnover. Furthermore, many of these steps are regulated by nuclear encoded factors that are often gene specific [39,40].

Early attempts to express foreign genes in *C. reinhardtii* chloroplasts generally resulted in non detectable or a very low level of protein accumulation [41-43]. All of these foreign genes were taken from other systems and were not optimized for expression in *C. reinhardtii* chloroplasts. It has been shown in a variety of organisms that certain codons are more frequently used than others. The chloroplast of *C. reinhardtii* displays such codon bias, with codons containing adenine or uracil nucleotides in the third position favoured over those with guanine or cytosine [44]. Franklin et al. (2002) have shown that there is a 80-fold increase in the expression of a codon-optimized *gfp* gene when compared to the expression of a non-codon optimized one. Subsequent studies for other genes have confirmed that codon bias plays a significant role in protein accumulation in *C. reinhardtii* chloroplasts [37,45].

In addition to codon optimization several other avenues have been examined to increase recombinant protein expression in the chloroplast, and much of this work has benefited from the current state of knowledge concerning endogenous gene expression in chloroplasts. Transcriptional efficiency, transcript stability, and translational efficiency have all been shown to impact the expression of endogenous proteins in *C. reinhardtii* chloroplasts, and the 5'UTR of chloroplast mRNAs plays a key role in each of these events. Transcriptional efficiency is regulated by both chloroplast gene promoters as well as sequences contained within the 5'UTR [46].

One of the most widely used promoters derives from the *psb*A gene. This promoter has been shown to yield the highest level of protein accumulation [31,43,47]. However, it is also tightly auto-regulated and controlled in a feedback manner; when the product of the gene, that is the D1 protein is present, translation of the *psb*A mRNA is down regulated. Since heterologous genes under the control of the *psb*A promoter carry the 5'UTR when transcribed, these are subjected to the same regulation as the *psb*A transcripts. It has been shown that when the D1 protein is absent, and hence the strains are not photosynthetic, the level of protein accumulation can be up to 5% of the total soluble protein TSP (12% of the total protein TP) [47]. Because *psb*A deficient strains are non-photosynthetic and therefore require the addition of a carbon source in the growth media, their use for commercial protein production would result in higher production costs and a greater risk of microbial contamination.

Because of the drawbacks of the *psb*A promoter, many laboratories around the world have been searching for alternative promoters to drive the expression of foreign genes. The promoters that have been tested include the ones from genes *psb*D [31], *atp*A [43], *rbc*L [32] and, *psa*A [48]. These promoters have yielded various levels of protein accumulation, *atp*A and *psb*D being the ones with the highest level of protein accumulation, although in a gene specific manner. That is, when these promoters have been used to drive the expression of genes other than the endogenous genes, the result has been less favourable. The influence of the transcription terminators from genes *atp*A, *rbc*L, *psb*A and tRNAarg has also been evaluated but the general conclusion seems to be that there is little effect on transcript accumulation and translation efficiency [49].

Looking for an efficient promoter/5' UTR combination that could boost heterologous protein production on a photosynthetic strain, Rasala et al. (2011) have examined the effect of various combinations of promoter and 5' UTR from the *rbc*L, *psb*D, *atp*A, *psb*A and 16S genes. The chimeric 16S promoter+*atp*A 5' UTR fusion allowed robust accumulation of *lux*Ct (a chloroplast optimized luciferase gene) and the therapeutic protein *14FN3* similar to those achieved using the endogenous *psb*A promoter/5' UTR on a D1-deficient strain to express the transgenes. This result indicates that there are combinations that could still be explored to increase protein accumulation in the chloroplast further. Curiously the combination of 16S promoter fused to the 5´UTR from gene 10 from the bacteriophage T7 has not been reported for the chloroplast of *C. reinhardtii*; for this promoter/5´UTR combination yields the highest level of protein accumulation for a recombinant protein in the chloroplast of higher plants [50].

In the case of plants, strategies for inducible expression of recombinant proteins have been reported. These approaches place the transgene under the transcriptional regulation of an inducible promoter (e.g., T7 polymerase transcribing genes downstream of the T7 promoter [51] or the lac repressor regulating the expression of genes under the control of the lac promoter [52]). Although these strategies have not been implemented in *C. reinhardtii* chloroplasts, there is no apparent biological reason why they would not work in this system. The ability to regulate recombinant protein expression in the chloroplast could be essential for the production of certain types of recombinant proteins, including those that would normally be lethal to the host cell.

Increasing heterologous protein expression levels is necessary for creating a truly competitive system against other recombinant platforms; this could be achieved by exploring factors affecting transgenes transcription, translation, and accumulation. Promoters play an important role in the expression of heterologous genes and since genetic data suggest the existence of nuclear encoded factors which interact specifically with the 5´UTR of chloroplast mRNA and play a major role in transcript stabilization, more studies are needed to examine these factors and understand how they can be used to boost recombinant protein production in the chloroplast of Chlamydomonas reinhardtii [53].

3. Recombinant Therapeutic Proteins Produced in the Chloroplast of *C. reinhardtii*

Since the first demonstration that genetic engineering of the chloroplast is possible, the technology has been applied to both plants and algae to evaluate its potential as an inexpensive platform for the production of recombinant proteins and metabolites for the biotechnological industry. However, unlike plants, where genetic engineering of the chloroplast has been used more widely to explore the capability of the chloroplast to produce oils, pigments, enzymes, resistant plants, plants capable of removing pollutants, etc. [54], genetic modification in *C. reinhardtii* has so far mainly been used to produce viral and bacterial antigens as well as mammalian antibodies and therapeutic proteins. Select examples of such cases will be presented in this section. A more extensive list can be found in Table 2 [55].

Table 2. Overview of recombinant proteins produces in the chloroplast of *C. Reinhardtii*

Recombinant therapeutic protein	Yield	Relevant information	Reference
VP1-CTB; Protein VP1 from foot and mouth disease virus (FMDV) fused to cholera toxin B (CTB)	3-4%Total Soluble Protein (TSP)	Demonstrated that the *C. reinhardtii* chloroplast derived VP1-CTB could bind to GM1-ganglioside receptor *in vitro*	[36]
HSV-lsc; Large single chain (lsc) antibody directed against glycoprotein D protein from Herpes simplex virus (HSV)	Not reported	First report to show that the *C. reinhardtii* chloroplast can efficiently fold antibodies and form disulfide bonds	[33]
TRAIL; Tumor necrosis factor-related apoptosis-inducing ligand	0.43% - 0.67% TSP		[81]
M-SAA; Mammary-associated serum amyloid	3%- 5% TSP	M-SAA was shown to generate mucin induction in a human intestinal epithelial cell line. Demonstrated that the *psb*A promoter yields high level or recombinant protein accumulation when the endogenous *psb*A gene is absent	[47]
CSFV-E2; Classical swine fever virus (CSFV) structural protein E2	1.5-2%TSP	Subcutaneous immunization of mice with E2 was shown to induced IgG antibodies	[35]
Human glutamic acid decarboxylase (hGAD65)	0.25 - 0.3% TSP	The protein was shown to immunoreact with sera from diabetic mice	[32]
IBDV-VP2; Infectious burial disease virus VP2 protein IHNV-G; Infectious haematopoietic necrosis virus IPNV-VP2; Infectious pancreatic necrosis virus VP2 protein IPNV-VP2 SBC; Infectious pancreatic necrosis virus Quorum sensing-regulated gene (LecA) p57 PCV2; Porcine circovirus type 2 VP-2C VP28	4-0.8%To- tal cell protein (TCP) < 0.5% TCP < 0.3% TCP 1-0.1%TCP 1-0.2%TCP < 0.5 TCP 0.9 - 0.2%TCP < 0.5% TCP 21 - 0.2% TCP	This report looked at the expression of 11 proteins. Nine proteins showed some level of accumulation, while the rest could not be detected. It showed that there are variations in the level of expression even amongst lines obtained with the transformation construct. Authors postulated the existence of the transformosome, a state in which particular genomic characteristics, induced incidentally with transformation, affect, negatively or positively, the expression of the transgene	[31]
HC-83K7C; Heavy chain human monoclonal antibody against anthrax protective antigen 83 (PA83) LC-83K7C; Light chain human monoclonal antibody against anthrax PA83	0.01% dwt	It was shown that the heavy and light chains expressed in trans could assembled into a fully-functional monoclonal antibody against PA83	[42]

Recombinant therapeutic protein	Yield	Relevant information	Reference
CTB-D2; D2 fibronectin-binding domain of *Staphylococcus aureus* fused to the cholera toxin B subunit	0.7% TSP	First report to show that an orally-administered alga expressing an antigen in the chloroplast triggers a mucosal and systemic immune response in mice	[38]
14FN3; Domain 14 of human fibronectin VEGF; Human vascular endothelial growth factor HMGB1; High mobility group protein B1	3%- 0.15% TSP 2%-0.1% TSP 2.5%-1% TSP	This report looked at the expression of seven therapeutic proteins. For three of the proteins, a level of accumulation above 1% was observed, whereas for the rest of the proteins, erythropoietin, interferon β, and proinsulin no protein was detected. Biological activity was evaluated for VEGF and HMGB1	[43]
acrV2 and vapA2; antigens from the fish pathogen *Aeromonas salmonicida*	0.8% and 0.3% TP respectively	Showed that the *psa*A promoter-exon1 element can be used to drive the expression of foreign genes in non-photosynthetic strains	[48]
Escherichia coli phytase gene (appA)	N.D.	This study showed that algae expressing a bacterial phytase gene in the chloroplast could be lyophilized and administered orally to broiler chicks. The enzyme was active in the gut and reduces the fecal excretion of phytate.	[82]
Pfs25 and Pfs28; surface proteins from *Plasmodium falciparum*	0.5% and 0.2% TSP respectively	First report to show that Pfs25 and Pfs28 can be produced without glycosilation and in a correct conformation recognized by monoclonal antibodies specific to conformational epitopes	[45]
α CD22PE40; monomeric immunotoxin consisting on the single chain antibody that recognizes the CD22 surface protein from B-cells, fused to domains II and III of exotoxin A (PE40) from *Pseudomonas aeruginosa* αCD22HCH23PE40; dimeric version of αCD22PE40	0.3% - 0.4% TSP 0.2% - 0.3% TSP	First report to show that immunotoxins can be produced in a eukaryotic system without being toxic to the cell.	[37]
CtxB-Pfs25; *Plasmodium falciparum* surface protein 25 fused to the β subunit of the cholera toxin from *Vibrio cholera*	0.09% TSP	Demonstrated that the fusion protein can induce IgA antibodies when administered orallyas part of a lyophilized powered. However, IgG antibodies could not be elicited with this route of administration	[64]
αCD22Gel; single chain antibody targeting the CD22 receptor from B-cells, fused to the eukaryotic ribosome inactivating protein, gelonin, from *Gelonium multiflorm* αCD22CH23Gel; dimeric version of αCD22Gel	0.2% - 0.3% TSP 0.1% - 0.2% TSP	Demonstrated that immunotoxin can efficiently bind to cancerous B-cells in vitro and kill them without affecting non B-cells	[58]

Some of the characteristics that make the *C. reinhardtii* chloroplast an outstanding system for the productions of recombinant proteins include: complex mammalian proteins are correctly folded and assemble, high level of protein accumulation can be achieved, there is no risk of contamination with human or animal pathogens, proteins that are toxic for eukaryotic cells can easily accumulate in the chloroplast, stable transformed lines can be generated in a short time (2-3 months), and microalgae can easily be grown in full containment reducing concerns about environmental release of the therapeutic protein. Furthermore, the Food and Drug Administration of the United States of America considers *C. reinhardtii* a Generally Regarded as Safe (GRAS) organism.

Taking advantage of these characteristics in *C. reinhardtii*, it has been possible to produce antibodies, immunotoxins, antigens intended for vaccinations and human therapeutic proteins. Examples of each of these will be given below.

Antibodies and Immunotoxins

Given their complexity, antibodies have been difficult to produce in prokaryotic systems and thus have to be produced in eukaryotic systems such as mammalian cell cultures, making the final product expensive and sometimes scarce [56]. If antibodies could be efficiently produced in a photosynthetic organism such as in *C. reinhardtii*, the production costs would certainly be lower and the product could become more accessible to developing countries at a lower price level.

Antibodies have been successfully expressed in the chloroplast of *C. reinhardtii*. In the first case, a monoclonal antibody directed against a glycoprotein of the herpes simplex virus D (*HSV8*) was expressed from a large single chain (*lsc*) coding sequence, which consisted on the entire IgA heavy chain fragment fused to the variable region of the light chain fragment by a flexible linker peptide. HSV8-lsc protein accumulated to approximately 0.5% TSP and was found completely in the soluble phase with no apparent accumulation as insoluble aggregates. Moreover, the HSV8-lsc antibody correctly assembled as a dimer and was capable of binding its target antigen, the herpes simplex viral coat protein [33]. A simplified version of this antibody, HSV8-scFv, has also been expressed in the *C. reinhardtii* chloroplast. The HSV8-scFv contained only the variable regions of the HSV8 light and heavy chains joined by a peptide linker. The HSV8-scFv accumulated to slightly higher levels than HSV8-lsc, and again was completely soluble and able to bind herpes viral coat protein in enzyme-linked immunosorbent assays (ELISA)[57].

In the second case, a monoclonal antibody directed against protein PA from *Bacillus anthracis* assembled into a functional tetrameric structure consisting of 2 heavy-chain and 2 light-chain subunits, capable of binding the PA83 antigen from the pathogen *in vitro*. Affinity was determined in parallel for the chloroplast produced antibody and an antibody expressed in a mammalian system and was shown to be similar in function. A comparison with antibody 83K7C, expressed in CHO cells, which had been demonstrated to provide protection from anthrax toxicity, in both cell-based assays and animal models [42], showed that, apart from the lack of glycosylation (inherent to chloroplasts), the chloroplast-produced antibody had the same PA83 binding properties than the antibody produced in the mammalian cell-line. Two remarkable features that were demonstrated for these strains are: 1) the antibody was soluble, an important feature at the time of purification and; 2) the strains were stable and continued to produce the antibody for more than a year. These findings established that algal chloroplasts have the capacity to synthesize fairly complex molecules in a soluble and active form.

Other types of molecules that are similar in structure and size to antibodies, but that are different in terms of function are immunotoxins. Immunotoxins are antibodies linked to a toxin of a protein nature. These chimeric molecules can be used in cancer treatment because they are highly specific, a characteristic conferred by the antibody they contain, and can be directed to one particular cell-type where the toxin binds to a target protein and kills the eukaryotic cell. Given their nature, immunotoxins are difficult to produce in eukaryotic systems. However, because of the prokaryotic origin of the chloroplast, they have been produced in the chloroplast of *C. reinhardtii* without much difficulty. As has been mentioned before, the chloroplast serves as the ideal compartment for the production of complex molecules, such as antibodies but at the same time it serves as a confined compartment where the toxin, or the immunotoxins for this matter, is sequestered from the cytosol, preventing it from killing the eukaryotic cell.

Two immunotoxins have been shown to accumulate in the chloroplast of *C. reinhardtii* without causing any harm to the cells. A single chain antibody recognizing the CD22 surface receptor from B-cells was fused either to domain II or III of Exotoxin A from *Pseudomonas aeruginosa* [37,58] or to the ribosome inactivating protein, gelonin, from *Gelonium multiflorm* [37,59]. In both cases it was demonstrated that the immunotoxins were capable of specifically binding to B-cells *in vitro* but not to Jurkat cells, which lack the CD22 receptor. In the case of the immunotoxin Exotoxin A, the life-span of mice implanted with a human B-cell tumour, was extended.

Antigens

In humans and animals, vaccination is the most effective method of inducing resistance to a certain pathogen. Traditional vaccines are commonly produced from an attenuated or killed form of the pathogenic organism. An alternative approach is to produce a recombinant antigenic protein from the pathogen, and to use it to elicit an immune response through vaccination (subunit vaccination). If antigens intended for subunit vaccination could be produced in GRAS organisms, such as *C. reinhardtii*, the purification process would not need to be concerned with the removal of toxins or remnants of a pathogen, making the process less expensive and the product safer than when it is obtained from microbial or mammalian cultures.

Subunit vaccination is currently used for the prevention of diseases like hepatitis A and B [60]. As algae contain the protein folding machinery that bacteria and other prokaryotes lack, they can be used to produce complex proteins that cannot be easily produced at large scale in bacterial cultures, without costly denaturation and refolding steps. Algae are also ideal for producing antigens, which exhibit little or no glycosylation, such as those from the parasite *Plasmodium falciparum* [61].

The first subunit viral vaccine produced in the chloroplast of *C. reinhardtii* was the translational fusion of the structural protein VP1, from the Foot and Mouth Disease Virus (FMDV), and the cholera toxin B subunit (*CtxB*) [36]. FMD in livestock can be prevented by subunit vaccination. The structural protein VP1 of FMDV carries critical epitopes, which can induce the generation of neutralizing antibodies, while the CtxB is a potent mucosal adjuvant that can bind to intestinal epithelial surfaces via GM1 ganglioside receptors. The CtxB-VP1 fusion protein produced in the chloroplast retained both GM1-ganglioside binding affinity and antigenicity. The CtxB-VP1 fusion protein accumulated up to 3% TSP, a 30-fold increase in comparison with the accumulation level obtained for the same antigen when produced from the nuclear genome of alfalfa and *Arabidopsis thaliana* [62,63].

Another viral protein, the classical swine fever virus E2 structural protein, was expressed in the chloroplast of *C. reinhardtii* [35]. ELISA quantification indicated that the accumulation of the E2 protein was 1.5-2% TSP, and retained immunological activity. Similarly, along with a list of other recombinant proteins, the white spot syndrome virus protein 28 (VP28) was expressed in the chloroplast and was reported to accumulate to a striking 10.5% TSP, although no data was presented to show how this level of expression was determined [31]. Recently, the D2 fibronectin-binding domain from a *Staphylococcus aureus* protein was fused to the B subunit of cholera toxin, and expressed in the chloroplast [38]. The transgenic algae were fed to mice, and shown to induce resistance against lethal doses of *S. aureus*, presumably by eliciting a systemic antigenic response to the *S. aureus* peptide. This is the first evidence that orally administrated transgenic algae are capable of inducing a protective immune response against common pathogens and opens the possibility of exploring the algae's immunogenicity capacity against deadly diseases such as tuberculosis or dengue.

The fact that chloroplasts do not glycosilate proteins might be considered a drawback. However, this characteristic has been exploited positively for certain proteins. The structurally complex non-glycosilated proteins Pfs25 and Pfs28 from *Plasmodium falciparum*, the parasite responsible for malaria have successfully been expressed in the chloroplast of *C. reinhardtii* [64]. Proteins accumulated to levels of 0.2- 0.5% TSP when expressed from the *psb*A promoter in photosynthetic-deficient strains. It has been suggested that Pfs25 and Pfs28 could be used to generate antibodies that bind, and neutralize them *in vivo*, thus preventing the formation of *P. falciparum* sporozoites in the mosquitos, which are ultimately responsible for the transmission of the parasite to humans. Producing these proteins in other systems has been difficult as diverse conformations occur and the proteins are glycosylated when produced in eukaryotic systems whilst native Pfs25 and Pfs28 from the parasite are not. When produced in the chloroplast, proteins folded correctly, lacked glycosylation, and generated antibodies that blocked the transmission of the parasite by preventing the formation of the sporozoites in the mosquito *in vitro*. This shows the fitness of the *C. reinhardtii* chloroplast for the expression of properly folded complex proteins.

Because *C. reinhardtii* is considered a GRAS organism, it is particularly suited for the production of oral vaccines. Once the guidelines for oral delivery of a subunit vaccine are established and approved, algae could be used to produce inexpensive oral vaccines, making vaccination accessible to developing countries, where vaccination is hindered by poor infrastructure. Furthermore, the oral delivery method and the option to store doses at ambient temperature would allow vaccines to be transported and administered to remote populations without the need for expensive refrigeration or highly trained medical personnel [65]. There is evidence that chloroplast-produced vaccines elicit similar immune responses as the actual pathogen when injected with standard vaccine adjuvants [65-68]. Furthermore, fusions of antigens with the cholera toxin B subunit show promise for eliciting immune response from mucosal delivery alone, as the cholera toxin B subunit allows a fused protein to penetrate the intestinal lining [36,69].

Mammalian Proteins

In the first report of a recombinant mammalian protein produced in algae, and one of the highest level of protein accumulation yet reported for the chloroplast of *C. reinhardtii*, it has been demonstrated that it is possible to produce a bovine mammary-associated serum amyloid (M-SAA) protein at levels above 5% of TSP in the chloroplasts of a photosynthetic strain [47]. M-SAA is found in the colostrum of mammals and can induce mucin synthesis in mammalian gut epithelial cells, which could offer protection against intestinal bacterial and viral infections in new-borns. M-SAA accumulated predominantly as a soluble protein and was shown to be bioactive when assessed by the induction of mucin in a human intestinal epithelial cell line (HT29) *in vitro*.

The second report for a human protein produced in the chloroplast of *C. reinhardtii* was for Glutamic Acid Decarboxylase 65 (hGAD65), which is a key auto antigen in insulin dependent diabetes mellitus (IDDM) [32]. hGAD65 has been identified as an important marker for the prediction and diagnosis of type 1 diabetes, as antibodies against it are present before the onset of the disease and it is believed that tolerance to this auto antigen can be induced using a vaccination-like scheme. The chloroplast of *C. reinhardtii* has been transformed with the native non codon-optimized hGAD65 gene under control of the chloroplast *rbc*L promoter [32]. The presence of hGAD65 was confirmed by western blotting using anti-GAD antibodies and the level of accumulation accounted for up to 0.3% TSP, a significantly higher value when compared to the 0.04% TSP obtained for a nuclear transgenic plant [70]. The antigenicity of algal-derived hGAD65 was determined by its immunoreactivity with sera from diabetic patients.

In an extensive study, Rasala et al. (2010) have evaluated the expression of a diverse group of therapeutic human proteins in *C. reinhardtii* chloroplast [43]. Four out of the seven proteins reported in the study accumulated to levels above 1% TSP. All four proteins 10FN3, 14FN3, HMGB1, and VEGF were expressed using the *psb*A and *atp*A promoters and were shown to be biologically active. In a follow up report, Rasala et al. (2011) reported the expression of 14FN3, a potential antibody mimic, using a line that lacked the *psb*A gene, and hence, was non-photosynthetic [71]. Lines expressing the 14FN3 from the wild-type *psb*A promoter and from a chimeric promoter (obtained by fusing the 16S rDNA promoter and the *psb*A 5'UTR) were generated. In parallel, using a photosynthetically active strain, the *psb*D, *atp*A, and the chimeric16S rDNA + *atp*A 5'UTR promoters driving the expression of 14FN3 were used to generate transplastomic lines. The wild-type *psb*A promoter allowed for the accumulation of 14FN3 to 0.5% TSP but only in the non-photosynthetic strain. The chimeric 16S+*atp*A and 16S+*psb*A promoters also yielded significant amounts of the heterologous protein in photosynthetic strains, 0.23% and 0.21% respectively. Finally, the wild-type *atp*A and *psb*D promoters supported only low levels of expression for 14FN3.

CONCLUSION

Despite the proven fact that *C. reinhardtii* chloroplasts can be used to produced diverse recombinant proteins, two problems still need to be solved: a) the development of an expression genetic tool that consistently allows for the expression of recombinant proteins at high level. These developments should include chimeric promoters, the use of sequences with a demonstrated ability to confer protein stability and the improvement in the algorithms to predict codon usage in highly expressed endogenous proteins; and, b) the development of an inducible system of expression by using systems that have a proven functionality in other systems or developing a specific system for *C. reinhardtii*.

As we have reviewed above, a series of endogenous and chimeric promoters have been tested with the aim of achieving high levels of expression. In some cases, recombinant protein accumulation has been reported to be 5% of the TSP, however this could only be achieved in non-photosynthetic strains, which results in slower growth and increases production costs, as it requires an external carbon source. The development of a chimeric promoter has already been shown to help solve this problem [71]. In plant chloroplasts high levels of protein accumulation have been achieved by fusing the 16S rDNA promoter to the 5´UTR of gene 10 from the bacteriophage T7. A similar chimeric promoter to drive the expression of recombinant proteins in *C. reinhardtii* chloroplast still needs to be developed.

Another potentially important characteristic still to be explored is the development of a strain able to grow on economic substrates such as starch, cellulose or even industrial substrates such as molasses. This would be important in cases in which photosynthetic strains cannot be used and thus a carbon source is required. Molasses are less expensive than glucose or even acetate, which is the substrate used in some culture media. The recent discovery that *Chlamydomonas reinhardtii* can degrade cellulose is important [72]. Non-photosynthetic cultures could be grown in less expensive substrates to reduce costs and keep the platform attractive and commercially viable.

Recombinant proteins are not the only products that could be produced in the chloroplast of *Chlamydomonas reinhardtii*. Secondary metabolites such as vitamins, carotenoids, biofuels (ranging from hydrogen to fatty acids and solvents), bioplastics, etc. could potentially be produced in algae. This would alleviate two problems, the ever-increasing worry of a decline in oil production and a reduction in atmospheric CO_2. Use in bioremediation, removal of toxic compounds from water, land and air has also been contemplated but poorly studied.

Genetic manipulation of algae is no longer limited to *Chlamydomonas reinhardtii*. Recent successes in generating transgenic algae are ever-growing in number. Exogenous genes have been expressed in the unicellular charophyte alga, *Closterium peracerosumstrigosumlittorale* complex [73]. The nuclear genome of volvocine alga *Gonium pectorale* has been stably transformed [74], as has the chlorophyceae *Haematococcus pluvialis* [75], by co-cultivation with *Agrobacterium* [76]. Some progress has been made by transient transformation of marine chlorarachniophyte *Lotharella amoebiformis* [77], chlorophyta alga *Ulva pertusa* [78], and red alga *Cyanidio schyzonmerolae* [79]. Methods have also been improved for previously transformed alga, such as *Dunaliella salina* [80], and cyanobacterial genetics have also been extensively explored. However, chloroplast transformation has so far been limited to *C. reinhardtii* and has been poorly studied in other species.

The chloroplast is a semiautonomous organelle; there is a tightly regulated mechanism for gene expression between the nucleus and the chloroplast. This mechanism has been suggested to function bi-directionally, that is, the nucleus sends signals to the chloroplast for the expression of chloroplast genes and, in turn, the chloroplast sends signals to the nucleus in a feedback manner. This fact reveals that even though manipulating the chloroplast genome can alter chloroplast gene expression, there are mechanisms controlled from the nucleus that have an effect on chloroplast gene expression. This is of particular importance because these mechanisms could also be altered to potentiate foreign gene expression in the chloroplast.

The *C. reinhardtii* chloroplast has been used to demonstrate that therapeutic recombinant proteins can be produced and it has been shown that it can compete with well-established production platforms. However, so far only a handful of proteins have been proven to retain biological activity, and none has made it to clinical trials yet. It is clear that the ultimate aim of producing proteins in the chloroplast of *C. reinhardtii* cannot be considered mature until there is an algae-produced product in the market.

REFERENCES

[1] Harris EH, Stern DB, Witman G. The Chlamydomonas sourcebook. *Cambridge Univ Press;* 2009.

[2] Fan B, Li Y, Zhai Z. The keratin-like intermediate filament network exists in the *Chlamydomonas* sp. pyrenoid. *Acta Botanica Sinica. 1*993;36(7):518–21.

[3] Stauber EJ, Fink A, Markert C, Kruse O, Johanningmeier U, Hippler M. Proteomics of *Chlamydomonas reinhardtii* light-harvesting proteins. *Eukaryotic Cell. Am Soc Microbiol;* 2003;2(5):978–94.

[4] Elrad D, Grossman AR. A genome's-eye view of the light-harvesting polypeptides of *Chlamydomonas reinhardtii. Curr Genet. Springer;* 2004;45(2):61–75.

[5] Rochaix J. *Chlamydomonas reinhardtii* as the photosynthetic yeast. *Annu rev genet;* 1995;29(1):209–30.

[6] Randolph-Anderson BL, Boynton JE, Gillham NW, Harris EH, Johnson AM, Dorthu M-P, et al. Further characterization of the respiratory deficient dum-1 mutation of *Chlamydomonas reinhardtii* and its use as a recipient for mitochondrial transformation. *Mol Gen Genet. Springer;* 1993;236(2-3):235–44.

[7] Yamasaki T, Kurokawa S, Watanabe KI, Ikuta K, Ohama T. Shared molecular characteristics of successfully transformed mitochondrial genomes in *Chlamydomonas reinhardtii. Plant Mol Biol. Springer;* 2005;58(4):515–27.

[8] Remacle C, Cardol P, Coosemans N, Gaisne M, Bonnefoy N. High-efficiency biolistic transformation of Chlamydomonas mitochondria can be used to insert mutations in complex I genes. P Natl Acad Sci USA. *National Acad Sciences;* 2006;103(12):4771–6.

[9] Purton S, Rochaix J-D. Characterisation of the ARG7 gene of *Chlamydomonas reinhardtii* and its application to nuclear transformation. *European Journal of Phycology. Taylor \& Francis;* 1995;30(2):141–8.

[10] Auchincloss AH, Loroch AI, Rochaix JD. The argininosuccinate lyase gene of *Chlamydomonas reinhardtii*: cloning of the cDNA and its characterization as a selectable shuttle marker. *Mol Gen Genet. Springer;* 1999;261(1):21–30.

[11] Fernández E, Schnell R, Ranum LP, Hussey SC, Silflow CD, Lefebvre PA. Isolation and characterization of the nitrate reductase structural gene of *Chlamydomonas reinhardtii. P Natl Acad Sci USA. National Acad Sciences;* 1989;86(17):6449–53.

[12] Kalakoutskii KL, Fernández E. *Chlamydomonas reinhardtii* nitrate reductase complex has 105 kDa subunits in the wild-type strain and a structural mutant. *Plant Science. Elsevier;* 1995;105(2):195–206.

[13] Mayfield SP, Kindle KL. Stable nuclear transformation of *Chlamydomonas reinhardtii* by using a *C. reinhardtii* gene as the selectable marker. *P Natl Acad Sci USA. National Acad Sciences;* 1990;87(6):2087–91.

[14] Diener DR, Curry AM, Johnson KA, Williams BD, Lefebvre PA, Kindle KL, et al. Rescue of a paralyzed-flagella mutant of Chlamydomonas by transformation. *P Natl Acad Sci USA. National Acad Sciences;* 1990;87(15):5739–43.

[15] Smart EJ, Selman BR. Complementation of a*Chlamydomonas reinhardtii* mutant defective in the nuclear gene encoding the chloroplast coupling factor 1 (CF1) γ-subunit (atpC). *J bioenerg biomembr. Springer;* 1993;25(3):275–84.

[16] Blankenship JE, Kindle KL. Expression of chimeric genes by the light-regulated cabII-1 promoter in *Chlamydomonas reinhardtii*: a cabII-1/nit1 gene functions as a dominant selectable marker in a nit1-nit2-strain. *Mol Cell Biol. Am Soc Microbiol;* 1992;12(11):5268–79.

[17] KozMINSKI KG, Johnson KA, Forscher P, Rosenbaum JL. A motility in the eukaryotic flagellum unrelated to flagellar beating. *P Natl Acad Sci USA. National Acad Sciences;* 1993;90(12):5519–23.

[18] Lodha M, Schroda M. Analysis of chromatin structure in the control regions of the Chlamydomonas HSP70A and RBCS2 genes. *Plant Mol Biol. Springer;* 2005; 59(3):501–13.

[19] León R, Fernández E. Nuclear transformation of eukaryotic microalgae. *Transgenic microalgae as green cell factories. Springer;* 2007. pp. 1–11.

[20] Coll JM. Methodologies for transferring DNA into eukaryotic microalgae: a review. *Spanish Journal of Agricultural Research.* 2006;4(4):316–30.

[21] Goldschmidt-Clermont M. Transgenic expression of aminoglycoside adenine transferase in the chloroplast: a selectable marker for site-directed transformation of Chlamydomonas. *Nucleic Acids Res. Oxford Univ Press;* 1991;19(15):4083–9.

[22] Bateman JM, Purton S. Tools for chloroplast transformation in Chlamydomonas: expression vectors and a new dominant selectable marker. *Mol Gen Genet. Springer;* 2000;263(3):404–10.

[23] Komine Y, Kikis E, Schuster G, Stern D. Evidence for in vivo modulation of chloroplast RNA stability by 3′-UTR homopolymeric tails in *Chlamydomonas reinhardtii. P Natl Acad Sci USA. National Acad Sciences;* 2002;99(6):4085–90.

[24] Franklin S, Ngo B, Efuet E, Mayfield SP. Development of a GFP reporter gene for *Chlamydomonas reinhardtii* chloroplast. *The Plant Journal. Wiley Online Library;* 2002;30(6):733–44.

[25] Sakamoto W, Kindle KL, Stern DB. In vivo analysis of Chlamydomonas chloroplast petD gene expression using stable transformation of beta-glucuronidase translational fusions. *P Natl Acad Sci USA. National Acad Sciences;* 1993;90(2):497–501.

[26] Fargo DC, Zhang M, Gillham NW, Boynton JE. Shine-Dalgarno-like sequences are not required for translation of chloroplast mRNAs in *Chlamydomonas reinhardtii* chloroplasts or in Escherichia coli. *Mol Gen Genet. Springer;* 1998;257(3):271–82.

[27] Ishikura K, Takaoka Y, Kato K, Sekine M, Yoshida K, Shinmyo A. Expression of a foreign gene in *Chlamydomonas reinhardtii* chloroplast. *J Biosci Bioeng. Elsevier;* 1999;87(3):307–14.

[28] Maul JE, Lilly JW, Cui L, Miller W, Harris EH, Stern DB, et al. The *Chlamydomonas reinhardtii* Plastid Chromosome Islands of Genes in a Sea of Repeats. *The plan Cell. Am Soc Plant Biol;* 2002;14(11):2659–79.

[29] Purton S. Tools and techniques for chloroplast transformation of Chlamydomonas. *Transgenic microalgae as green cell factories. Springer;* 2007. pp. 34–45.

[30] Newman SM, Boynton JE, Gillham NW, Randolph-Anderson BL, Johnson AM, Harris EH. Transformation of chloroplast ribosomal RNA genes in Chlamydomonas: molecular and genetic characterization of integration events. *Genetics. Genetics Soc America;* 1990;126(4):875–88.

[31] Surzycki R, Greenham K, Kitayama K, Dibal F, Wagner R, Rochaix J-D, et al. Factors effecting expression of vaccines in microalgae. *Biologicals. Elsevier;* 2009;37(3):133–8.

[32] Wang X, Brandsma M, Tremblay R, Maxwell D, Jevnikar AM, Huner N, et al. A novel expression platform for the production of diabetes-associated autoantigen human glutamic acid decarboxylase (hGAD65). *BMC Biotechnol. BioMed Central Ltd;* 2008;8(1):87.

[33] Mayfield SP, Franklin SE, Lerner RA. Expression and assembly of a fully active antibody in algae. *P Natl Acad Sci USA. National Acad Sciences;* 2003;100(2):438–42.

[34] Mayfield SP, Schultz J. Development of a luciferase reporter gene, luxCt, for *Chlamydomonas reinhardtii* chloroplast. *The Plant Journal. Wiley Online Library;* 2004;37(3):449–58.

[35] He D-M, Qian K-X, Shen G-F, Zhang Z-F, Li Y-N, Su Z-L, et al. Recombination and expression of classical swine fever virus (CSFV) structural protein E2 gene in *Chlamydomonas reinhardtii* chroloplasts. *Colloids and Surfaces B: Biointerfaces. Elsevier;* 2007;55(1):26–30.

[36] Sun M, Qian K, Su N, Chang H, Liu J, Shen G. Foot-and-mouth disease virus VP1 protein fused with cholera toxin B subunit expressed in *Chlamydomonas reinhardtii* chloroplast. *Biotechnol Lett. Springer;* 2003;25(13):1087–92.

[37] Tran M, Henry RE, Siefker D, Van C, Newkirk G, Kim J, et al. Production of anti-cancer immunotoxins in algae: Ribosome inactivating proteins as fusion partners. *Biotechnol Bioeng. Wiley Online Library;* 2013;110(11):2826–35.

[38] Dreesen IA, Hamri GC-E, Fussenegger M. Heat-stable oral alga-based vaccine protects mice from *Staphylococcus aureus* infection. *J Biotechnol. Elsevier;* 2010;145(3):273–80.

[39] Marin-Navarro J, Manuell AL, Wu J, Mayfield SP. Chloroplast translation regulation. *Photosynth Res. Springer;* 2007;94(2-3):359–74.

[40] Stern DB, Goldschmidt-Clermont M, Hanson MR. Chloroplast RNA metabolism. *Annu Rev Plant Biol. Annual Reviews;* 2010;61:125–55.

[41] Zhang G-H, Shen G-F, RU B-G. Survival of Human Metallothionein-2 Transplastomic *Chlamydomonas reinhardtii* to Ultraviolet B Exposure. *Acta Biochim Biophys Sinica. Wiley Online Library;* 2006;38(3):187–93.

[42] Tran M, Zhou B, Pettersson PL, Gonzalez MJ, Mayfield SP. Synthesis and assembly of a full-length human monoclonal antibody in algal chloroplasts. *Biotechnol Bioeng. Wiley Online Library;* 2009;104(4):663–73.

[43] Rasala BA, Muto M, Lee PA, Jager M, Cardoso RM, Behnke CA, et al. Production of therapeutic proteins in algae, analysis of expression of seven human proteins in the

chloroplast of *Chlamydomonas reinhardtii. Plant Biotechnol J. Wiley Online Library;* 2010;8(6):719–33.

[44] Nakamura S, Ikehara T. Mutant male *Chlamydomonas reinhardtii* cells with few chloroplast nucleoids and exceptional chloroplast gene transmission. *J Plant Res. Springer;* 2000;113(2):165–70.

[45] Gregory JA, Li F, Tomosada LM, Cox CJ, Topol AB, Vinetz JM, et al. Algae-produced Pfs25 elicits antibodies that inhibit malaria transmission. *PLoS ONE. Public Library of Science;* 2012;7(5):e37179.

[46] Klein U, Salvador ML, Bogorad L. Activity of the Chlamydomonas chloroplast rbcL gene promoter is enhanced by a remote sequence element. *P Natl Acad Sci USA. National Acad Sciences;* 1994;91(23):10819–23.

[47] Manuell AL, Beligni MV, Elder JH, Siefker DT, Tran M, Weber A, et al. Robust expression of a bioactive mammalian protein in Chlamydomonas chloroplast. *Plant Biotechnol J. Wiley Online Library;* 2007;5(3):402–12.

[48] Michelet L, Lefebvre-Legendre L, Burr SE, Rochaix J-D, Goldschmidt-Clermont M. Enhanced chloroplast transgene expression in a nuclear mutant of Chlamydomonas. *Plant Biotechnol J. Wiley Online Library;* 2011;9(5):565–74.

[49] Barnes D, Franklin S, Schultz J, Henry R, Brown E, Coragliotti A, et al. Contribution of 5′-and 3′-untranslated regions of plastid mRNAs to the expression of Chlamydomonas reinhardtii chloroplast genes. *Mol Genet Genomics. Springer;* 2005;274(6):625–36.

[50] Oey M, Lohse M, Kreikemeyer B, Bock R. Exhaustion of the chloroplast protein synthesis capacity by massive expression of a highly stable protein antibiotic. *The Plant Journal. Wiley Online Library;* 2009;57(3):436–45.

[51] Magee AM, Coyne S, Murphy D, Horvath EM, Medgyesy P, Kavanagh TA. T7 RNA polymerase-directed expression of an antibody fragment transgene in plastids causes a semi-lethal pale-green seedling phenotype. *Transgenic Res. Springer;* 2004;13(4):325–37.

[52] Mühlbauer SK, Koop H-U. External control of transgene expression in tobacco plastids using the bacterial lac repressor. *The Plant Journal. Wiley Online Library;* 2005;43(6):941–6.

[53] Rochaix JD. Post-transcriptional regulation of chloroplast gene expression in *Chlamydomonas reinhardtii. Post-Transcriptional Control of Gene Expression in Plants. Springer;* 1996. pp. 327–41.

[54] Maliga P, Bock R. Plastid biotechnology: food, fuel, and medicine for the 21st century. Plant Physiol. *American Society of Plant Biologists;* 2011;155(4):1501–10.

[55] Almaraz-Delgado AL, Flores-Uribe J, Perez-España VH, Salgado-Manjarrez E, Badillo-Corona JA. Production of therapeutic proteins in the chloroplast of *Chlamydomonas reinhardtii. AMB express.* 2014 Aug 2;4:57–65.

[56] Leader B, Baca QJ, Golan DE. Protein therapeutics: a summary and pharmacological classification. *Nature Reviews Drug Discovery. Nature Publishing Group;* 2008;7(1):21–39.

[57] Mayfield SP, Franklin SE. Expression of human antibodies in eukaryotic micro-algae. *Vaccine. Elsevier;* 2005;23(15):1828–32.

[58] Tran M, Van C, Barrera DJ, Pettersson PL, Peinado CD, Bui J, et al. Production of unique immunotoxin cancer therapeutics in algal chloroplasts. *P Natl Acad Sci USA. National Acad Sciences;* 2013;110(1):E15–E22.

[59] Tran M, Henry RE, Siefker D, Van C, Newkirk G, Kim J, et al. Production of anti-cancer immunotoxins in algae: Ribosome inactivating proteins as fusion partners. *Biotechnol Bioeng. Wiley Online Library;* 2013;110(11):2826–35.

[60] Powdrill TF, Johnston JM. Immunologic priming with recombinant hepatitis A virus capsid proteins produced in Escherichia coli. *J Virol. Am Soc Microbiol;* 1991;65(5):2686–90.

[61] Gowda DC, Davidson EA. Protein glycosylation in the malaria parasite. *Parasitol today. Elsevier;* 1999;15(4):147–52.

[62] Carrillo C, Wigdorovitz A, Oliveros JC, Zamorano PI, Sadir AM, Gomez N, et al. Protective immune response to foot-and-mouth disease virus with VP1 expressed in transgenic plants. *J Virol. Am Soc Microbiol;* 1998;72(2):1688–90.

[63] Wigdorovitz A, Carrillo C, Dus Santos MJ, Trono K, Peralta A, Gomez MC, et al. Induction of a protective antibody response to foot and mouth disease virus in mice following oral or parenteral immunization with alfalfa transgenic plants expressing the viral structural protein VP1. *Virology. Elsevier;* 1999;255(2):347–53.

[64] Gregory JA, Topol AB, Doerner DZ, Mayfield S. Alga-produced cholera toxin-Pfs25 fusion proteins as oral vaccines. *Appl Environment Microbiol. Am Soc Microbiol;* 2013;79(13):3917–25.

[65] Chebolu S, Daniell H. Chloroplast-derived vaccine antigens and biopharmaceuticals: expression, folding, assembly and functionality. *Plant-Produced Microbial Vaccines. Springer;* 2009. pp. 33–54.

[66] Tregoning JS, Nixon P, Kuroda H, Svab Z, Clare S, Bowe F, et al. Expression of tetanus toxin fragment C in tobacco chloroplasts. *Nucleic Acids Res. Oxford Univ Press;* 2003;31(4):1174–9.

[67] Koya V, Moayeri M, Leppla SH, Daniell H. Plant-based vaccine: mice immunized with chloroplast-derived anthrax protective antigen survive anthrax lethal toxin challenge. *Infection and Immunity. Am Soc Microbiol;* 2005;73(12):8266–74.

[68] Molina V, Shoenfeld Y. Infection, vaccines and other environmental triggers of autoimmunity. *Autoimmunity;* 2005;38(3):235–45.

[69] Harakuni T, Sugawa H, Komesu A, Tadano M, Arakawa T. Heteropentameric cholera toxin B subunit chimeric molecules genetically fused to a vaccine antigen induce systemic and mucosal immune responses: a potential new strategy to target recombinant vaccine antigens to mucosal immune systems. I*nfection and Immunity. Am Soc Microbiol;* 2005;73(9):5654–65.

[70] Avesani L, Vitale A, Pedrazzini E, DeVirgilio M, Pompa A, Barbante A, et al. Recombinant human GAD65 accumulates to high levels in transgenic tobacco plants when expressed as an enzymatically inactive mutant. *Plant Biotechnol J. Wiley Online Library;* 2010;8(8):862–72.

[71] Rasala BA, Muto M, Sullivan J, Mayfield SP. Improved heterologous protein expression in the chloroplast of *Chlamydomonas reinhardtii* through promoter and 5' untranslated region optimization. *Plant Biotechnol J. Wiley Online Library;* 2011;9(6):674–83.

[72] Blifernez-Klassen O, Klassen V, Doebbe A, Kersting K, Grimm P, Wobbe L, et al. Cellulose degradation and assimilation by the unicellular phototrophic eukaryote *Chlamydomonas reinhardtii. Nature communications. Nature Publishing Group;* 2012;3:1214.

[73] Abe J, Hiwatashi Y, Ito M, Hasebe M, Sekimoto H. Expression of exogenous genes under the control of endogenous HSP70 and CAB promoters in the Closterium peracerosum--strigosum--littorale complex. *Plant Cell Physiol. Jpn Soc Plant Physiol;* 2008;49(4):625–32.

[74] Lerche K, Hallmann A. Stable nuclear transformation of Gonium pectorale. *BMC Biotechnol. BioMed Central Ltd;* 2009;9(1):64.

[75] Kathiresan S, Sarada R. Towards genetic improvement of commercially important microalga Haematococcus pluvialis for biotech applications. *J Appl Phycol. Springer;* 2009;21(5):553–8.

[76] Kathiresan S, Chandrashekar A, Ravishankar GA, Sarada R. Agrobacterium-mediated transformation in the green alga *Haematococcus pluvialis* (Chlorophyceae, volvocales) 1. *J Phycol. Wiley Online Library;* 2009;45(3):642–9.

[77] Hirakawa Y, Kofuji R, Ishida K-I. Transient transformation of a Chlorarachniophyte alga, Lotharella amoebiformis (Chlorarachniophyceae), with uidA and egfp reporter genes. *J Phycol. Wiley Online Library;* 2008;44(3):814–20.

[78] Kakinuma M, Ikeda M, Coury DA, Tominaga H, Kobayashi I, Amano H. Isolation and characterization of the rbcS genes from a sterile mutant of Ulva pertusa (Ulvales, Chlorophyta) and transient gene expression using the rbcS gene promoter. *Fisheries Science. Springer;* 2009;75(4):1015–28.

[79] Ohnuma M, Yokoyama T, Inouye T, Sekine Y, Tanaka K. Polyethylene glycol (PEG)-mediated transient gene expression in a red alga, Cyanidioschyzon merolae 10D. Plant and Cell Physiology. *Jpn Soc Plant Physiol;* 2008;49(1):117–20.

[80] Feng S, Xue L, Liu H, Lu P. Improvement of efficiency of genetic transformation for Dunaliella salina by glass beads method. *Mol Biol Rep. Springer;* 2009;36(6):1433–9.

[81] Yang Z, Chen F, Li D, Zhang Z, Liu Y, Zheng D, et al. Expression of human soluble TRAIL in *Chlamydomonas reinhardtii* chloroplast. *Chinese Science Bulletin. Springer;* 2006;51(14):1703–9.

[82] Yoon S-M, Kim SY, Li KF, Yoon BH, Choe S, Kuo MM-C. Transgenic microalgae expressing Escherichia coli AppA phytase as feed additive to reduce phytate excretion in the manure of young broiler chicks. *Appl Microbiol Biot. Springer;* 2011;91(3):553–63.

In: Microalgae and Other Phototrophic Bacteria
Editor: Luis G. Torres Bustillos
ISBN: 978-1-63482-078-3

Chapter 2

REVIEW ON INSTRUMENTATION, MODELLING AND AUTOMATIC CONTROL FOR MICROALGAE CULTURE

I. Chairez
Bioprocesses Department, UPIBI-IPN, Mexico

Abstract

Microalgae have been recognized as versatile microorganisms that are used in various applications. Their capacity of converting sunlight and carbon dioxide in diverse metabolites makes them an efficient microbiological factory. Despite these characteristics, there is a general trend on optimizing the microalgae cultures considering the potential industrial and economical benefits that can be obtained from them. This task is closely related to the instruments used to monitor and regulate the reactor activity where the microalgae are growing. Moreover, these instruments are useless if they are not associated to an automatic control law that can force the optimal performance in the reactor operation. This chapter reviews the recent results on instrumentation strategies as well as the novel automatic control methods that have been used to optimize reaction operation to improve microalgae productivity either increasing biomass or their corresponding metabolites concentrations.

PACS 05.45-a, 52.35.Mw, 96.50.Fm
Keywords: Microalgae culture, bioreactor instrumentation, mathematical model of bioprocesses, automatic control for photobioreactors
AMS Subject Classification: 53D, 37C, 65P

1. Introduction

For many years, society has faced catastrophic predictions regarding oil production level. However, the actual production of energetics from oil is on its highest historic levels. It is reasonable to assume that fossil fuels would be the main energetic supplies to the modern societies for several decades [1, 2]. Nevertheless, the increasing demands of energetics and the non-renewable condition of fossil oils demand preventive actions that can ensure the survival of the human race for the following generations. Therefore, the real challenge is sustainability: smoothing the transition from fossil fuels in the distant future, stabilizing and reversing global climate change and minimizing political and economic energy volatility.

The relative importance of these three aspects has been weighted in different ways for the governments of developed and in-developing countries. However, at this moment the environmental issue has gained more and more attention due to the effects on health of all the vegetal and animal species. A different face of this aspect concerns the reduction of crude oil reserves and the increasing difficulties in their extraction and processing, leading to cost increments [3]. This situation is particularly acute in the transportation sector. Today, there are no reliable alternatives to fossil fuels. Transportation and energy sectors are the major anthropogenic pollution sources, responsible for 20% and 60% of greenhouse gas (GHG) emissions in Europe, respectively [4].

Agriculture sector is considered the third most important anthropogenic source of GHG (about 9%) emissions. Among other gases, the most important are nitrous oxide (N_2O) and methane (CH_4) [5]. Moreover, in the emerging economies, such as India and China (where billions of persons are living with population in billions), the global consumption of energy is rapidly increasing and is exerting relevant environmental damage [6]. This condition has been reached considering poor regulations on environmental aspects that are still applied on the majority of the countries all over the world.

Additionally, GHG contributes to global warming (GW) as well as to other effects on the environment and health of human race. The increasing amount of carbon dioxide absorbed (one-third of the CO2 emitted each year comes from human activities) from the atmosphere that is dissolved in oceans decreases the water pH gradually. This pH variation has been shown to be one of the main factors on the quick loss of coral reefs and of marine ecosystem biodiversity. These huge implications in ocean life will have consequently a remarkable and maybe irreversible effect in earth life [7].

The different effects of irrational utilization of fossil oils over environment represent a real and relevant problem that must be considered as major priority. Therefore, not only a single but a host of solutions is needed to address it. The number of options studied and implemented today in practice and academy, with different degrees of success, is growing every day. Solar energy, either thermal or photovoltaic, hydroelectric, geothermal, wind, biofuels, and carbon sequestration are the most remarkable examples of renewable energy generation methods [8].

Which one of these methods represents the most adequate method to produce energy depends on the area of application. Different options consider sequential or/and parallel combinations of individual methods.

Despite the economical profits obtained by renewable energy methods, the main aspect to consider is the transportation emissions reduction. Therefore, carbon neutral renewable processes are needed to eventually displace petroleum-derived fuels that greatly contribute to global warming and compromise the world's energetic security. The urgent necessity of reducing carbon emissions, and the declining reserves of crude oil, liquid or gaseous fuels derived from plant or microbiological materials (also termed biofuels) seem to be reliable alternative sources of renewable energy.

Biofuels have the major advantage of storing energy within microorganisms. Moreover, biofuels can be used in existing engines and transportation infrastructures after blending them at various degrees with petroleum diesel [9, 10]. Among the scientific and industrial communities, there are many open discussions regarding the potential of biofuels to provide a complete displacement of petroleum-derived transport fuels, such as gasoline, jet fuel and diesel [11]. However, biofuels are considered as real (maybe not total) contributors to reach that goal, particularly in the short term. Ethanol, butanol, biodiesel and hydrogen have been the most developed biotechnologies to produce renewable energy. Each one of these methods has energetic efficiency as the major challenge. Even when there are other options that can use chemical or physical processes to generate these alternative fuels, they are less efficient than biotechnological options. Furthermore, biofuels production is expected to offer new economical opportunities by diversifying income and fuel supply sources. To become a more viable alternative fuel and to impact the energetics market, biodiesel shall compete economically with oil based diesel. The final cost of biodiesel mainly depends on the price of feedstocks that represent 60–75% of the total cost of biodiesel fuel.

What is the best biofuel is still a matter of discussion among the scientific and technological specialists. The main alternative to fossil diesel fuel in all over the world is biodiesel, representing 82% of total biofuels production and is still growing based on political and economic objectives [12]. Biodiesel is produced by biological and non-biological processes. No biological biofuels are generated from vegetable oils (edible or non-edible) or animal fats. Since vegetable oils are used for human consumption, an increment for price of food-grade oils can be expected. As a consequence, the cost of biodiesel may increase and preventing its usage, even if it has advantages comparing with oil based diesel fuel. Moreover, the available quantities of wasted oils and animal fats are clearly not enough to supply the current demands for alternative diesel from biological sources.

Biodiesel has two major advantages: a) postprocessing biological material is relatively cheap compared to all other biofuels sources and b) biodiesel can be obtained by a very efficient biological organisms, microalgae. Actually, it is known that producing biodiesel from oil crops and bioethanol from sugarcane is the most popular method to obtain large amounts of renewable biofuels, but at that level, their production is not economically sustainable [13]. This fact confirms that microalgae are an alternative to produce biodiesel in large scale. Microalgae are photosynthetic microorganisms that convert sunlight, water and carbon dioxide to the so-called algal biomass. In dramatic contrast with the best oil-producing crops, microalgae biodiesel has the potential to be able to completely displace petroleum-derived transport fuels without adversely impacting supplies of food and other agricultural products [14]. It is further demonstrated that microalgae biodiesel is a better alternative than bioethanol from sugarcane, which is currently the most widely used transport biofuel [15].

The term algae has no formal understanding of taxonomy, as this term is commonly used to indicate a set of photosynthetic organisms artificially attached polyphyletic and not cohesive. This classification includes organisms that do not share a common origin, but continue multiple or independent lines of evolution. There are several exceptions, as some colorless organisms. Based on this definition, plants can be considered a division of algae, producing the same reserve components, using similar methods of defense against predators or parasites. Moreover, there are relevant morphological similarities between them [16].

On the other hand, there are also remarkable differences between algae and plants. Therefore, microalgae are considered as primitive plants (thallophytes), i.e. lacking roots, stems and leaves, have no sterile covering of cells around the reproductive cells and have chlorophyll as primary photosynthetic pigment. Their structures are primarily for energy conversion without any development beyond cells, and their simple development allows them to prevail over harmful environmental conditions and prosper in the long term [17].

Recently, microalgae have been exploited in different areas, such as, biofuels production, waste treatment, food supplements, carbon dioxide (CO_2) abatement, production of fine chemicals, immobilization systems for production of some extracellular compounds or bio-absorption of heavy metals [18]. Despite the large number of applications where microalgae are applied, there are some challenges that must be solved before they can represent a true solution in many industrial processes. It is estimated that there are between one and ten million species of algae, and most of them are microalgae.

There are several microalgae species that can be induced to accumulate relevant internal lipids concentrations [19]. Despite the average lipid contents vary between 1% and 70%, some species may reach 90% (w/wDW) under certain reaction conditions [10]. The most common microalgae species such as Chlorella, Dunaliella, Isochrysis, Nannochloris, Nannochloropsis, Neochloris, Nitzschia, Phaeodactylum and Porphyridium spp. possess oil levels between 20% and 50%, along with interesting productivity [20].

There is a general trend among diverse scientific communities regarding which one is the most adequate species for biodiesel production. In general, it is considered that many other factors should be taken into account, for example, the ability to uptake available nutrients (nitrogen, CO_2, salts, etc.) or grow under specific environmental conditions (type of bioreactor, reaction regimen, etc.). A relevant fatty acid production or higher productivity is not the only way to consider a successful microalgae culture. Fatty acid profile of microalgae biomass is also a key factor: the heating power of the resulting biodiesel relies upon the lipids distribution; most of said moieties are saturated and unsaturated fatty acids containing 12–22 carbon atoms, often of the $\omega3$ and $\omega6$ types [21]. It is well known that different nutritional and processing factors, cultivation conditions and growth phases severely affect the fatty acid composition of microalgae biomass [22].

2. Microalgae culture and photobioreactors

To accomplish the growing needs of each microalgae, it is important to develop a particular system required to keep regulated the conditions to promote their growth. These devices are known as photobioreactors which are considered a specific class of bioreactors which are devices designed and constructed to support a biologically active environment. In particular, these photobioreactors consider the light as a variable to be controlled in order to optimize microalgae growing.

Although the term photobioreactor has been applied to open algal ponds and channels, it is best reserved for devices that allow monoseptic culture which is fully isolated from a potentially contaminating environment. These devices can be further categorized according to the orientation of tubes or panels, the mechanism to circulate culture media, the method used to provide light, the type of gas exchange system, the arrangement of the individual growth units, and the materials of construction employed [23].

Generally, photobioreactors can be classified as open or closed. Their mass transfer rates are very poor yielding to low biomass productivity. Different forms of open photobioreactors can be distinguished: these may be natural water forms such as lakes and ponds, waterfall type inclined, circular ponds, and rough raceways systems.

Closed systems reduce contamination risks and increase the possibility of implementing control systems on the culture conditions that yields to better mass transfer conditions [24]. It can be found flat plate systems, bubble column, air-lift, stirred tank, toroidal, among others [17]. In these photobioreactors, several different instruments can be set-up to promote the culture medium recirculation, the exchange of CO_2, etc. [25, 26, 27]. Since the illumination is essential for growing microalgae, closed photobioreactors are designed with transparent containers. Indeed, light penetration is the key factor that affect the reactor design: a high superficial area and volume ratio are required [23].

Tubular photobioreactors remain as the most popular options to microalgae culture outdoors. Actually, hydrodynamics regime is easily improved in this type of systems (mixture periods and low cutting efforts), they can be sterilized without high resource investments, microalgae strain can be acclimated in short periods of time and the photo-oxidation effect is no longer observed [28].

Despite all their aforementioned benefits, closed systems have the major disadvantages of the high cost of construction, operation and maintenance [29]. Even when the economical aspect is neglected, microalgae form dense films and can be affected by cytotoxic gradients of pH, dissolved oxygen and carbon dioxide [30].

On the other hand, open systems can be easily constructed and operated. These are the two major reasons because they are the most widely used commercially. These reactors are classified into shallow lagoons, waterfalls systems, circular ponds, stirred ponds and raceways [25]. These last devices are stirred by a mechanical driver [14].

However, they also have important problems such as low light utilization cell efficiency, evaporative losses of liquid medium, diffusion of CO_2 to the atmosphere, requirement of large areas of land, contamination by predators and other fast growing heterotrophs. The most extended type of open systems is raceway. Consequently, raceway reactors are restricted to microalgae species that can handle wide range of temperatures that can grow in specific environments with high salinity or/and alkalinity [3].

In the previous years, there is a new trend on developing the so-called semi-closed photobioreactors where the main characteristics of the open and closed systems are gathered. These systems look in general like raceways but are partially covered (also known as closed raceway pond). This mixed scheme tries to reduce the contamination and evaporation losses. Considering that these systems were introduced just recently, there are many open question regarding its operation.

3. Automatization in microalgae culture

Automatization of bioreactors is an active field within the bioprocess disciplines. Automatization is a general concept that includes the application of fundamental principles of several disciplines. Among others, electrical and electronic engineering, systems theory and image and signal processing methods have been combined to produce reliable systems to implement automatic systems that can monitor and regulate the bioreactor behavior.

These automatic systems are composed of electrical/electronic sensors that provide online information of the internal bioreactor environment. This information can be used to evaluate the health of the microorganism culture. Among others, pH level, carbon dioxide and oxygen concentrations, light intensity, temperature, alkalinity and conductivity are the most regular examples of sensors used within a microalgae culture. All these variables need to be monitored considering the relative importance that each one has over the microorganism growing dynamics.

Output of these sensors must be adjusted into an electronic board which is designed to supply the information into a digital or analog electronic device. This device is the section of the automatization system where the sensors information can be displayed, processed and recorded. One additional process includes the utilization of such information as input of some automatic control design.

Automatic controller is a result of predefined rules to adjust the devices that can modulate the bioreactor operation conditions. Some of these automatic controller designs are presented in the section (5.) of this chapter. The output of these automatic controllers shall be transformed into useful signals for the so-called bioreactor actuators. These devices are responsible of modulating the aforementioned bioreactor operation conditions. Among others, these devices can include peristaltic pumps (used to adjust the pH, alkalinity, etc.), gases pumps which provide the possibility of regulating carbon dioxide or oxygen concentration, thermal actuators to set the reactor temperature, etc.

The core section of all automatization modules in bioreactor is the automatic controller algorithm. This section can be designed using two possible frameworks: a) based on a suitable and feasible mathematical model used to describe the microalgae growing and b) without considering the model but using robust and adaptive strategies. The first one has proven to be more accepted because it takes into account the nature of the microorganism culture. The following section reviews some mathematical models used to describe the microalgae growing dynamics.

4. Modelling photobioreactors

Understanding the mechanisms of algae growth, the utilization of nutrients and developing models in order to predict biomass formation are essential to enhance the photobioreactors operation. Kinetic modeling of microalgae growth has become significant because an accurate model can be considered a prerequisite for designing an efficient photobioreactor, predicting process performance and overall, optimizing operational conditions.

Mathematical models of biological processes in photobioreactors have considered two main factors: a) when the light affects the reaction and b) when the light is not modifying the biomass growing dynamics. The first step of the model development is to derive mass and energy balances on the photobioreactor. Mass balances must be considered relating all the input and output materials feed or extracted from the reactor (biomass, nitrogen, phosphorus, inorganic carbon, etc.). An energy balance is also needed for assessing the change in energy associated to the culture, including biochemical and thermodynamic conversions.

The experimental setup considered to construct the models under the assumption of batch or continuous regimens (with equal input and output flows). Therefore, the bioreactor keeps a constant volume. This approach considers a close photobioreactor which is a convenient device to control the growth conditions of microalgae and to mimic conditions encountered in natural ecosystems like lakes. These conditions have been extensively used to study the growth of microorganisms populations, and competition or cooperation between them [31].

Modelling microalgae growth requires a deep understanding on the different phases used by the cell to achieve its mature stage. These phases are:

1. A lag phase, where a delay in growth initially occurs due to the presence of nonviable cells or spores in the inoculum or physiological adjustments to change in nutrient concentration or culture conditions.

2. An exponential phase, where cells grow and divide as an exponential function of time, as long as mineral substrates and light intensity are saturated.

3. A linear growth phase, where growth rate is linear as a function of time.

4. A stationary growth phase, where the growth rate remains constant. However, an increment of nutrient concentration may lead to luxury storage of nutrients by algae during this phase.

5. A decline or death phase, where the decrease in the concentration of nutrients and/or accumulation of toxic waste products leads to microorganisms' death.

4.1. Droop Model

The Droop model [32] is a simple and widely used model that can represent the growth of microalgae. This model describes the dynamic evolution of biomass, the internal quota of nitrogen and the substrate used to feed microalgae. The model is formally presented as:

$$\frac{d}{dt}X(t) = X(t)\left(\mu\left(Q_N(t)\right) - D\right)$$

$$\frac{d}{dt}Q_N(t) = \rho\left(S(t)\right) - \mu\left(Q_N(t)\right)Q_N(t) \tag{1}$$

$$\frac{d}{dt}S(t) = D\left(S^{in}(t) - S(t)\right) - \rho\left(S(t)\right)X(t)$$

where the time varying parameters $\rho(S(t))$ and $\mu(Q_N(t))$ are:

$$\rho(s(t)) = \rho_m \frac{S(t)}{S(t) + K_s}$$

$$\mu(Q_N(t)) = \bar{\mu}\left(1 - \frac{K_Q}{Q_N(t)}\right)$$

The biomass concentration is denoted as $X\left[\mu mol L^{-1}\right]$, the internal quota $Q_N\left[\mu mol L^{-1}\right]$ which is defined as the quantity of intracellular nitrogen per unit of biomass and the substrate concentration is denoted by $S\left[\mu mol L^{-1}\right]$. The substrate uptake rate is represented as $\rho\left[L\,\mu m^{-3} d^{-1}\right]$, while $\mu\left[L\,\mu m^{-3} d^{-1}\right]$ is the specific growth rate and $D\left[d^{-1}\right]$ is the constant dilution rate.

4.2. Extended Droop Model

When the carbon metabolism is introduced into the model structure, the so-called extended Droop model arises. This model uses the assumption that organic carbon is split into a functional pool (proteins, nucleic acids, membranes, etc.) and two main storage pools: sugar and neutral lipids. It has been demonstrated that carbon intake from CO_2 is firstly incorporated into the internal algal sugar metabolism. At a second stage, these carbohydrates are metabolized to produce functional carbon included in enzymes used to interiorize nitrogen.

At the same metabolic level, the same carbohydrates are used to produce neutral lipids, labeled L, that can be accumulated or mobilized to generate membranes. Under the assumption that carbon is mainly used for these two substances (lipids and enzymes), the Droop model is completed with one additional equation that describes the neutral lipid quota Q_L. If all the hydrodynamics into the reactor satisfy perfect mixture conditions (ideally), the model now includes

$$\frac{d}{dt}Q_L(t) = \mu\left(Q_L(t)\right)\left(\beta Q_N(t) - Q_L(t)\right) - \gamma\rho\left(s(t)\right) \tag{2}$$

The parameters β and γ represent the maximum velocity of lipid accumulation and the rate of transformation from lipids to nitrogenated compounds in the microalgae cells.

4.3. Modelling of inorganic carbon consumption

A specific model based on the one proposed by Nouals [33, 34], by neglecting the substrate inhibition term, implies the association of the Monod model for the light effect and the Contois model for limitation effect by a substrate as the total inorganic carbon. The dynamics of the bioreactor is described by a set of two differential equations.

Biomass evolution is obtained by a standard mass balance in the photobioreactor assuming perfectly-stirred conditions under continuous mode (CSTR). Total inorganic carbon (TIC, $mole/L$) is modeled to include carbon dioxide (CO_2), bicarbonate (HCO_3^-) and carbonate ion (CO_3^{-2}) species in the culture medium. Partition coefficient associated to the pH is determined by the pH of the culture. Consumption of TIC is dynamically modelled by a regular mass balance procedure:

$$\frac{d}{dt}\left[TIC(t)\right] = -\mu\left(Q_N(t)\right)\frac{X(t)}{Y} - D\left[TIC(t)\right] + K_{L_a}a\left(\left[CO_2^* - CO_2(t)\right]\right)$$

In this equation, the parameters Y and K_{L_a} are the biomass conversion yields which represent the amount of biomass produced to the amount of total inorganic carbon consumed and the gas-liquid transfer coefficient of carbon dioxide, respectively.

The equilibrium carbon dioxide concentration CO_2^* is defined as $CO_2^* = P_{CO_2}H$. The variable P_{CO_2} is the partial pressure of CO_2 ($0.05\ atm$) under the majority of reported experimental conditions. The parameter H is the Henry's constant for CO_2 at the experimental temperature used to develop the microalgae culture. The CO_2 concentration in the culture is calculated by the following expression:

$$CO_2(t) = \frac{TIC(t)}{1 + \dfrac{K_1}{H^+(t)} + \dfrac{K_1K_2}{\left[H^+(t)\right]^2}}$$

where K_1 and K_2 are the dissociation constants of the chemical equilibriums between CO_2/HCO_3^- as well as CO_2/CO_3^{-2}. respectively. The hydrogen ion concentration H^+ is given by $H^+(t) = 10^{-pH(t)}$.

4.4. Light intensity effect on biomass growing dynamics

The light intensity effect on the biomass growing dynamics has been analyzed in just a few of works. The following model presents one example of specific growth rate for microalgae as a function of light intensity limitation and substrate limitation effect (see [33] for more details):

$$\mu\left(Q_N, E, TIC, X\right) = \bar{\mu}\left(\frac{E}{K_E + E}\right)\left(1 - \frac{K_Q}{Q_N}\right)\left(\frac{[TIC]}{K_{CL}X + [TIC]}\right)$$

where K_E and K_{CL} are the half saturation constants for light intensity available per cell (denoted by E) and the half saturation constants for TIC, respectively.

The light intensity accessible per cell is described as follows:

$$E = \frac{(I_{in} - I_{out})}{VX}A_r$$

Light intensities at the input and output of the reactor are I_{in} and I_{out} that can be measured with simple light sensors. The bioreactor illuminated area is described by A_r $\left[m^2\right]$. The variable V $[L]$ is the reactor's volume.

When the outgoing light intensity cannot be measured on-line, this variable can be calculated by an analytical expression as a function of biomass and the incident light intensity according to the following relation

$$I_{out} = C_1 I_{in} X^{C_2}$$

with the parameters C_1 and C_2 are constants that regularly depend on the reactor geometry and its material.

4.5. Complex interactions in photobioreactors

Just few papers have considered the effect of more complex interactions between the variables involved in the photobioreactor dynamics. The results presented in [35] described the effect of some additional interactions between biomass, product and substrate into the reactor. The minimal model describing the relationship between reactor variables is as follows:

$$\frac{d}{dt}x\left(t\right) = -Dx\left(t\right) + \varphi_1(P\left(\left(t\right)\right), x\left(t\right))$$

$$\frac{d}{dt}P\left(t\right) = -DP\left(t\right) - \varphi_1(s\left(t\right), x\left(t\right)) + \varphi_2(P\left(t\right), x\left(t\right))$$

$$\frac{d}{dt}s\left(t\right) = D\left(s^{in}(t) - s\left(t\right)\right) + \varphi_2(s\left(t\right), x\left(t\right))$$

The variable P represents the variable related to the internal product accumulated within the microalgae biomass. The functions labeled φ_1 and φ_2 are related to the growth and uptake rates that may include effects of biomass, substrate or and product inhibition, light intensity, nitrogen limitation or excess, etc. These functions may include nonlinear relationships between reactor variables that can help to understand more complex metabolic relationships and hydrodynamic limitations.

4.6. Kinetic models of microalgae growing

The most studied complex relationship for microalgae growing is the nature of the specific growing rate. These studies have been conducted for the so-called kinetic models of microalgae culture. The specific models presented above are the most studied complete forms to describe the interaction between different variables in the reactor. However, all of them can be modified by adjusting the kinetic model associated to the growth of microalgae. Any possible variation of this parameter can be helpful to describe the effect of many factors over the culture evolution (light intensity, nitrogen harvesting, etc.). The following subsection describe some of these kinetic models:

4.6.1. Carbon, Nitrogen and Phosphorus concentrations

The models that have considered the effect of inorganic carbon (from carbon dioxide), nitrogen (using ammonia or nitrate) or phosphorus concentrations effect over the microalgae growing dynamics have considered the following couple of mathematical relationships:

- Modified Monod model

$$\mu(t) = \mu_{\max,Y} \left(\frac{Y(t)}{K_{S,Y} + Y(t)} \right)$$

where Y is either the carbon [36, 37, 38], nitrogen [39, 40, 41] and phosphorus [39, 42] concentrations while $K_{S,Y}$ is their corresponding saturation constant.

- Hybrid Monod model

$$\mu(t) = \begin{cases} \mu_{\max,Y} \left(\dfrac{Y(t) - Y_\mu}{K_{S,Y} + Y(t) - Y_\mu} \right) & if \quad Y(t) \geq Y_\mu \\ 0 & otherwise \end{cases}$$

where Y_μ is the finite concentration of carbon, nitrogen or phosphorus. The only one study regarding this model has primarily analyzed the effect of nitrogen [43].

More complex models have been proposed when phosphorus is considered as the limiting factor on microalgae growing. They can be consulted in [44] and [45] for example.

4.6.2. Light Intensity

- Monod model

$$\mu(t) = \mu_{\max}\left(\frac{I(t)}{K_{S,I} + I(t)}\right)$$

where I is the light intensity and $K_{S,I}$ is the saturation light intensity [46, 47, 48].

- Exponential model

$$\mu(t) = \mu_{\max}\left(1 - e^{\frac{I(t)}{K_{E,I}}}\right)$$

where $K_{E,I}$ is a variation of the saturation light intensity parameter [47].

- Distributed model

$$\mu(t) = K\left(\frac{\varepsilon \alpha_l X I_0}{XV} - I_m\left(1 - V_m\right)\right)$$

where K is a proportionality constant (kg/mol), ε a constant dependent on the reactor geometry, α_l is the effective light absorption surface area of each cell (m^2), X is the biomass concentration expressed on (kg/m^3), V is the liquid volume in the reactor (m^3), I_0 is the incident light intensity $(mol/m^2/d)$, I_m is the maintenance rate $(mol/m^2/d)$ and V_m is the illuminated fraction of the reactor [49].

4.6.3. Nutrient effects

- Nitrogen and carbon dioxide effect

$$\mu(t) = \mu_{\max}\left(\frac{S_N(t)}{K_{S,N} + S_N(t)}\right)\left(\frac{S_{CO_2}(t)}{K_{S,CO_2} + S_{CO_2}(t)}\right)\left(1 - e^{\frac{I(t)}{K_{E,I}}}\right)\frac{I_a}{I_o}$$

where S_N is the nitrogen concentration, S_{CO_2} is the carbon dioxide concentration, $K_{S,N}$ and K_{S,CO_2} are the corresponding saturation constants for nitrogen and carbon dioxide while I_a is the average light intensity. The rest of the variables and parameters have already been described in [50].

- Maximum concentrations allowed in microalgae affect

$$\mu(t) = \mu_{\max}\left(\frac{S(t)}{K_S + S(t) + K_{I,S}^{-1} S^2(t)}\right)\left(\frac{I(t)}{K_{S,I} + I(t)}\right)\left(1 - \frac{C_x(t)}{C_m}\right)\left(1 - \frac{C_p(t)}{C_{pm}}\right)$$

where C_x is the cell concentration, C_m is the maximum allowed cell concentration, C_p is the product concentration, C_{pm} is the maximum allowed product concentration. The constant $K_{I,S}$ referred to as the substrate inhibition constant. As one can notice, this model considers the eventual effect of inhibition by many usual factors [51, 52].

- Maximum population growth rate

$$\mu_{\max}(t) = \mu'_{\max}\left(\frac{S_N(t)}{K_{S,N} + S_N(t)}\right)\left(1 - \frac{C_p(t)}{C_{pm}}\right)$$

This model considers that $\mu_{\max}$ is not longer a constant but a time varying parameter depending on current conditions of nitrogen concentrations and product inhibition. $\mu'_{\max}$ is considered the value of $\mu_{\max}$ when no lipids are accumulated within the microalgae and there is no nitrogen limitations [53].

4.6.4. Temperature

Temperature has shown to be a very relevant factor in microalgae culture. However, in literature, only two models have considered the effect of such important operation condition:

- Modified Monod model

$$\mu(t) = T\mu_{\max,T}\left(\frac{S(t)}{K_S + S(t)}\right)$$

where T is the current temperature in the reactor and $\mu_{\max,T}$ is the maximum biomass growing rate at the given temperature T [54].

- Andrews Model with pH and temperature effect

$$\mu(t) = Ae^{-(E/RT)}\left(\frac{H^+(t)}{H^+(t) + K_{OH} + K_{I,H}^{-1}\left[H^+(t)\right]^2}\right) \tag{3}$$

where H^+ is the concentration of hydrogen ions, E is the activation energy of the growth limiting reaction, R is the universal gas constant, K_{OH} and $K_{I,H}$ are the pH saturation constants and the hydrogen ions inhibition parameter [55].

5. Automatic control design for photobioreactors

The last step prior to obtaining optimal or at least suboptimal performance on photobioreactors is to design a close loop control action. This control action usually refers to an algorithm implemented in digital or and analogical systems. These control designs usually are proposed to optimize some predefined performance indexes. These indexes consider the energetic cost to produce high concentrations of biomass or lipid contents within microalgae cells. In order to increase the productivity and profitability of bioprocesses, many research efforts have been devoted to their improvement via process systems engineering approaches. In this way, mathematical modeling, optimization and control have become fundamental tools to optimally design and operate production facilities in the bioprocess industries.[56]

Consequently, keeping the photobioreactor under regulated conditions at a desired value or predefined profile is a relevant objective in microalgae culture. Which control method is the more adequate option to regulate the reactor conditions is still a matter of discussion. This is the main reason because the applications of automatic controllers have been popularized for microalgae culture [57]. It is regularly accepted that controlling bioprocesses has a large number of challenging problems. For instance, the presence of living organisms and their interaction yields to interdependent complicated mathematical models [58, 59].

In order to compensate the presence of high degree of nonlinearities and the complexity of this kind of processes, several nonlinear control strategies have been developed such as optimization-based approaches [60], adaptive approaches [61, 62, 63], sliding mode control [64, 65, 66, 67], exact linearization approach [68], backstepping approach.[69] and model predictive controllers [70, 71]. In the specific case of microalgae culture, the study presented in [68] used the input-output linearization technique [72] to regulate the biomass at a constant value when the photobioreactor is operated in continuous regimen. Also, in [60] a nonlinear model predictive controller was applied to maintain the culture at optimal population density (priory defined) that corresponds to a constant high biomass density mode. This condition has been selected to obtain high biomass final concentration. In [73], an output-based model predictive control designed to fulfill an interior point optimization, the controller was supplied with a moving-horizon observer. This controller was successfully applied to maximize and regulate the lipid production in fed-batch microalgae cultivation. Both, estimator and controller were designed using a set of linearized models associated to the microalgae growth process.

One additional problem when photobioreactors are proposed to be controlled, is that only a few measurements can be obtained. Moreover, measurement devices can provide unreliable or inaccurate measurements [74, 75, 76]. In order to circumvent this problem with measuring devices, observers (also called state estimator or software sensor) can be used . They are dynamic systems which are used to estimate important process variables by means of accessible measured variables. Their design and application in process control have been an active research area over the past decades, especially in bioprocess applications [77, 78, 79].

In the context of the microalgae applications, in [80] a high gain observer based on the Droop model to monitor phytoplankton was designed. They applied the proposed software sensor to a real experimental setup, and they showed the validity and the efficiency of the observer. Continuous-discrete interval observers for this culture were developed by [81, 82]. Moreover, other studies have been carried out such as the design of an extended Kalman filter to estimate biomass from dissolved oxygen measurements and a simple unstructured process model or the design of a moving horizon estimator in the context of lipid production optimization [83].

Even when the state estimator offers a remarkable option to obtain the information of immeasurable variables, their major utility relies on providing information for the automatic controllers. This class of solutions is usually known as output-based feedback controllers. These controllers have been used in bioreactor engineering for many years. In particular, when only simple measurements are needed to recover relevant information from the reactor, state observers become an invaluable section of the reactor operation. For example, in [84], a state observer used only dissolved oxygen to recover the microalgae biomass concentration that was further used to control the reactor. Moreover, an adequate output-based controller may produce suboptimal reaction regimens as the case presented in [61] where the Extremum Seeking Method was applied to optimize the accumulation of secondary metabolites. In literature, there are many examples where similar results have been obtained [63, 85].

Conclusion

Today, there is a long way before the potential offered by microalgae as source of biodiesel becomes feasible. The actual productive processes based on photobioreactors (open, closed or semiclosed) are still debatable if net energy balance and global warming contribution are factors to consider. Without evident positive energy balances, there is a relevant risk that many proposed academic and even industrial processes would be inappropriate from the point of view of sustainability. The scenarios available for microalgae culture still depend on many factors. One of the these factors relies on the photobioreactor reaction conditions which indeed may have a significant effect on biodiesel productivity.

Useful biodiesel productivity likely to be achieved in practice strongly depends on an accurate regulation of the reactor conditions. It is clearly known that major breakthroughs remain to be overtaken towards design and development of technologies that can reduce costs while increasing yields. Among others, automatic controllers working together with modern and reliable instrumentation devices may solve the problem of the optimization process which surely constitutes an actual and urgent contribution to bring innovation into the microalgae cultures even in continuous regiments running in outdoor reactors. This combination will eventually meet with success of producing alternative energy based on biodiesel produced by microalgae.

Without any doubt, automatic control can help the urgent necessity of providing responses to climate and environmental issues. Human race cannot longer systematically delay its responsibility on attenuating its impact on the environment, specially if energy security cannot be jeopardized. This review tries to make a call to consider new aspects of engineering that can serve as a key factor on avoiding an undesirable future scenario for the humanity.

References

[1] J. W. Tester, *Sustainable Energy: Choosing among Options*. Cambridge, MA.: MIT Press, 2005.

[2] D. F. Savage, J. Way, , and P. A. Silver, "Defossiling fuel: How synthetic biology can transform biofuel production," *ACS Chemical Biology*, vol. 3, no. 1, pp. 13–16, 2008.

[3] T. Mata, A. A. Martins, and N. S. Caetano, "Microalgae for biodiesel production and other applications: A review," *Renewable and Sustainable Energy Reviews*, vol. 14, pp. 217–232, 2010.

[4] "Greenhouse gas emission trends and projections in europe 2004: progress by the eu and its member states towards achieving their kyoto protocol targets," European Environmental Agency (EEA), Copenhagen, Denmark, Tech. Rep. Report N85, 2004.

[5] "Greenhouse gas emission trends and projections in europe 2007: tracking progress towards kyoto targets." European Environmental Agency (EEA), Copenhagen, Denmark, Tech. Rep. Report N85, 2007.

[6] "World energy outlook 2007. china and india insights," International Energy Agency (IEA), Paris, France, Tech. Rep., 2007.

[7] W. Ormerod, P. Freund, A. Smith, and J. Davison, "Ocean storage of co2. iea greenhouse gas r&d programme," International Energy Agency, United Kingdom, Tech. Rep., 2002.

[8] J. Dewulf and H. Van Langenhove, *Renewables-based technology: sustainability assessment*. John Wiley & Sons Ltd., 2006.

[9] S. J. and S. Gu, "Commercialization potential of microalgae for biofuels production," *Renewable Sustainable Energy Reviews*, vol. 14, pp. 2569–2610, 2010.

[10] H. M. Amaro, A. C. Guedes, and F. X. Malcata, "Advances and perspectives in using microalgae to produce biodiesel," *Applied energy*, vol. 88, pp. 3402–3410, 2011.

[11] Y. Li, M. Horsman, N. Wu, C. Q. Lan, and N. Dubois-Calero, "Biofuels from microalgae," *Biotechnol. Prog.*, vol. 24, pp. 815–820, 2008.

[12] K. Bozbas, "Biodiesel as an alternative motor fuel: production and policies in the european union," *Renewable and Sustainable Energy Reviews*, vol. 12, pp. 542–552, 2008.

[13] "The gallagher review of the indirect effects of biofuels production," Renewable Fuel Agency (RFA), Tech. Rep., 2008.

[14] Y. Chisti, "Biodiesel from microalgae," *Biotechnol. Adv.*, vol. 25, pp. 294–306, 2007.

[15] M. Canakci and H. Sanli, "Biodiesel production from various feedstocks and their effects on the fuel properties," *Journal of Industrial Microbiology and Biotechnology*, vol. 35, pp. 431–441, 2008.

[16] M. Borowitzka and N. Moheimani, Eds., *Algae for biofuels and energy.* Springer, 2013.

[17] L. Brennan and P. Owende, "Biofuels from microalgae - a review of technologies for production, processing, and extractions of biofuels and co-products," *Renewable and sustainable energy reviews*, vol. 14, pp. 557–577, 2010.

[18] C. Ugwu, H. Aoyagi, and H. Uchiyama, "Photobioreactors for mass cultivation of algae," *Biore*, vol. 99, pp. 4021–4028, 2008.

[19] J. Sheehan, T. Dunahay, J. Benemann, and P. Roessler, "A look back at the us department of energys aquatic species program: biodiesel from algae," USA: National Renewable Energy Laboratory, Tech. Rep. NREL/TP-580-24190, 1998.

[20] Q. Hu, . Sommerfeld, M, E. Jarvis, M. Ghirardi, M. Posewitz, M. Seibert, and et al., "Microalgal triacylglycerols as feedstocks for biofuel production: perspectives and advances," *Plant Journal*, vol. 54, pp. 621–639, 2008.

[21] W. H. Thomas, T. G. Tornabene, and J. Weissman, "Screening for lipid yielding microalgae: activities for 1983," SERI/STR-231-2207, Tech. Rep., 1984.

[22] C. Yoo, S. Y. Jun, J. Y. Lee, C. Y. Ahn, and H. M. Oh, "Selection of microalgae for lipid production under high levels of carbon dioxide," *Bioresource Technology*, vol. 101, no. S71-4, 2010.

[23] E. Molina, F. Acien, F. Garcia, and Y. Chisti, "Photobioreactors: light regime, mass transfer, and scaleup," *Journal of Biotechnology*, vol. 70, pp. 231–247, 1999.

[24] Y. Chisti, "Biodiesel from microalgae beats bioethanol," *Trends in Biotechnology*, vol. 26, no. 3, pp. 126–131, 2008.

[25] M. A. Borowitzka, "Commercial production of microalgae: ponds, tanks, tubes and fermenters," *Journal of Biotechnology*, vol. 70, no. 1-3, pp. 313–321, 1999.

[26] M. Balat and H. Balat, "Progress in biodiesel processing," *Applied Energy*, vol. 87, no. 6, pp. 1815–1835, 2010.

[27] N. Eriksen, "The technology of microalgal culturing," *Biotechnology Letters*, vol. 30, no. 9, pp. 1525–1536, 2008.

[28] L. Brennan and P. Owende, "Biofuels from microalgaea review of technologies for production, processing, and extractions of biofuels and co-products," *Renewable and Sustainable Energy Reviews*, vol. 14, no. 2, pp. 557–577, 2010.

[29] F. Acien, J. Fernendez, and E. Molina, *Microalgal Biotechnology: Potential and Production*, Gruyter. Alemania, 2012, ch. Principles of photobioreactor design.

[30] A. Singh, P. S. Nigam, and J. D. Murphy, "Mechanism and challenges in commercialisation of algal biofuels," *Bioresource Technology*, vol. 102, no. 1, pp. 26–34, 2011.

[31] H. L. Smith and P. Waltman, *The theory of the chemostat. Dynamics of microbial competition.* Cambridge Studies in Mathematical Biology, Cambridge University Press, 1995.

[32] M. Khaksar Toroghi, G. Goffaux, and M. Perrier, "Output feedback passivity-based controller for microalgae cultivation," in *Preprints of the 12th IFAC Symposium on Computer Applications in Biotechnology, The International Federation of Automatic Control*, Mumbai, India, 2013.

[33] R. Filali, S. Tebbani, D. Dumur, A. Isambert, P. D., and F. Lopes, "Identification of the growth model parameters for a culture of chlorella vulgaris in a photobioreactor," in *Proceedings of 11th Computer Applications In Biotechnology*, Leuven, Belgium, July 2010.

[34] W. Su, J. Li, and N. S. Xu, "State and parameter estimation of microalgal photobioreactor cultures based on local irradiance measurement," *Journal of Biotechnology*, vol. 105, pp. 165–178, 2003.

[35] E. Rocha-Cotzal and A. Vande, "State and input estimation in phytoplanktonic cultures using quasi-unknown input observers," *Chemical Engineering Journal*, vol. 175, pp. 39–48, 2011.

[36] H. T. Hsueh, W. J. Li, H. H. Chen, and H. Chu, "Carbon bio-fixation by photosynthesis of thermosynechococcus sp. cl-1 and nannochloropsis oculta," *Journal of photochemistry and photobiology. B, Biology*, vol. 95, no. 1, pp. 33–39, 2009.

[37] M. G. de Morais and J. A. V. Costa, "Biofixation of carbon dioxide by spirulina sp. and scenedesmus obliquus cultivated in a three-stage serial tubular photobioreactor," *Journal of biotechnology*, vol. 129, no. 3, pp. 439–445, 2007.

[38] D. Tang, W. Han, P. Li, X. Miao, and J. Zhong, "Co2 biofixation and fatty acid composition of scenedesmus obliquus and chlorella pyrenoidosa in response to different co2 levels," *Bioresource technology*, vol. 102, no. 3, pp. 3071–3076, 2011.

[39] S. Aslan and I. Kapdan, "Batch kinetics of nitrogen and phosphorus removal from synthetic wastewater by algae," *Ecological Engineering*, vol. 28, no. 1, pp. 64–70, 2006.

[40] S. F. Baldia, T. Nishijima, Y. Hata, and K. Fukami, "Growth characteristics of a blue-green alga spirulina platensis for nitrogen utilization," *Nippon Suisan Gakkaishi*, vol. 57, no. 4, pp. 645–654, 1991.

[41] N. F. Tam and Y. Wong, "Effect of ammonia concentration s on growth of chlorella vulgaris and nitrogen removal from media," *Bioresource Technology*, vol. 57, pp. 45–50, 1996.

[42] U. Sommer, "A comparison of the droop and the monod models of nutrient limited growth applied to natural populations of phytoplankton," *British Ecology Society*, vol. 5, no. 4, pp. 535–544, 2011.

[43] W. Sunda, K. Shertzer, and D. Hardison, "Ammonium uptake and growth models in marine diatoms: Monod and droop revisited," *Marine Ecology Progress Series*, vol. 386, pp. 29–41, 2009.

[44] K. J. Flynn, "How critical is the critical n/ p ratio?" *Phycology*, vol. 38, no. 5, pp. 961–970, 2002.

[45] B. Yao, B. Xi, C. Hu, S. Huo, J. Su, and H. Liu, "A model and experimental study of phosphate uptake kinetics in algae: Considering surface adsorption and p-stress." *Journal of Environmental Sciences*, vol. 23, no. 2, pp. 189–198, 2011.

[46] S. R. Chae, E. J. Hwang, and H. S. Shin, "Single cell protein production of euglena gracilis and carbon dioxide fixation in an innovative photo-bioreactor," *Bioresource technology*, vol. 97, no. 2, pp. 322–329, 2006.

[47] E. Martinez, F. Camacho, J. M. Jimenez, and J. B. Espinola, "Influence of light intensity on the kinetic and yield parameters of chlorella pyrenoidosa mixotrophic growth," *Process Biochemistry*, vol. 32, no. 2, pp. 93–98, 1997.

[48] D. Sasi, P. Mitra, A. Vigueras, and G. Hill, "Growth kinetics and lipid production using chlorella vulgaris in a circulating loop photobioreactor," *Journal of Chemical Technology & Biotechnology*, vol. 86, no. 6, pp. 875–880, 2011.

[49] J. C. Ogbonna, H. Yada, and H. Tanaka, "Kinetic study on light-limited batch cultivation of photosynthetic cells," *Fermentation and Bioengineering*, vol. 80, no. 3, pp. 259–264, 1995.

[50] A. Yang, "Modeling and evaluation of co 2 supply and utilization in algal ponds," *American Chemical Society*, vol. 50, no. 19, pp. 11 181–11 192, 2011.

[51] X. W. Zhang, X. D. Gong, and F. Chen, "Kinetic models for astaxanthin production by high cell density mixotrophic culture of the microalga haematococcus pluvialis," *Journal of Industrial Microbiology & Biotechnology*, vol. 23, pp. 691–696, 1999.

[52] B. Magbanua, Y. Lu, and G. Leslie, "A technique for obtaining representative biokinetic parameter values from replicate sets of parameter estimates," *Water Research*, vol. 32, no. 3, pp. 849–855, 1998.

[53] L. Xin, H. Hu, G. Ke, and Y. Sun, "Effects of different nitrogen and phosphorus concentrations on the growth, nutrient uptake, and lipid accumulation of a freshwater microalga scenedesmus sp." *Bioresource technology*, vol. 101, no. 14, pp. 5494–5500, 2010.

[54] R. W. Sterner and J. P. Grover, "Algal growth in warm temperature reservoirs: Kinetic examination of nitrogen, temperature, light and other nutrients," *Science*, vol. 32, no. 12, 1998.

[55] A. W. Mayo, "Effects kinetic of growth cultures temperature of unialga containing ph chlorella bacteria and on the vulgaris," *Water environment federation*, vol. 69, no. 1, pp. 64–72, 1997.

[56] J. R. E. Banga, B.-C. C. G., A. Moles, and A. Alonso, "Dynamic optimization of biorreactors: a review," *Proceedings of the Indian Acaddemy of Sciences*, vol. 13, pp. 1–21, 2003.

[57] M. K. Toroghi, G. Goffaux, and M. Perrier, "Observer-based backstepping controller for microalgae cultivation," *Industrial and Engineering Chemistry Research*, vol. 52, pp. 7482–7491, 2013.

[58] R. A. Khandelwal, B. G. Olivier, F. M. Rling, B. Teusink, and F. J. Bruggeman, "Community flux balance analysis for microbial consortia at balanced growth," *Plos one*, vol. 10, p. DOI: 10.1371/journal.pone.0064567, 2013.

[59] W. F. M. Roling, M. Ferrer, and P. N. Golyshin, "Systems approaches to microbial communities and their functioning," *Current Opinion in Biotechnology*, vol. 21, pp. 532–538, 2010.

[60] G. Becerra-Celis, G. Hafidi, S. Tebbani, and D. Dumur, "Nonlinear predictive control for continuous microalgae cultivation process in a photobioreactor," in *10th International Conference on Control, Automation, Robotics and Vision*, 2008.

[61] N. I. Marcos, M. Guay, and D. Dochain, "Output feedback adaptive extremum seeking control of a continuous stirred tank bioreactor with monods kinetics," *Journal of Process Control*, vol. 14, pp. 807–818, 2004.

[62] P. Cougnon, D. Dochain, M. Guay, and M. Perrier, "On-line optimization of fedbatch bioreactors by adaptive extremum seeking control," *Journal of Process Control*, vol. 21, pp. 1526–1532, 2011.

[63] L. Mailleret, O. Bernard, and J. P. Steyer, "Nonlinear adaptive control for bioreactors with unknown kinetics," *Automatica*, vol. 40, pp. 1379–1385, 2004.

[64] D. Selisteanu, E. Petre, and V. Rasvan, "Sliding mode and adaptive sliding-mode control of a class of nonlinear bioprocesses," *International Journal of Adaptive Control and Signal Processing*, vol. 21, pp. 795–822, 2007.

[65] J. Gonzalez, G. Fernandez, R. Aguilar, M. Barron, and J. Alvarez-Ramirez, "Sliding mode observer-based control for a class of bioreactors," *Chemical Engineering Journal*, vol. 83, pp. 25–32, 2001.

[66] J. Moreno, J. Alvarez-Ramirez, E. Rocha-Cotzal, and J. Diaz-Salgado, "Super-twisting observer-based output feedback control of a class of continuous exothermic chemical reactors," in *Proceedings of the 9th International Symposium on Dynamics and Control of Process Systems (DYCOPS 2010)*, 2010.

[67] A. Vargas, J. Moreno, and I. Mendoza, "Time-optimal output feedback controller for toxic wastewater treatment in a fed-batch bioreactor," in *Preprints of the 18th IFAC World Congress*, 2011.

[68] G. Becerra-Celis, C. Joannis-Cassan, A. Isambert, and H. Siguerdidjane, "Control strategy for continuous microalgae cultivation process in a photobioreactor," in *IEEE International Conference on Control Applications, CCA.*, 2008.

[69] D. Dochain and M. Perrier, "Adaptive backstepping nonlinear control of bioprocesses," in *7th Advanced control of chemical processes and International Symposium, Advanced control of chemical processes*, 2004.

[70] M. Berenguel, F. Rodriguez, F. Acien, and J. Garcia, "Model predictive control of ph in tubular photobioreactors," *Journal of Process Control*, vol. 14, pp. 377–387, 2004.

[71] S. Celikovsky, S. Papacek, A. C. Herrera, and J. R. Leon, "Singular perturbation based solution to optimal microalgal growth problem and its infinite time horizon analysis," in *47th IEEE Conference on Decision and Control*, 2008.

[72] A. Isidori, *Nonlinear control systems*, ser. Communications and control engineering series. Springer-Verlag, 1995, vol. 1.

[73] J. Abdollahi and S. Dubljevic, "Lipid production optimization and optimal control of heterotrophic microalgae fed-batch bioreactor," *Chemical Engineering Science*, vol. 84, pp. 619–627, 2012.

[74] J. P. Gauthier, H. Hammouri, and S. Othman, "A simple observer for nonlinear systems applications to bioreactors," *IEEE Transactions on Automatic Control*, vol. 37, no. 6, pp. 875–880, 1992.

[75] F. Lafont, E. Busvelle, and J. P. Gauthier, "An adaptive high-gain observer for wastewater treatment systems," *Journal of Process Control*, vol. 21, no. 6, pp. 893–900, 2011.

[76] F. Muoz-Palacios and C. Ben-Youssef, "Biomass and phenol estimation using dissolved oxygen measurement," *Proceedings of the Electronics, Robotics and Automotive Mechanics*, vol. 2, pp. 206–214, 2006.

[77] D. Theilliol, J. C. Ponsart, J. Harmand, C. Join, and P. Gras, "On-line estimation of unmeasured inputs for anaerobic wastewater treatment processes," *Control Engineering Practice*, vol. 11, no. 9, pp. 1007–1019, 2003.

[78] T. Zhang and M. Guay, "Adaptive nonlinear observers of microbial growth processes," *Journal of Process Control*, vol. 12, no. 5, pp. 633–643, 2002.

[79] D. Dochain, "State and parameter estimation in chemical and biochemical processes: a tutorial." *Journal of Process Control*, vol. 13, pp. 801–818, 2003.

[80] O. Bernard, A. Sciandra, and O. G. Sallet, "A non-linear software sensor to monitor the internal nitrogen quota of phytoplanktonic cells," *Oceanography Acta*, vol. 24, pp. 435–442, 2001.

[81] G. Goffaux, A. Vande-Wouwer, and O. Bernard, "Continuous - discrete interval observers for monitoring microalgae cultures," *Biotechnology Progress*, vol. 3, no. 3, pp. 667–675, 2009.

[82] ——, "Improving continuous discrete interval observers with application to microalgaebased bioprocesses," *Journal of Process Control*, vol. 19, no. 7, pp. 1182–1190, 2009.

[83] H. Khodadadi and H. Jazayeri-Rad, "Applying a dual extended kalman filter for the nonlinear state and parameter estimations of a continuous stirred tank reactor," *Computers and Chemical Engineering*, vol. 35, pp. 2426–2436, 2011.

[84] J. Li, N. S. Xu, and W. W. Su, "Online estimation of stirred-tank microalgal photobioreactor cultures based on dissolved oxygen measurement," *Biochemical Engineering Journal*, vol. 14, pp. 51–65, 2003.

[85] N. Martnez-Fonseca, I. Chairez, and A. Poznyak, "Uniform step by step pbserver for aerobic bioreactor based on super-twisting algorithm," *Bioprocess and Biosystems Engineering*, 2014.

In: Microalgae and Other Phototrophic Bacteria
Editor: Luis G. Torres Bustillos
ISBN: 978-1-63482-078-3

Chapter 3

THE BIOFUEL POTENTIAL OF THE GREEN COLONIAL MICROALGA *BOTRYOCOCCUS BRAUNII*

Ivette Cornejo-Corona[1*], Hem R. Thapa[2†], Timothy P. Devarenne[2‡] and Edmundo Lozoya-Gloria[1#].
[1]Departamento de Ingeniería Genética CINVESTAV-IPN Irapuato, Gto., México
[2]Department of Biochemistry and Biophysics, Texas A&M University, College Station, Texas, US

ABSTRACT

Ever since the 1970's algae have been proposed as a source of oils for transportation fuel production, and this interest has been renewed in recent years. Algae fix CO_2 very efficiently and are capable of producing large amounts of oils that can be used for combustion engine fuel production. The freshwater green alga *Botryococcus braunii* produces long-chain liquid hydrocarbon oils in large quantities that can be converted into petroleum-equivalent fuels suitable for combustion engines. Additionally, *B. braunii* hydrocarbon oils have been found as major constituents of currently used petroleum and coal deposits. Recently, genes and enzymes for the biosynthesis of *B. braunii* oils known as botryococcenes have been identified, and many studies have assayed different culture conditions for effectively growing *B. braunii*. Business evaluations of a *B. braunii* oil production system suggest that it may become competitive in the fuel market by the mid-twenty-first century. This chapter summarizes current knowledge on the biosynthesis, biochemistry and molecular biology, as well as cost analysis of hydrocarbon production in *B. braunii*.

[*] Email: icornejo@ira.cinvestav.mx.
[†] E-mail: hrt2369@tamu.edu.
[‡] E-mail: tpd8@tamu.edu.
[#] Email: elozoya@ira.cinvestav.mx.

INTRODUCTION

In the past several decades there has been growing concern over buildup of greenhouse gases such as carbon dioxide (CO_2) in the Earth's atmosphere, which causes the greenhouse effect leading to increased global temperatures, altered weather patterns, glacier melting, and rising sea level. There is a consensus that these greenhouse gases are from man-made actions, principally the burning of coal for electricity generation and from transportation fuels derived from petroleum. Both petroleum and coal are carbon-rich fossil fuels that have been stored underground for hundreds of millions of years. Thus, burning these fuels increases atmospheric CO_2, leading to the greenhouse effect. Continued use of fossil fuel sources will only exacerbate the problem, and atmospheric CO_2 levels must be reduced or at least leveled off to curb global climate change.

One solution to this problem is to use fuel and energy sources that are at least carbon neutral; implying that the same amount of carbon released during burning of a fuel is used to make new fuel. Thus, atmospheric CO_2 levels remain constant. Carbon neutral sources include fuels derived from photosynthetic organisms. Since these organisms fix atmospheric CO_2 during photosynthesis the use of fuels, or biofuels, derived from photosynthetic material creates a carbon cycle of fixation, conversion to biofuel, and release back to the atmosphere when burning fuel.

There are several different types of biofuels that can be produced from photosynthetic organisms. First generation biofuels are currently in use and include ethanol produced through yeast fermentation of photosynthetic sugars, and biodiesel in the form of fatty acid methyl esters derived from triacylglycerols (TAGs) produced by oil crops such as soybean and oil palm. Second generation biofuels include ethanol produced from plant cell wall cellulose, a glucose polymer, which is currently available but requires further development for full economic feasibility. All of these fuels have associated problems such as high-energy inputs for biofuel generation, low CO_2 emission savings, inability to produce enough fuel to meet demand, or directly competing with food sources. It is expected that, most of these issues will be solved through the use of third generation biofuels: microalgae-derived biodiesel or petroleum-equivalent fuels.

Microalgae are considered to be a promising source of biofuels that can help reduce the problem of CO_2 emissions because of their photosynthetic efficiency and oil productivity [1, 2]. Algae are capable of attaining oil contents up to 80% of their dry weight and usually range between 20% and 50% [3]. It is important to note that most oil producing microalgae accumulate TAG oils used for biodiesel production. Other positive aspects of algae biofuels are that algae can attain large amounts of biomass, have a very quick cellular doubling time, can be grown on non-agricultural land, do not compete with food crops, and calculations suggest an equivalent amount of land equaling only 3% of total US agricultural land is needed to produce enough algal based fuels to meet 50% of the annual US transportation fuel needs [1, 3]. However, there are many limitations to the use of microalgae that currently place the price of algal biofuel too high to be economically competitive. But, these roadblocks will be overcome with continued research directed at resolving these issues [4].

One oleaginous alga the green colonial microalga *Botryococcus braunii*, has received significant attention for many years as it can produce up to 86% of its dry weight as oils. These oils are not TAGs, but rather liquid hydrocarbons, the majority of which are found in the colony extracellular matrix [5, 6]. The presence of large amounts of *B. braunii* hydrocarbons and derived molecules in petroleum and oil shale deposits indicates the direct contribution of this microalga to the formation of existing fossil oil deposits [7, 8]. Furthermore, caustic hydrolysis of liquid hydrocarbons extracted from *B. braunii* results in fuels that are highly compatible with the existing petroleum infrastructure [9]. In this chapter we will give a synopsis of the characteristics of *B. braunii*, its hydrocarbons, and its potential as a biofuel source.

CHARACTERISTICS OF *B. BRAUNII*

The Trebouxiophyte green alga *B. braunii* was first identified in 1849 [10]. It is widely distributed across Earth and is found in fresh and brackish water, reservoirs, ponds and seasonal lakes in many climatic zones including temperate, alpine and tropical [5]. A lack of phylogenetic relationships between strains and locality suggests that *B. braunii* spreads between bodies of water via air currents or animals such as birds [11]. Based on the hydrocarbons present in this alga, there are three different races of *B. braunii*. Race A produces fatty acid derived alkadienes and alkatrienes, race B produces the triterpenoid hydrocarbons botryococcene, squalene and their methylated derivatives, and race L produces the tetraterpenoid hydrocarbon known as lycopadiene (Figure 1) [5].

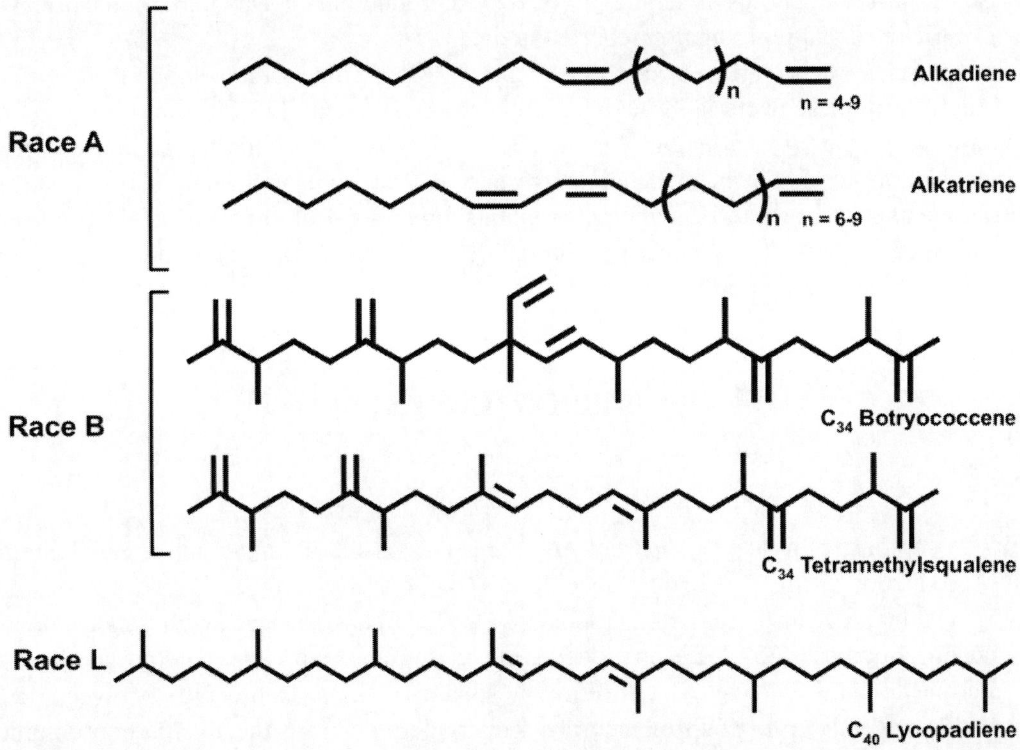

Figure 1. Hydrocarbons produced by three chemical races of *B. braunii*.

The *B. braunii* colony is an amorphous group of individual pyriform-shaped cells embedded within an extracellular matrix (ECM) [12]. The cells of race L are smaller (8 to 9 μm × 5 μm) compared to race A and race B (13 μm × 8 to 9 μm) [13, 14]. Cells are protected by a -1, 4 and/or -1, 3-glucan-containing cell wall and sometimes partially enclosed by a lipophilic cuticle [15, 16]. Colonies of cells are held together by the ECM, which is made up of cross-linked hydrocarbon polymers specific to each race. The hydrocarbons are produced inside the cell, are found in numerous intracellular oil bodies, and are excreted into the ECM, which is permeated by the hydrocarbons [17-19]. Externally, a polysaccharide fiber sheath mainly composed of arabinose-galactose units wraps *B. braunii* colonies, and between the ECM and polysaccharide sheath is a retaining wall that keeps the liquid hydrocarbons within the ECM [15, 16].

Many ultrastructural studies of *B. braunii* race B have been conducted to analyze cell and colony structure. While these studies did not provide evidence to suggest how the hydrocarbons may be secreted from cells, a recent study did show intracellular oil bodies associated with the endoplasmic reticulum (ER) and chloroplast, a phenomenon also seen in *Chlamydomonas reinhardtii* [16]. Each cell contains a single large "U-shaped" chloroplast enveloping cytoplasmic organelles, except the ER, including a centrally located nucleus [20]. The ER of *B. braunii* is found as a continuous layer just below the cell membrane and contains many fenestrations along its surface and the retaining wall contacts the cell wall at the vertex of each cell [16, 21]. Presumably the retaining wall and hydrocarbons guard the *B. braunii* cells from various stresses and the hydrocarbons allow the colonies to float for increased exposure to photosynthetic light [6, 16]. Reproduction appears to be autosporic with no zoosporic or sexual reproduction reported to date [22].

The presence of bacteria associated with the ECM of some strains suggests *B. braunii* may benefit from these bacteria through a mutualistic relationship [23]. It is known that some algae are incapable of producing vitamin B_{12} and thus obtain this vital nutrient from associated bacteria. However, evidence suggests that *B. braunii* is a vitamin B_{12} autotroph [24] and the reason for bacterial symbionts remains unknown. But, inclusion of these bacteria in culture media may help to increase oil production and growth rates and improve the value of *B. braunii* as a source of biofuel.

HYDROCARBON BIOSYNTHESIS

Race B

Hydrocarbons from race B in particular have been considered an attractive candidate for biofuel production as this race can produce up to 86% of its dry weight as liquid hydrocarbon oils [5]. The majority of these hydrocarbons are C_{30}-C_{37} botryococcenes with minor amounts of tetramethylsqualene [25, 26]. Botryococcenes are isoprenoid in nature and are products of the non-mevalonate or methylerythritol phosphate (MEP) isoprenoid pathway [27]. Isoprenoids are a diverse group of natural products derived from the five-carbon common precursors isopentenyl diphosphate (IPP) and dimethylallyl diphosphate (DMAPP). Depending on the organism, IPP and DMAPP biosynthesis occurs via two different pathways, the mevalonate (MVA) pathway and the MEP pathway [28-31]. Green algae such as *B.*

braunii are known to utilize only the MEP pathway for isoprenoid biosynthesis [27, 28, 32, 33].

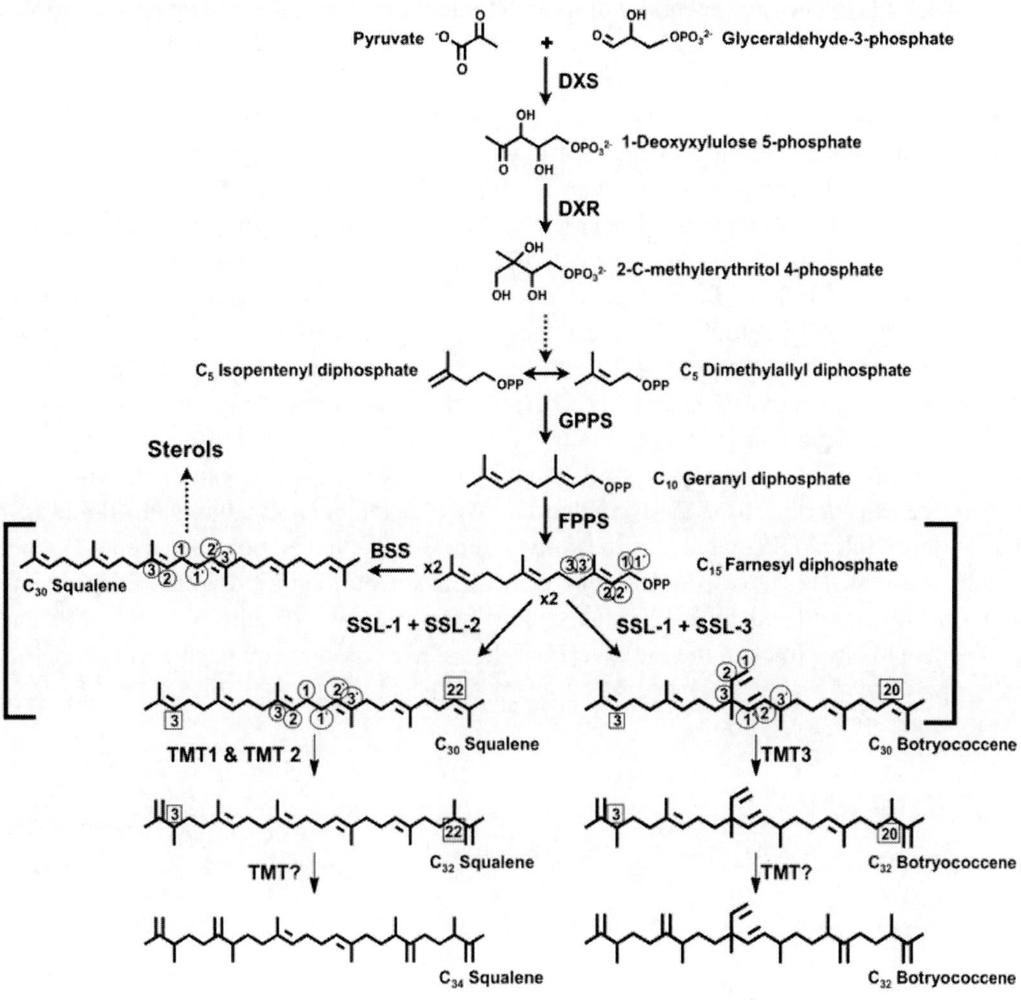

Figure 2. Biosynthesis of methylated triterpenes in race B of *B. braunii*. Enzyme name abbreviations are: (DXS), 1-deoxy-xylulose-5-phosphate synthase; (DXR), 1-deoxy-xylulose-5-phosphate reductoisomerase; (GPPS), geranyl diphosphate synthase; (FPPS), farnesyl diphosphate synthase; (BSS), *Botryococcus* squalene synthase; (SSL), squalene synthase-like; (TMT), triterpene methyltransferases. Carbon position numbers in circles indicate carbon numbering for FPP and the linkage of the two FPP molecules in squalene and botryococcene. Carbon position numbers in squares indicate methylation positions for squalene and botryococcene. Dashed arrows indicate multiple reactions. Brackets indicate reaction detail highlighted in Figure 3. From experimental work described in references 35, 36, and 40.

The biosynthesis of botryococcenes through the MEP pathway in race B is shown in Figure 2. The first step involves the condensation of pyruvate and glyceraldyhyde-3-phosphate to 1-deoxy-xylulose-5-phosphate (DOXP) by DOXP synthase (DXS). The DXS reaction is irreversible and is considered the committed step for carbon flux into the MEP pathway [32]. Race B expresses three DXS paralogs simultaneously coinciding with botryococcene production, suggesting their critical role in the production of large amounts of triterpenes [34]. DOXP is then reduced by DXP reductoisomerase (DXR) to 2-C-methylerythritol 4-phosphate (MEP), which after several enzymatic steps is converted to IPP and DMAPP [31]. The next steps involve the condensation of IPP and DMAPP to form C_{10} geranyl diphosphate (GPP) and addition of another IPP to form C_{15} farnesyl diphosphate (FPP) [33]. This process of adding IPP can continue to make diphosphate intermediates of hundreds of carbons long.

The production of botryococcenes starts with the use of FPP and follows an enzymatic process similar to the production of squalene (Figure 2). C_{30} botryococcene and C_{30} squalene are similar in structure as both are produced by condensation of two FPP molecules, but differ in how the farnesyl molecules are connected [27, 35]. For squalene production, the enzyme squalene synthase (BSS for *Botryococcus* squalene synthase) carries out a two-step reaction. In the first step two FPP molecules are condensed to produce the cyclopropyl intermediate presqualene diphosphate (PSPP; Figure 3). The second step involves reductive rearrangement of PSPP in the presence of NADPH to form a C1–C1' bond between the two FPP molecules yielding C_{30} squalene (Figure 3). In contrast, C_{30} botryococcene has a C3–C1' connection between the two FPP molecules with the reaction catalyzed by two squalene synthase-like (SSL) enzymes [26, 35, 36]. SSL-1 condenses two FPP molecules to yield PSPP and SSL-3 in the presence of NADPH catalyzes the conversion of PSPP to C_{30} botryococcene (Figure 3). An additional mechanism of squalene biosynthesis has been identified where PSPP from the SSL-1 reaction is utilized by SSL-2 to produce squalene [36] (Figure 3). This squalene is hypothesized to have roles in the biogenesis of the ECM and other squalene derivatives [36].

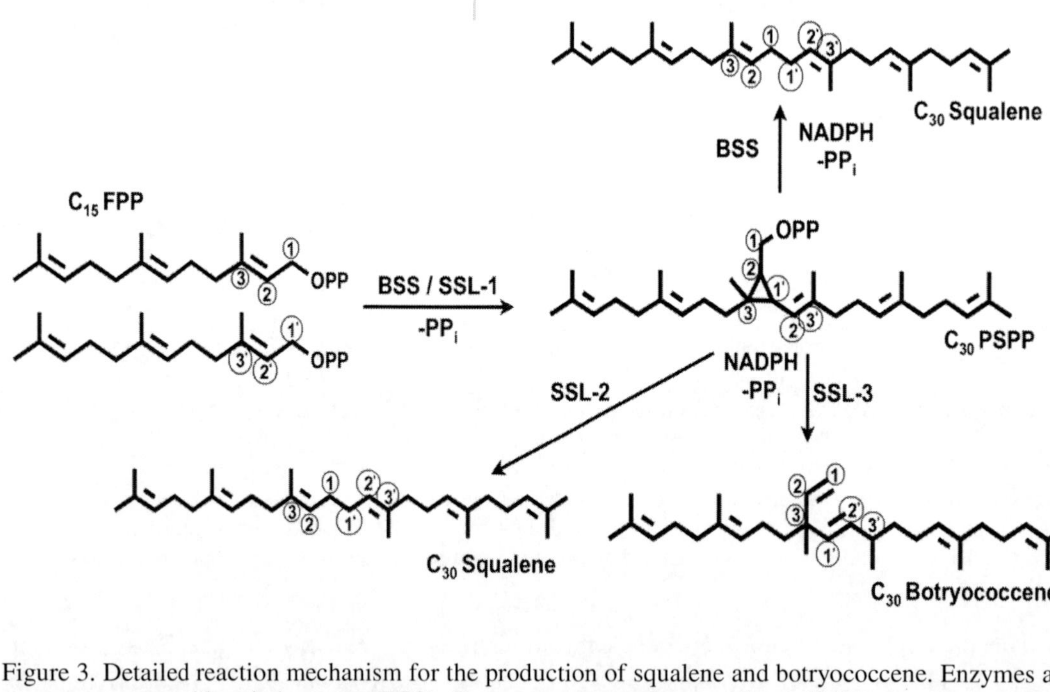

Figure 3. Detailed reaction mechanism for the production of squalene and botryococcene. Enzymes and precursors name abbreviations are: (BSS), *Botryococcus* squalene synthase; (SSL), squalene synthase-like. (PSPP), presqualene diphosphate. Carbon position numbers in circles indicate carbon numbering for FPP and the linkage of the two FPP molecules in squalene and botryococcene. From experimental work described in references 35 and 36.

Depending on strains and culture conditions, C_{30} botryococcene is methylated to produce C_{31}-C_{37} botryococcenes (Figure 2) with C_{34} botryococcene as the predominant hydrocarbon produced by race B [37, 38]. Similarly, C_{30} squalene is also methylated to C_{31}-C_{34} squalene with C_{34} tetramethylsqualene as the predominant methylated squalene [39]. Three methyltransferases S-adenosyl methionine (SAM) dependent related to methylated botryococcene and squalene synthesis, have been identified in race B and are designated as triterpene methyltransferases (TMTs). TMT-1 and TMT-2 catalyze the methylation of squalene at carbons 3 and 22 to produce C_{32} squalene, whereas TMT-3 is involved in two successive methylations of C_{30} botryococcene at positions 3 and 20 to produce C_{32} botryococcene (Figure 2) [40]. Enzymes catalyzing further methylation of C_{32} botryococcene and C_{32} squalene have not been identified (Figure 2).

In C_{34} botryococcene-producing strains of race B such as the Berkeley (Showa) strain, the majority (~90-95%) of C_{34} botryococcene exists in the ECM, while the lower carbon number botryococcenes are found in numerous intracellular oil bodies [37, 41]. The intracellular botryococcenes are excreted to the extracellular matrix as they become methylated to C_{34} botryococcene. However, methylation reactions can take place both inside cells and in the ECM [37]. Radiolabeling studies with [^{14}C]-methionine suggest that low molecular weight botryococcenes are more rapidly methylated inside cells compared to outside [37]. In order to fill in gaps in the understanding of botryococcene biosynthesis and identify additional genes involved, a high quality transcriptome has been produced [42] and genome sequencing is being carried out at the Joint Genome Institute, both using the Berkeley strain of race B.

Race L

The biosynthetic pathway leading to the tetraterpenoid lycopadiene in the L race of *B. braunii* remains virtually unknown. Lycopadiene is the predominant hydrocarbon in the L race [43], however three minor isomers of lycopadiene and small amounts of lycopatriene have been reported in two Indian strains of the L race [44, 45]. Thus, two possible biosynthetic routes have been proposed; condensation of two phytyldiphosphate molecules or the reductive conversion of lycopersene to lycopadiene [43]. The latter route could account for the presence of lycopatriene as a reductive intermediate. Alternatively, lycopadiene biosynthesis may occur via reduction of a carotenoid like phytoene or lycopene and could also account for a lycopatriene intermediate [45]. The L race accumulates the lowest amount of hydrocarbons with lycopadiene levels in the range of 5-10% of dry weight.

Race A

Unlike race B and L, hydrocarbons from race A are non-isoprenoid in origin and are derived from fatty acids. These hydrocarbons are odd numbered and primarily constitute C_{23}-C_{33} alkadienes and alkatrienes, but can have one to four double bonds including a terminal double bond [46]. Feeding experiments with radiolabeled fatty acids showed oleic acid as the direct precursor for alkadienes/trienes, and their biosynthesis is suggested to occur via elongation of oleic acid to a very long chain fatty acid (VLCFA) followed by decarboxylation [47-49]. However, the detection of fatty acyl-CoA reductase and decarbonylase enzyme activities from race A microsomes suggests an alternate mechanism where VLCFAs of appropriate length are first reduced to an aldehyde and then decarbonylated to produce alkadienes/trienes [50-52]. Total hydrocarbon content in race A can vary widely between strains with a Bolivian and a French strain accumulating hydrocarbons at 0.4% and 61% of dry weight, respectively [5]. Despite similar culture conditions, hydrocarbon makeup differs between strains of race A. For instance, the Yamanaka strain synthesizes only alkadienes, whereas the Bolivian strain produces alkamonoenes and alkatetraenes plus C_{26} and C_{28} even-numbered hydrocarbons [46].

CULTURING OF *B. BRAUNII*

Like other microalgae, *B. braunii* requires CO_2, light, inorganic elements and water for growth. Modified Chu 13 media has been developed and is found to be favorable for standard growth for all *B. braunii* races [5]. In order to exploit *B. braunii* as a potential renewable biofuel source, many studies have been conducted in the last two decades to determine optimal culture conditions for maximum biomass and hydrocarbon production [5]. Studies on nitrogen metabolism in *B. braunii* using NO_3^-, NO_2^- and NH_3 showed that NO_3^- is the primary source of nitrogen and hydrocarbon production was optimal when KNO_3 was maintained at 0.2 g/L (5). However, the replacement of NO_3^- with NO_2^- increased the growth of *B. braunii* race A [53]. Use of ammonia inhibited botryococcene biosynthesis as the cells diverted their metabolic flux toward amino acid biosynthesis [54]. Phosphorus is an essential nutrient for the growth of *B. braunii* and is supplied in the form of dibasic potassium phosphate (K_2HPO_4). An increase in phosphate level (K_2HPO_4 at 43.6 mg/L) over that in the modified Chu13 medium (10 mg/L) did not affect total growth or nature and location of hydrocarbons, but showed a slight increase in total hydrocarbon production from 1.48 g/L to 1.82 g/L [55]. Using a mathematical and statistical approach known as response surface methodology, potassium nitrate and ferric citrate showed an important effect on biomass and hydrocarbon yield of race A. Concentrations of 19.5 mg/L KH_2PO_4, 50 mg/L KNO_3, 20 mg/L $MgSO_4$ and 18.5 mg/L $C_6H_5FeO_7$ were optimum for biomass (650 mg/L) and hydrocarbon (50.65%) production. Overall results indicated that an N:P ratio of 1:4 should be used for biomass and 1:0.5 is better for hydrocarbon yield [56].

Air enriched with 1% CO_2 can enhance *B. braunii* growth by doubling algal biomass and achieving a 5-fold increase in hydrocarbon production compared to aeration without CO_2 [57]. More extensive studies have been done to determine the effect of CO_2 on algal growth where cultures were grown in 0-50% CO_2 enrichment [58]. Consistent with previous results, the specific growth rate remained maximum at 0.2-5% CO_2, but decreased over 5% CO_2 and came to halt at 50% CO_2 [58]. An increase in CO_2 concentration decreased media pH from 7.5 (ambient air) to 4.54 (50% CO_2) [58]. Optimal growth was observed when the pH of the medium was in the range of 5.44 to 6.5 [58]. Interestingly, the best yield ever reported for hydrocarbon production from race B (22.5 mg oil/L/photo-h) was obtained with high-density cells (20 g dry weight/L) maintained under continuous cultures for 3 months [59].

Table 1. Yields and culture conditions of several parameters of *B. braunii* race B cultures

Culture condition	BP	CD	HA	HP	RD	TL	CS	VP	RY
Shake flask		0.86	15.00						[13] 2007
				19.00					[80] 2010
Raceway-outdoor	0.14	2.00							[81] 2010
Shake flask	0.28	3.11			0.24	0.61			[82] 2010
Shake flask		1.70	13.00						[83] 2011
Open raceway		1.80	11.00						[83] 2011
Conical Fernbach	0.14		18.00						[72] 2011
Shake flask	0.06								[82] 2011
					0.16				[65] 2011
Indoor				40.00					[83] 2011
Wastewater	0.07	2.92		17.41	0.30				[84] 2011
	0.22								[85] 2011
Panel reactor-outdoor	0.02			2.50	0.09				[86] 2012
Air stirring	0.10								[87] 2012
Shake flask	0.10	2.12							[88] 2012
Shake flask				18.20		39.00			[89] 2012
Raceway-outdoor	0.11	2.00	28.00						[90] 2012
Shake flask			27.00						[91] 2012
Tube reactor	0.05	1.45			0.09				[92] 2012
Shake flask		1.30	27.40	110.00					[63] 2012
Symbiotic consortia		2.61	41.60		0.10		0.03		[93] 2012
Brewery wastewater		8.50		48.00					[94] 2012
Airlift bioreactor	0.12		64.30						[82] 2012
Continuous culture		20.00	22.50			5.00	0.34	22.50	[59] 2014
Airlift bioreactor	0.04			13.10					[95] 2014

(BP), Biomass productivity (g/L /day).
(CD), Cell densities (g DW/L).
(HA), Hydrocarbons accumulation (% DW).
(HP), Hydrocarbon productivity (mg/L/day).
(RD), Rate/day.
(TL), Triterpene oil (g/L).
(CS), Colony size (mm).
(VP), Volumetric productivity (mg oil/L/photo-h).
(RY), Reference year to show progress along time.

Studies have also analyzed effects of the light/dark cycle and light intensity on *B. braunii* growth and oil production. When the combined effect of temperature and photoactive radiation was tested with a 14:10 light:dark cycle, maximum growth rate occurred at 30°C and 850 $\mu E \cdot m^{-2} s^{-1}$ [58]. In another study, 64 different light conditions were tested using a high-throughput microfluidic photobioreactor array to determine the effect of light intensity and cycles on *B. braunii* growth and oil production [60]. For the 16 different light intensities tested using a 12:12 light: dark cycle, maximum growth and oil production was observed at 113 $\mu E \cdot m^{-2} s^{-1}$ [60]. Similarly, when using eight different light/dark cycles at a light intensity of 120 $\mu E \cdot m^{-2} s^{-1}$, a 2:2 light: dark cycle showed 1.8-fold increase in oil production per unit area of the colony when compared to the standard 12:12 light:dark cycle [60].

Micronutrients are essential for optimal algal biomass and oil production. Iron has been shown to be very important for algal growth as it has functions in numerous basic cellular processes such as mitochondrial electron transport and nitrate and nitrite reduction [61]. The requirements for natural blooms of *B. braunii* have been dificult to explain, but may depend on the micronutrient nickel. High concentration of this metal seems to be required for blooms, but laboratory experiments have not been done to confirm this result [62]. Additionally, a recent study showed that a mixture of 0.2 mM iron, 0.70 mM manganese, 0.62 mM molybdenum and 3.38 mM nickel were found to be optimal for biomass production. Interestingly, these concentrations are 40 and 10 times higher for iron and manganese, respectively, and 60 and 2 times lower for molybdenum and nickel, respectively, than the one required for optimal hydrocarbon yield [63]. Table 1 summarizes reported yields and culture conditions of several parameters for *B. braunii* race B cultures from the last seven years.

B. braunii has also been shown to be able to grow mixotrophically or heterotrophically under a variety of conditions and carbon sources [64, 65]. Mixotrophic conditions with glucose produced the highest increase in biomass accumulation. However, under heterotrophic conditions using glucose or mannose as the carbon source and a complete lack of light, *B. braunii* greatly increased the amount of hydrocarbons produced as well as the cell/colony size [65, 66]. Wastewater from sources such as swine lagoons, municipal sewage, breweries, and soybean curd processing has been used to effectively grow *B. braunii* [67-70] suggesting that *B. braunii* could be used to clean up these wastewaters before returning them to the environment.

CONVERSION OF *B. BRAUNII* HYDROCARBONS TO FUELS

As mentioned above, hydrocarbons from all three races of *B. braunii* have been found as constituents of currently used petroleum and coal deposits suggesting that *B. braunii* has been the main contributor to these deposits over geologic time [7, 8, 68-71]. *B. braunii* hydrocarbons are well suited for fuel production using standard petroleum hydrocracking procedures and they do not contain sulfur or nitrogen, which can affect the hydrocracking process and produce air pollutants when combusted [6]. The B race botryococcenes appear to be the most suitable for biofuel production for several reasons. Compared to the A race, botryococcenes can be easily separated from the biomass without rupturing cells, which can cause contamination with pigments and other cellular material [72]. Botryococcenes have a high energy density and a high degree of unsaturation leading to the hydrocracking production of cyclic alkanes/alkenes and aromatics, which are major constituents of petroleum-derived gasoline. The highly hydrophobic nature of botryococcenes affords efficient separation from other oils such as TAGs, which require different methods for conversion to fuels. Most importantly, botryococcenes under standard hydrocracking and distillation practices can be converted into 67% gasoline, 15% aviation fuel, 15% diesel fuel and 3% residual oil [9]. The gasoline from *B. braunii* hydrocarbons has been shown to have an octane rating of ~95 making it appropriate for current combustion engines [73].

COST-BENEFIT OF BIOFUEL FROM *B. BRAUNII*

One of the most important aspects of biofuels is the cost-benefit of their production. It is necessary to consider a cost-benefit analysis in order to have a realistic idea of the economic impact biofuel production will have for the companies and governments involved. Despite several reports on production costs of third generation biofuels, the costs are quite variable depending on the reviewed literature or production preferences. Ranges between $0.88 - $24.60/L and $0.17 - $2.09/L were proposed for the cost of algae biofuel [74, 75]. This variation resulted mainly from production procedures and cost of supplies like algae biomass calculated between $0.35/kg - $7.32/kg [76].

The use of *B. braunii* for biofuel production has some limitations, but also strong potential. Mostly, *B. braunii* is hindered by slow growth rate with cells doubling every 6-7 days, likely due to commitment of large amounts of energy for hydrocarbon production [77]. This growth rate is much slower compared to other oleaginous algae, will increase the cost of producing the large amounts of biomass needed, and limits the feasibility of growing *B. braunii* in open raceway ponds [6]. Also, contamination with other microorganisms can decrease biomass yield and hydrocarbon production. Growth of algal competitors could be overcome by the use of herbicide resistant mutants of *B. braunii* [78, 79]. The economic feasibility of biofuel production using *B. braunii* was studied employing a 19-hectare semi-open pond system. Assuming the balance of operation costs and the total sale of fuel from this alga, the cost of *B. braunii* derived oil was calculated to be $240/barrel, which is projected to be competitive with petroleum costs by the middle of the current century [80]. However, future work on harvesting algae, separation of hydrocarbons, and processing into transportation fuels is required for *B. braunii* to be used as a production host for biofuels.

CONCLUSION

There is no doubt about the urgent necessity to decrease greenhouse gas levels in the atmosphere. The challenge is to do so without losing increases in and development of economic, cultural and social conditions around the world, which requires large energy inputs. Producing this energy in the form of fuels derived from photosynthetic organisms would be an efficient way to begin reducing atmospheric greenhouse gases. After many years of studies, algae are considered one of the best photosynthetic organisms to produce oils that can be converted into fuels. The green alga *Botryococcus braunii* has a unique place because the hydrocarbons produced can be easily converted into petroleum-equivalent fuels. Of course there are some challenging bottlenecks to overcome with *B. braunii*, which with additional research should be overcome in the near future. The main problem for *B. braunii* is the slow cell doubling time, even under the best growing conditions. However, in spite of these problems, business evaluations of a *B. braunii*-based oil and fuel production system suggest that *B. braunii* fuels may become competitive in the fuel market by the mid-twenty-first century [80]. It seems that there is some light at the end of the tunnel and light means energy.

ACKNOWLEDGMENTS

We thank CONACYT for a PhD scholarship to IC-C and a grant from the 2012 Texas A&M University - CONACYT Collaborative Research Grant Program to EL-G and TPD.

REFERENCES

[1] Chisti, Y. (2007). Biodiesel from microalgae. *Biotechnol. Adv., 25*, 294-306.

[2] Haag, A. L. (2007). Algae bloom again. *Nature, 447*, 520-521.

[3] Chisti, Y. (2008). Biodiesel from microalgae beats bioethanol. *Trends Biotech., 26*, 126-131.

[4] Chisti, Y. (2013). Constraints to commercialization of algal fuels. *J. Biotech., 167*, 201-214.

[5] Metzger, P. & Largeau, C. (2005). *Botryococcus braunii*: a rich source for hydrocarbons and related ether lipids. *Appl. Microbiol. Biotechnol., 66,* 486-496.

[6] Banerjee, A., Sharma, R., Chisti, Y. & Banerjee, U.C. (2002). *Botryococcus braunii*: a renewable source of hydrocarbons and other chemicals. *Crit. Rev.Biotechnol., 22,* 245-279.

[7] Moldowan, J. & Seifert, W.K. (1980). First discovery of Botryococcane in petroleum. *J.C.S. Chem. Comm., 19*, 912-914.

[8] Mckirdy, D.M., Volkman, J.K., & Howell, V.J. (1986). Botryococcane in a new class of Australian non-marine crude oils. *Nature, 320*, 57-59.

[9] Hillen, L.W., Pollard, G., Wake, L.V. & White, N. (1982). Hydrocracking of the oils of *Botryococcus braunii* to transport fuels. *Biotechnol. Bioeng., 24*, 193-205.

[10] Kützing, F.T. (1849). Species Algarum. *Lipsiae* (Leipzig) FA. Brockhaus. pp. 1-922.

[11] Kawachi, M., Tanoi, T., Demura, M., Kaya, K. & Watanabe, M.M. (2012). Relationship between hydrocarbons and molecular phylogeny of *Botryococcus braunii. Algal Res., 1*, 114-119.

[12] Blackburn, K.B. (1936). *Botryococcus* and the algal coals. I. A reinvestigation of the alga *Botryococcus braunii* Kützing. *Trans. Roy. Soc. Edinb., 58,* 841-854.

[13] Dayananda, C., Sarada, R. Kumar, V. & Ravishankar, G.A. (2007). Isolation and characterization of hydrocarbon producing green alga *Botryococcus braunii* from Indian freshwater bodies. *Electron. J. Biotechnol., 10*, 78-91.

[14] Metzger, P., Casadevall, E., & Coute, A. (1988). Botryococcene distribution in strains of the green-alga *Botryococcus braunii*. *Phytochemistry, 27*, 1383-1388.

[15] Wolf, F.R. & Cox, E.R. (1981). Ultrastructure of active and resting colonies of *Botryococcus braunii* (Chlorophyceae). *J. Phycol., 17,* 395-405.

[16] Weiss, T.L., Roth, R., Goodson, C., Vitha, S., Black, I., Azadi, P., Rusch, J., Holzenburg, A., Devarenne, T.P. & Goodenough, U. (2012). Colony organization in

the green alga *Botryococcus braunii* (Race B) is specified by a complex extracellular matrix. *Eukaryot. Cell, 11*, 1424-1440.

[17] Maxwell, J.R., Douglas, A.G., Eglinton, G. &McCornick, A. (1968). Botryococcenes-hydrocarbons of novel structure from the alga *Botryococcus braunii* Kützing. *Phytochemistry, 7*, 2157-2171.

[18] Knights, B.A., Brown, A.C., Conway, E. & Middleditch, B.S. (1970). Hydrocarbons from the green form of the freshwater alga *Botryococcus braunii. Phytochemistry, 9*, 1317-1324.

[19] Metzger, P., Rager, M.N. & Fosse, C. (2008). Braunicetals: acetals from condensation of macrocyclic aldehydes and terpene diols in *Botryococcus braunii. Phytochemistry, 69*, 2380-2386.

[20] Beakes, C.W & Cleary, A.L. (1999). Visualization of plastids and lipophilic components in living colonies of a wild strain of the hydrocarbon-forming green alga *Botryococcus* by laser scanning confocal microscopy. *Appl. Phycol., 10*, 435-446.

[21] Noguchi, T. & Kakami, F. (1999). Transformation of trans-Golgi network during the cell cycle in a green alga, *Botryococcus braunii. J. Plant Res., 112*, 175-186.

[22] Senousy, H.H., Beakes, G.W. & Hack, E. (2004). Phylogenetic placement of *Botryococcus braunii* (Trebouxiophyceae) and *Botryococcus sudeticus* isolate UTEX 2629 (Chlorophyceae). *J. Phycol., 40,* 412-423.

[23] Tanabe, Y., Kato, S., Matsuura, H. & Watanabe, W.M. (2012). A *Botryococcus* strain with bacterial ectosymbionts grows fast and produces high amount of hydrocarbons. *Procedia Environ. Sci., 15*, 22-26.

[24] Tanabe, Y., Ioki, M. & Watanabe, M.M. (2014). The fast-growing strain of hydrocarbon-rich green alga *Botryococcus braunii,* BOT-22, is a vitamin B12 autotroph. *J. Appl. Phycol., 26*, 9-13.

[25] Metzger, P., Casadevall, E., Pouet, M.J. & Pouet, Y. (1985). Structures of some Botryococcenes - branched hydrocarbons from the B-race of the green-alga *Botryococcus braunii. Phytochemistry, 24*, 2995-3002.

[26] Huang, Z., & Poulter, C.D. (1989). Tetramethylsqualene, a triterpene from *Botryococcus braunii* Var Showa. *Phytochemistry, 28*, 1467-1470.

[27] Sato, Y., Okada, S., Murakami, M. & Abe, H. (2003). Biosynthesis of the triterpenoids, botryococcenes and tetramethylsqualene in the B race of *Botryococcus braunii* via the non-mevalonate pathway. *Tetrahedron Lett., 44*, 7035-7037.

[28] Schwender, J., Gemünden, C. & Lichtenthaler, H.K. (2001). Chlorophyta exclusively use the 1-deoxyxylulose 5-phosphate/2-C-methylerythritol 4-phosphate pathway for the biosynthesis of isoprenoids. *Planta, 212*, 416-423.

[29] Rodríguez-Concepción, M & Boronat, A. (2002). Elucidation of the methylerythritol phosphate pathway for isoprenoid biosynthesis in bacteria and plastids. A metabolic milestone achieved through genomics. *Plant Physiol., 130,* 1079-1089.

[30] Miziorko, H.M. (2011). Enzymes of the mevalonate pathway of isoprenoid biosynthesis. *Arch. Biochem. Biophys., 505*, 131-143.

[31] Phillips, M.A., León, P., Boronat, A. & Rodríguez-Concepción, M. (2008). The plastidial MEP pathway: unified nomenclature and resources. *Trends Plant Sci., 13*, 619-623.

[32] Vranová, E., Coman, D. & Gruissem, W. (2013). Network analysis of the MVA and MEP pathways for isoprenoid synthesis. *Annu. Rev. Plant Biol., 64*, 665-700.

[33] Lohr, M., Schwender, J. & Polle, J.E. (2012). Isoprenoid biosynthesis in eukaryotic phototrophs: A spotlight on algae. *Plant Sci., 185-186*, 9-22.

[34] Matsushima, D., Jenke-Kodama, H., Sato, Y., Fukunaga, Y., Sumimoto, K., Kuzuyama, T., Matsunaga & S., Okada, S. (2012). The single cellular green microalga *Botryococcus braunii*, race B possesses three distinct 1-deoxy-D-xylulose 5-phosphate synthases. *Plant Sci., 185*, 309-320.

[35] Okada, S., Devarenne, T.P., Murakami, M., Abe, H. & Chappell, J. (2004). Characterization of botryococcene synthase enzyme activity, a squalene synthase-like activity from the green microalga *Botryococcus braunii*, Race B. *Arch. Biochem. Biophys., 422,* 110-118.

[36] Niehaus, T.D., Okada, S., Devarenne, T.P., Watt, D.S., Sviripa, V. & Chappell, J. (2011). Identification of unique mechanisms for triterpene biosynthesis in *Botryococcus braunii. Proc. Natl. Acad. Sci. USA, 108*, 12260-12265.

[37] Metzger, P., David, M. & Casadevall, E. (1987). Biosynthesis of triterpenoid hydrocarbons in the B-race of the green-alga *Botryococcus braunii* - Sites of production and nature of the methylating agent. *Phytochemistry, 26*, 129-134.

[38] Metzger, P., Casadevall, E. & Coute, A. (1988). Botryococcene distribution in strains of the green-alga *Botryococcus braunii*. *Phytochemistry,* 27, 1383-1388.

[39] Achitouv, E., Metzger, P., Rager, M.N. & Largeau C. (2004). C31-C34 methylated squalenes from a Bolivian strain of *Botryococcus braunii*. *Phytochemistry, 65*, 3159-3165.

[40] Niehaus, T.D., Kinison, S., Okada, S., Yeo, Y., Bell, S.A., Cui, P., Devarenne, T.P., & Chappell, J. (2012). Functional identification of triterpene methyltransferases from *Botryococcus braunii* race B. *J. Biol. Chem., 287,* 8163-8173.

[41] Weiss, T.L., Johnston, J.S., Fujisawa, K., Okada, S. & Devarenne, T.P. (2010). Genome size and phylogenetic analysis of the A and L races of *Botryococcus braunii. J. Appl. Phycol., 23*, 833-839.

[42] Molnár, I., Lopez, D., Wisecaver, J.H., Devarenne, T.P., Weiss, T.L., Pellegrini, M. & Hackett, J.D. (2012). Bio-crude transcriptomics: gene discovery and metabolic network reconstruction for the biosynthesis of the terpenome of the hydrocarbon oil-producing green alga, *Botryococcus braunii* race B (Showa). *BMC Genomics, 13*, 576-603.

[43] Metzger, P. & Casadevall, E. (1987). Lycopadiene, a tetraterpenoid hydrocarbon from new strains of the green-alga *Botryococcus braunii*. *Tetrahedron Lett., 28*, 3931-3934.

[44] Metzger, P., Pouet, Y. & Summons, R. (1997). Chemotaxonomic evidence for the similarity between *Botryococcus braunii* L race and *Botryococcus neglectus. Phytochemistry, 44*, 1071-1075.

[45] Zhang, Z. & Sachs, J.P. (2007). Biomarker evidence for the co-occurrence of three races (A, B and L) of *Botryococcus braunii* in El Junco Lake, Galápagos. *Org. Geochem., 38*, 1459-1478.

[46] Metzger, P. & Largeau, C. (1999). Chemicals of *Botryococcus braunii*. In: Cohen, Z. (Ed.), *Chemicals from Microalgae*. Taylor and Francis Ltd., London, pp. 205-260.

[47] Templier, J., Largeau, C. & Casadevall, E. (1984). Hydrocarbon formation in the green-alga *Botryococcus braunii.* 4 Mechanism of non-isoprenoid hydrocarbon biosynthesis in *Botryococcus braunii*. *Phytochemistry, 23*, 1017-1028.

[48] Chan Yong T.P., Largeau, C. & Casadevall, E. (1986). Biosynthesis of non-isoprenoid hydrocarbons by the microalga *Botryococcus braunii* evidence for an elongation

decarboxylation mechanism; activation of decarboxylation. *Nouv. J. Chim.,10*, 701-707.

[49] Templier, J., Largeau, C. & Casadevall, E. (1987). Hydrocarbon formation in the green-alga *Botryococcus braunii*. 5 Effect of various inhibitors on biosynthesis of non-isoprenoid hydrocarbons in *Botryococcus braunii*. *Phytochemistry, 26*, 377-383.

[50] Dennis, M.W. & Kolattukudy, P. E. (1991). Alkane biosynthesis by decarbonylation of aldehyde catalyzed by a microsomal preparation from *Botryococcus braunii*. *Arch. Biochem. Biophys., 287*, 268-275.

[51] Dennis, M. & Kolattukudy, P.E. (1992). A cobalt-porphyrin enzyme converts a fatty aldehyde to a hydrocarbon and CO. *Proc. Nat. Acad. Sci. USA, 89*, 5306-5310.

[52] Wang, X. & Kolattukudy, P.E. (1995). Solubilization and purification of aldehyde-generating fatty Acyl-Coa reductase from green-alga *Botryococcus braunii*. *FEBS Lett., 370*, 15-18.

[53] Yang, S.L., Wang, J., Cong, W., Cai, Z.L. & Fan, O.Y. (2004). Utilization of nitrite as a nitrogen source by *Botryococcus braunii*. *Biotechnol. Lett., 26,* 239-243.

[54] Ohmori, M., Wolf, F.R. & Bassham, J.A. (1984). *Botryococcus braunii*: carbon/nitrogen metabolism as affected by ammonia addition. *Arch. Microbiol., 140*, 101-106.

[55] Casadevall, E., Dif, D., Largeau, C., Gudin, C., Chaumont, D. & Desanti, O. (1985). Studies on batch and continuous cultures of *Botryococcus braunii*: hydrocarbon production in relation to physiological state, cell ultrastructure, and phosphate nutrition. *Biotechnol. Bioeng., 27*, 286-295.

[56] Dayananda, C., Sarada, R., Usha Rani, M., Shamala, T.R. & Ravishankar, G.A. (2007). Autotrophic cultivation of *Botryococcus braunii* for the production of hydrocarbons and exopolysaccharides in various media. *Biomass. Bioenerg., 31*, 87-93.

[57] Chirac, C., Casadevall, E., Largeau, C. & Metzger, P. (1985). Bacterial influence upon growth and hydrocarbon production of the green alga *Botryococcus braunii*. *J. Phycol., 21*, 380-387.

[58] Yoshimura, T., Okada, S. & Honda, M. (2013). Culture of the hydrocarbon producing microalga *Botryococcus braunii* strain Showa: optimal CO_2, salinity, temperature, and irradiance conditions. *Bioresource Technol., 133*, 232-239.

[59] Khatri, W., Hendrix, R., Niehaus, T., Chappell, J. & Curtis, W.R. (2014). Hydrocarbon production in high density *Botryococcus braunii* race B continuous culture. *Biotechnol. Bioeng., 111*, 493-503.

[60] Kim, H.S., Weiss, T.L., Thapa, H.R., Devarenne, T.P. & Han, A. (2014). A microfluidic photobioreactor array demonstrating high-throughput screening for microalgal oil production. *Lab Chip, 14*, 1415-1425.

[61] Sunda, W.G & Huntsman, S.A. (2004). Relationships among photoperiod, carbon fixation, growth, chlorophyll *a*, and cellular iron and zinc in a coastal diatom. *Limnol. Oceanogr., 49*, 1742-1753.

[62] Wake, L.V & Hillen, L.W. (1980). Study of a ''bloom'' of the oil-rich alga *Botryococcus braunii* in the Darwin River reservoir. *Biotechnol. Bioeng., 22,* 1637-1656.

[63] Song, L., Qin, J.G., Su, S., Xu, J., Clarke, S. & Shan, Y. (2012). Micronutrient requirements for growth and hydrocarbon production in the oil producing green alga *Botryococcus braunii* (Chlorophyta). *PLoS ONE, 7*, e41459.

[64] Brennan, L. & Owende, P. (2010). Biofuels from microalgae- a review of technologies for production, processing, and extractions of biofuels and co-products. *Renew. Sust. Energ. Rev., 14,* 557-577.

[65] Tanoi, T., Kawachi, M. & Watanabe, M.M. (2011). Effects of carbon source on growth and morphology of *Botryococcus braunii. J. Appl. Phycol., 23,* 25-33.

[66] Gim, G.H., Kim, J.K., Kim, H.S., Kathiravan, M.N., Yang, H., Jeong, S.H. & Kim, S.W. (2014). Comparison of biomass production and total lipid content of freshwater green microalgae cultivated under various culture conditions. *Bioprocess Biosyst. Eng.*, *37*, 99-106.

[67] Liu, J., Ge, Y., Cheng, H., Wu, L. & Tian, G. (2013). Aerated swine lagoon wastewater: A promising alternative medium for *Botryococcus braunii* cultivation in open system. *Bioresource Technol., 139,* 190-194.

[68] Kong, W.B., Song, H., Hua, S.F., Yang, H., Yang, Q. & Xia, C.G. (2012). Enhancement of biomass and hydrocarbon productivities of *Botryococcus braunii* by mixotrophic cultivation and its application in brewery wastewater treatment. *Afr. J. Microbiol. Res., 6,* 1489-1496.

[69] Yonezawa, N., Matsuura, H., Shiho, M., Kaya, K. & Watanabe, M.K. (2012). Effects of soybean curd wastewater on the growth and hydrocarbon production of *Botryococcus braunii* strain BOT-22. *Bioresource Technol., 109*, 304-307

[70] Testa, M., Gerbaudo, S. & Andri, E., (2001). Data report: *Botryococcus* colonies in miocene sediments in theWestern Woodlark Basin, Southwest Pacific (ODP Leg 180). *Proc. ODP, Sci. Results*, *180*, 1-6.

[71] Audino, M., Grice, K., Alexander, R. & Kagi, R.I. (2002). Macrocyclic alkanes in crude oils from the algaenan of *Botryococcus braunii*. *Org. Geochem.*, *33*, 979-984.

[72] Eroglu, E., Okada, S. & Melis, A. (2011). Hydrocarbon productivities in different *Botryococcus* strains: comparative methods in product quantification. *J. Appl. Phycol.*, *23*, 763-775.

[73] Kitazato, H., Asaoka, S. & Iwamoto, H. (1989). Catalytic cracking of hydrocarbons from microalgae. *Sekiyu Gakkaishi, 32*, 28-34.

[74] Alabi, A.O., Tampier, M. & Bibeau, E. (2007). Microalgae technologies and processes for biofuels and bioenergy production in British Columbia: Current technology, suitability, and barriers to implementation, *Aquacult. Update, 101,* 1-9.

[75] Lundquist, T.J., Woertz, I.C., Quinn, N.W.T. & Benemann, J.R. (2010). A realistic technology and engineering assessment of algae biofuel production. *Energy Biosciences Institute, 1,* 1-178.

[76] Thilakaratne, R., Wright, M.M. & Brown, R.C. (2014). A techno-economic analysis of microalgae remnant catalytic pyrolysis and upgrading to fuels. *Fuel, 128,* 104-112.

[77] Wolf, F.R., Nemethy, E.K., Blanding, J.H. & Bassham, J. A. (1985). Biosynthesis of unusual acyclic isoprenoids in the alga *Botryococcus braunii*. *Phytochemistry, 24*, 733-737.

[78] Deng, L., Senseman, S.A., Gentry, T.J., Zuberer, A.D., Weiss, T.L., Devarenne, T.P. & Camargo, E.R. (2012). Effect of selected herbicides on growth and hydrocarbon content of *Botryococcus braunii* (Race B). *Ind. Crop. Prod., 39,* 154-161.

[79] Ioki, M., Baba, M., Nakajima, N., Shiraiwa, Y. & Watanabe, M.M. (2012). Transcriptome analysis of an oil-rich race B strain of *Botryococcus braunii* (BOT-22)

by de novo assembly of pyrosequencing cDNA reads. *Bioresour. Technol., 109*, 292-296.

[80] Shiho, M., Kawachi, M., Horioka, K., Nishita, Y., Ohashi, K., Kaya, K. & Watanabe, M.M. (2012). Business evaluation of a green microalga *Botryococcus braunii* oil production system. *Procedia Environ. Sci., 15*, 90-109.

[81] Matsuura, H; Watanabe, MM; & Kaya, K. (2012). Echinenone production of dark red-coloured strain of *Botryococcus braunii. J. App. Phycol., 24*, 973-977.

[82] Xu, Li., Wang, F., Guo, C. & Liu, C.Z. (2012). Improved algal oil production from *Botryococcus braunii* by feeding nitrat and phosphate in anairlift bioreactor. *Eng. Life Sci., 12*, 171-177.

[83] Yeesang, C. & Cheirsilp, B. (2011). Effect of nitrogen, salt, and iron content in the growth medium and light intensity on lipid production by microalgae isolated from fresh water sources in Thailand. *Bioresource Technol., 102*, 3034-3040.

[84] Yonezawa, N., Matsuura, H., Shiho, M., Kaya, K. & Watanabe, M.K. (2011). Effects of soybean curd wastewater on the growth and hydrocarbon production of *Botryococcus braunii* strain BOT-22. *Bioresource Technol., 109*, 304-307.

[85] Zhang, F., Cheng, L.H., Gao, W.L., Xu, X.H., Zhang, L. & Chen, H.L. (2011). Mechanism of lipid extraction from *Botryococcus braunii* FACHB 357 in biphasic bioreactor. *J. Biotechnol., 154,* 281-284.

[86] Bazaes, J., Sepulveda, C., Acién, F.G., Morales, J., Gonzales, L., Rivas, M. & Riquelme, C. (2012). Outdoor pilot-scale production of *Botryococcus braunii* in panel reactors. *J Appl Phycol., 24*, 1353-1360.

[87] Bouno, S., Langellotti, A.L., Martello, A., Bimonte, M., Tito, A., Carola, A., Apone, F., Colucci, G. & Fogliano, V. (2012). Biological activities of dermatological interest by the water extract of the microalga *Botryococcus braunii. Arch. Dermatol. Res., 304*, 755-764.

[88] Kim, B.H., Ramanan, R., Cho, D.H., Choi, G.G., La, H.J., Ahn, C.Y., Oh, H.M. & Kim, H.S. (2012). Simple, rapid and cost-effective method for high quality nucleic acids extraction from different strains of *Botryococcus braunii. PLoS ONE 7*, e37770.

[89] Liu, J., Ge, Y. & Tian, G. (2012). Enhancement of hydrocarbon productivity of *Botryococcus braunii* by an adenine type phytohormone. *Adv. Mat. Res., 515,* 397-400.

[90] Rao, A.R., Racishankar, G.A. & Sarada, R. (2012). Cultivation of green alga *Botryococcus braunii* in raceway, circular ponds under outdoor conditions and its growth, hydrocarbon production. *Bioresource Technol., 123*, 528-533.

[91] Sakamoto, K., Baba, M., Suzuki, I., Watanabe, M.M. & Shiraiwa, Y. (2012). Optimization of light for growth, photosynthesis, and hydrocarbon production by the colonial microalga *Botryococcus braunii* BOT-22. *Bioresource Technol., 110*, 474-479.

[92] Shimamura, R., Watanabe, S., Sakakura, Y., Shiho, M., Kaya, K. & Watanabe, M.M. (2012). Development of *Botryococcus* seed culture system for future mass culture. *Procedia Environ. Sci., 15,* 80-89.

[93] Tanabe, Y., Kato, S., Matsuura, H. & Watanabe, W.M. (2012). A *Botryococcus* strain with bacterial ectosymbionts grows fast and produces high amount of hydrocarbons. *Procedia Environ. Sci., 15*, 22-26.

[94] Kong, W.B., Song, H., Hua, S.F., Yang, H., Yang, Q. & Xia, C.G. (2012). Enhancement of biomass and hydrocarbon productivities of *Botryococcus braunii* by mixotrophic cultivation and its application in brewery wastewater treatment. *Afr. J. Microbiol. Res., 6*, 1489-1496.

[95] Xu, Li., Wang, S.K., Wang, F., Guo, C. & Liu, C.Z. (2014). Improved biomass and hydrocarbon productivity of *Botryococcus braunii* by periodic ultrasound stimulation. *Bio Energy Res., 7*, 986-992.

In: Microalgae and Other Phototrophic Bacteria
Editor: Luis G. Torres Bustillos
ISBN: 978-1-63482-078-3

Chapter 4

EFFECT OF CULTURE CONDITIONS ON THE QUANTITY AND QUALITY OF LIPIDS: ITS IMPORTANCE ON BIODIESEL SYNTHESIS

Luis Fernández Linares**, Eder Gutiérrez Márquez and Aarón Millán Oropeza
Unidad Profesional Interdisciplinaria de Biotecnología
del Instituto Politécnico Nacional, Mexico

ABSTRACT

Microorganisms such as bacteria and microalgae are the most viable sources for the production of sustainable energy since they lack the major drawbacks of first and second generation biofuels. Microalgae have the capacity to store large quantities of lipids that can be used in the production of biodiesel. Due to the advantages microalgae present vis-a-vis other raw material sources in the production of biodiesel, there are great efforts to overcome technological and economical obstacles to develop this technology on an industrial scale. In this chapter the effect of nitrogen limitation and CO_2 supply in the fatty acids profiles of microalgal oils from *Dunaliella tertiolecta* and *Nannochloropsis oculata* was determined. *Nannochloropsis oculata* and *Dunaliella tertiolecta* were cultured under different nutrient conditions and culture systems. In the photobioreactor culture, D. *tertiolecta* accumulated mainly C16:0 (42%), C18:0 (32%), and C18:3 (10%), whereas N. *oculata* accumulated C16:0 (41%), followed by C16:1 (21%) and C18:0 (10%).Combining the variation in the NO_3 and CO_2 concentrations did not modify importantly the fatty acids composition of *D. tertiolecta* in any of the analyzed levels. With sudden nitrogen limitation, the proportion of C18:3 in D. *tertiolecta* increased from 10 to 30% of the fatty acids. In *N. oculata* no important change occurred in the fatty acids composition up to 72 h of limitation. In the semicontinuous raceway culture, *N. oculata* presented predominantly C16:0 (37%), followed by C16:1 and C18:0. The C16:1 and C18:0 proportion was very different between the photobioreactor and the raceway cultures. C16:1 proportion was of 21% in the photobioreactor and of 6.7% in the raceway, whereas the proportion of C18:0 was of 10.2% in the photobioreactor and

* E mail: lfernand36@gmail.com

of 30.5% in the raceway. Addition of 2 L/min during 30 min/day of CO_2 did not affect the fatty acids composition of the culture. The raceway culture presented a very homogeneous composition; high in saturated lipids C16:0 and C18:0, despite the lack of control of the system as it is an open system.

INTRODUCTION

Currently, fossil fuel resources are not considered sustainable and their usage as fuels is questioned from an economic, ecologic and environmental viewpoint. Fossil fuel combustion contributes in great measure to an increase in atmospheric carbon dioxide and thus to the global warming observed in the last decades (Mabee et al., 2005). The need to find alternative energy sources that minimize the utilization of fossil fuels and limit the production of carbon dioxide and other harmful gases has led to an increased effort in the development of fuels of biological origin, derived from organic waste, plants, and, more recently, microalgae.

First generation biofuels such as biodiesel obtained from vegetable oil, bioethanol, and biogas are more eco-friendly than petrol-derived diesel. However, first generation biofuels possess the disadvantage of competing for agricultural soil and can generate ecological damage stemming from intensive farming. Thus, developing more efficient alternatives based on conventional and renewable technologies is recommended (Eisberg, 2006).

Second generation biofuels are obtained from non-food vegetable biomass, ideally agricultural waste. Although it is a more advantageous technology than first generation fuels in terms of meeting the demand for liquid fuels, it would require large quantities of crops and farmland, entirely devoted to biofuel production (Chisti, 2007). Furthermore there are still noteworthy technical limitations to the pretreatment of raw materials allowing large-scale production

Microorganisms such as bacteria and microalgae are the most viable sources for the production of sustainable energy since they lack the major drawbacks of first and second generation biofuels (Texeira, 2010). In particular, algae oil has sparked great interest in recent years because of its capacity to store large quantities of lipids that can be used in the production of biodiesel. Due to the advantages microalgae present vis-a-vis other raw material sources in the production of biodiesel, there are great efforts to overcome technological and economical obstacles to develop this technology on an industrial scale.

It has been demonstrated that the quality of biodiesel depends on the type of fatty acids that compose it (Dorado, 2008) Therefore, requirements are not only microalgal strain with high lipid content but also a consideration of the specific fatty acids that it produces. In this regard, it is of the utmost importance to analyze the microalgae fatty acid content to provide a key evaluation criterion as a potential candidate for large-scale culture.

MICROALGAE FOR THE PRODUCTION OF BIODIESEL

Of all biofuel sources, biodiesel produced from microalgal oils is the only one with the potential to entirely displace fossil diesel (Chisti, 2007). The advantages of microalgae include (Scott et al., 2010; Texeira, 2010):

1. Capacity to store large amounts of lipids, far exceeding the productivities of the best vegetable oils
2. Capacity to store large amounts of triacylglycerols.
3. Higher growth and production rates.
4. Allowing year round cultivation.
5. Require less water than land crops.
6. Microalgae do not require the use of herbicides or pesticides.
7. Microalgae do not require farmland for production.

However, an array of technological and economic limitations must be overcome in order to achieve industrial-scale production. Selecting a robust microalgal strain that can be cultivated on a large-scale exposed to the environment, has a high content of neutral lipids, a high growth rate, and is also resistant to contamination remains a challenge. On the other hand, for the downstream processing to be successful, the development of an effective and energy efficient lipid extraction process is required. Despite the widespread and routine use of extraction protocols for lipid content determination, the variables that affect the extraction of microalgal lipids are not well known and there is no established method for industrial scale production (Halim et al., 2011).

RELEVANT FATTY ACIDS FOR THE PRODUCTION OF BIODIESEL

The most common type of lipids are fatty acids. Fatty acids are molecules composed of a hydrophilic carboxyl group attached to a hydrophobic hydrocarbon chain. Fatty acids are named based on the number of carbons and double bonds in the hydrocarbon chain. For example, cis-9-octadecenoic acid or oleic acid has 18 carbons in the aliphatic chain and the double bond is 9 carbons away counting from the distal methyl group; its simplified denomination is C18:1(n9). Saturated fatty acids are those without double bonds, while unsaturated fatty acids have one or more double bonds along the chain.

Lipids produced by microalgae can be categorized in two groups: structural lipids that are generally polar and storage lipids that are generally neutral. Structural lipids typically contain a larger content of polyunsaturated fatty acids (PUFAs), and include phospholipids and sphingolipids. Storage lipids are mainly found as acylglycerols, the most abundant being triacylglycerols (TAGs), predominantly of the saturated and monounsaturated kind (MUFAs). Many microalgae species produce large quantities of TAGs under conditions of stress (Chen et al., 2010). These TAGs are the most desirable for the large-scale production of biodiesel. However, other neutral lipids that do not contain fatty acids exist, namely, hydrocarbons, sterols, ketones, and pigments such as chlorophylls and carotenoids. While these lipid fractions are soluble in organic solvents, they cannot be converted into biodiesel (Halim et al., 2012).

Generally, microalgae lipids are extracted utilizing organic solvents or supercritical fluids, once extracted they undergo a transesterification reaction for conversion to biodiesel.

TRANSESTERIFICATION OF OILS

Given their structural similarity with the hydrocarbons that compose fossil diesel, the long hydrocarbon chains of acylglycerols comprising an oil (animal, plant or microalgae) are usable as fuel. However, the direct use of oil on motors is problematic: it is between 11 and 17 times more viscose than diesel and much less volatile so it is not fully combusted and forms deposits in the engine nozzle (Matheus et al., 1998).

Different alternatives to adapt oils for motor use exist: dilution (25%) with diesel, forming microemulsions with short chain alcohols, thermal decomposition, catalytic cracking, and transesterification with alcohols. Among these alternatives, transesterification is the best because the physical properties of the fatty acid esters are very similar to those of petrol diesel and because the process is relatively uncomplicated (Matheus et al., 1998).

In the production of biodiesel, transesterification is carried out by reacting the lipid extract with an alcohol, usually methanol or ethanol. When methanol is used, fatty acid methylesters (FAMEs) are produced. Acid catalysts, such as H_2SO_4 and HCl, or alkaline catalysts, such as NaOH and KOH, are used in the reaction. As alkaline catalysts possess higher reaction rates and increased conversion capacity than acid catalysts for the transesterification of triacylglycerols, these catalysts are the most commonly used on an industrial scale for biodiesel production from vegetable and animal oils (Huang et al., 2010). However, alkaline transesterification has limited efficacies for lipid compounds other than acylglycerol lipids and free fatty acids.

BIODIESEL

According to the ASTM D6751, biodiesel is defined as the set of mono-alkyl esters of long chain fatty acids derived from vegetable oils or animal fats, for use in compression-ignition engines. The properties of biodiesel are strongly determined by the fatty acid esters that compose it (Knothe, 2005).

In general, the quality of biodiesel is influenced by various factors, which include: quality of raw materials, fatty acid composition of the oil, production materials and processes, and post-production parameters (Van Gerpen et al., 2004). There are various parameters that characterize a biodiesel and determine its quality. The oil composition determines the following parameters:

1. *Oxidative stability. Reflects the degree of total unsaturation of biodiesel. Unsaturated fatty acids are more prone to oxidation (regardless of factors such as air, light, and presence of metals)* which results in the formation of degradation products that can affect the operability of the engine. FAMEs of saturated and monounsaturated fatty acids are considered better than *esters from* polyunsaturated *fatty acids (PUFAs),* in terms of oxidative stability *(Imahara et al., 2006). It is even recommended to avoid the presence of PUFAs since they have a disproportionately large effect on biodiesel auto-oxidation, which creates problems under conditions of prolonged storage (Dunn et al., 2002).*

2. Cetane number (CN). CN is a measure of the quality of combustion during ignition. Generally, biodiesel has a larger CN value compared to fossil diesel, which is an advantage in terms of engine performance and emissions (Knothe et al., 2003). The longer the fatty acid carbon chain, the greater the value of CN. Additionally, higher CN values have been observed in saturated FAMES than in highly unsaturated FAMES (Knothe et al., 2003).
3. Caloric value. The quantity of energy that a fuel can generate depends on the length of the chain (number of carbons and hydrogens). From this viewpoint, oil sources with saturated long chain components are more suitable (Mittelbach, 2004).
4. Kinematic viscosity. This is an important parameter, particularly when motor operate at low temperatures. A higher viscosity fuel requires more pressure and volume to be injected by the injection pump. With longer fatty acids, viscosity increases and decreases with the degree of unsaturation (Allen et al., 2005).
5. Per*formanc*e in cold environments. One of the major drawbacks of biodiesel is that it has poor flow properties at low temperatures. Biodiesel produced using saturated lipids tends to form waxes at room temperature, which can clog fuel lines and filters. An elevated content of polyunsaturated fatty acids is recommended in this regard (Knothe et al., 2005).

In the search for an ideal biodiesel composition, the high presence of monounsaturated fatty acids, low presence of polyunsaturated fatty acids, and a controlled amount of saturated fatty acids are highly recommended. In this regard, C18:1 and C16:1 are the most appropriate fatty acids in terms of oxidative stability, cold flow properties, and other features (Dorado, 2008). Meng et al. (2009) suggested a high content of C18:0 and C18:1 in microorganisms as a prerequisite for industrial production. Palmitic acid C16:0 has also been proposed as suitable for the production of biodiesel, as it is a common fatty acid in both microalgae and oleaginous plants. However, if the obtained oil is high in polyunsaturated fatty acids, oil refining through hydrogenation may account for more flexibility since the process allows the removal of double bonds. However, it is also feasible to add antioxidants to biodiesel or oils containing a greater amount of saturated fatty acids.

VARIATIONS IN CULTURE SYSTEM CONDITIONS FOR LIPID ACCUMULATION

Lipid content in different microalgae species varies considerably. In terms of dry biomass, the lipid fraction can vary between 5 and 80%. Some species are richer in neutral lipids than others (Lv et al., 2010). In general, lipid accumulation occurs when a key nutrient, usually nitrogen, is depleted. When a nutrient is depleted, carbon assimilation continues and it accumulates in energy storage molecules. Generally, this reaction is probably a survival response to conditions of nutritional stress (Bashan, 2011).

Moreover, the type of fatty acids that each particular species accumulates is affected by culture medium age, diet and conditions, including substrate composition, temperature, pH, lighting, aeration, and nutrient imbalances (Ratledge & Winn, 2002). It has been observed that in heterotrophic cultures, when sugar concentration is high, accumulated fatty acids are more saturated: the molecules commonly formed are triglycerides, which provide more energy from oxidation than polyunsaturated fatty acids and thus provide a higher energy storage source. On the other hand, autotrophic cultures form a higher number of highly unsaturated fatty acids such as C16:3 and C18:3 (Day et al., 1991).

In microalgae cultures, lipid content can be manipulated by varying culture conditions (Chisti, 2007; Mata et al., 2010). This is due to the fact that autotrophic metabolism is influenced by various factors such as the incidence of light on the culture, system temperature, and supply or limitation of nutrients.

Light incidence has an important role in culture productivity, as well as in the sustainability and economic feasibility of a bioproduction process of microalgae biomass (Borowitzka and Moheimani, 2013). The purpose of utilizing a microalgae culture with autotrophic metabolism is to take advantage of natural light energy and convert CO_2 into organic compounds via photosynthesis.

Several studies have focused on irradiance in culture systems. Yeesang and Cheirsilp (2011) compared the effect of three different light intensities (33, 49.5 and 82.5 $\mu E/m^2$ s) on the production of lipids in different species of *Botryococcus,* concluding that the highest lipid content was obtained at 49.5 $\mu E/m^2$ s; Chrismadha and Borowitzka (1994) observed that a higher light intensity yielded a higher cell density in a semi-continuous culture of *Phaeodactylum tricornutum*, comparing three light intensities (56, 212, and 286 $\mu E/m^2 \bullet$ s). Furthermore, Cheirsilp and Torpee (2012) evaluated different light intensities (27 at 135 $\mu E/m^2 \bullet$ s) in *Chlorella sp.* and *Nannochloropsis sp.* cultures and obtained the highest quantity of lipids at different light intensities depending on the species, 397 mg/L at 108 $\mu E/m^2 \bullet$ s for *Chlorella sp.* and 572 mg/L at 67.5 $\mu E/m^2 \bullet$ s for *Nannochloropsis sp.* The previous point indicates that the largest production of lipids is reached when cultures are exposed to an intermediate light incidence; however, the optimal irradiance value is unique to each species.

Another factor that influences the production of biomass and microalgae lipid content is the duration of culture photoperiods. Regarding this factor, Wahidin et al. (2013) evaluated cell growth and lipid accumulation in *Nannochloropsis sp.* using three light intensities (50, 100, and 200 $\mu E/m^2 \bullet$ s) and three photoperiods (12:12, 18:6, and 24:0 of light:darkness) and observed that photoperiod cycles of 18:6 at a light intensity of 100 $\mu E/m^2 \bullet$ s yielded the highest cellular density and maximum lipid content; 6.5×10^7 cells/mL and 31.1% based on dry weight, respectively.

As a general formula, the higher the light intensity, the higher the growth rates until photoinhibition levels are reached. However, the increase in light intensity reduces lipid accumulation in the cell structure since metabolism directs itself towards replication (Cheirsilp and Torpee, 2012; Chrismadha and Borowitzka, 1994) or towards the excretion of exopolysaccharides (Rebolloso Fuentes et al., 1999). The excess of irradiance in crops can cause a state of physiological stress that inhibits growth during light photoperiods (Wahidin et al., 2013). This affects outdoor crops, where light intensity peaks at noon can reach 2000$\mu E/m^2 \bullet$ s (Molina Grima et al., 1999). This value may be up to 10-times higher than saturation irradiance, causing crop photoinhibition.

With regards to the effect of temperature on lipid accumulation, Kalacheva et al. (2002) evaluated *Botryococcus braunii* lipid production using three different temperatures (18, 25, and 32°C). The authors observed that the highest TAGs production rate was obtained at 18°C and showed a higher content of palmitic (16:0) and linolenic acid (18:3). On the other hand, Converti et al. (2009) studied lipid productivity in *Chlorella vulgaris* strains at25, 30, 35, and

38°C and *Nannochloropsis oculata* at 15, 20, and 25°C; at 25°C, the highest lipid productivities were obtained, yielding 20.22 and 10.10 mg/L d for *C. vulgaris* and *N. oculata*, respectively. In addition, the profiles of fatty acid methylesters (FAMEs) showed that *C. vulgaris* predominantly produced palmitic acid (16:0) representing 66% of overall lipid fraction, while *N. oculata* generated 58% palmitic acid (16:0) and 19% oleic acid (18:1) relative to the total content of FAMEs. Although culture temperature cannot be controlled in open systems or on a large scale, the use of a microalgal species that can survive under a wide range of temperature variation is recommended, considering diurnal and seasonal temperature fluctuations.

Another strategy to increase lipid content is two-stage cultivation. The first stage focuses on cell growth using a culture medium with sufficient nutrients while the second stage focuses on lipid accumulation by using another medium that promotes nutrient limitation or another type of physiological stress (Courchesne et al., 2009). The nutrient restriction strategy can be performed by: progressive limitation, where the nutrient limitation phase is achieved through cell growth and progressive consumption of substrates in the culture; and sudden limitation in which the culture is maintained with an abundance of substrates until obtaining high cell concentrations. Subsequently, the culture medium is replaced by another with nutritional deficiency. In this regard, VanVooren et al. (2012) demonstrated higher TAGs productivity in *N. oculata* culture with sudden limitation employing nitrogen, accounting for a 28% increase compared to progressive limitation. Notwithstanding, sudden limitation processes presents the inconvenience of requiring an additional separation step during the bioprocess. This step includes separating culture medium cells for another medium lacking a nitrogen source.

Furthermore, the use of culture media with low substrate concentration allows for: rapid induction of the limiting step, a beneficial effect on lipid accumulation, controlling pollution in open systems, and is environmentally friendly (Courchesne et al., 2009).

The effect of specific nutrient limitation or starvation on lipid production has been studied. The elements with a higher positive influence on lipid accumulation are N and Fe (Converti et al., 2009; Courchesne et al., 2009; Chen et al., 2011b). In this regard, Rodolfi et al. (2009) studied nitrogen limitation in *Chlorella sp.*, *Scenedesmus sp.*, *Tetraselmis suecica,* and *Nannochloropsis sp.* cultures. As lipid accumulation increased, biomass productivity progressively decreased, however, there was no significant difference between lipid content with nitrogen restriction and controls with an abundance of nitrogen in freshwater strains of *Chlorella sp.* and *Scenedesmus sp.* While in *Tetraselmis suecica* and *Nannochloropsis sp.* strains, nitrogen limitation increased lipid accumulation, doubling lipid content after 9 and 4 days in the low nitrogen-containing medium, respectively.

Moreover, Lin and Lin (2011) analyzed the composition of FAMEs in *Scenedesmus rubescens* under the limitation of different nitrogen sources: urea, $NaNO_3$, and $(NH_4)_2CO_3$. The largest production of biomass was obtained using $NaNO_3$, whereas the highest production of FAMEs was 0.133 g/L • d and was obtained after 12 days of $(NH_4)_2CO_3$ limitation. Additionally, the ratio of fatty acids of the C16 and C18 series increased at the end of the culture (from 83.62% to 90.06%), with fatty acidsC18:0, C18:1n-7, 9, 11, C18:2n-6, and C18:3n-3 accounting for64% of total FAMEs. This points to the fact that the type of nitrogen source influences the production of biomass; it also has a dual effect on the accumulation of lipids, both qualitatively and quantitatively.

The strategy to limit nutrients involves a compromise between lipid productivity and biomass productivity since the limitation of nitrogen stops cell replication during lipid accumulation. Nitrogen stress reduces CO_2 due to the overload in the photosynthetic electron transport chain, where the increase in O_2 concentration results in photo-oxidative and photo-inhibiting stress to the culture. To release the excess electrons, cells synthesize compounds that demand large quantities of NADPH, such as TAGs; in this way, metabolism is directed towards *de novo* synthesis of TAGs, mainly C18 since approximately 24 molecules of NADPH are required; this value is double the required amount for the synthesis of proteins or carbohydrates of the same molecular weight (Hu et al., 2008; Stephenson et al., 2009). Simultaneously, the reduction in photosynthetic activity caused by the limitation of nitrogen results in a drastic reduction of cellular chlorophyll and an increase in carotenoid content as a defense mechanism to oxidative damage of chlorophyll (Li et al., 2008b; Zhekisheva et al., 2002).

Different complementary strategies to overcome the challenge to cellular replication during nitrogen source limiting periods have been proposed. One option is the addition of inorganic carbon; in this regard, Lin et al.(2012) concluded the highest biomass and FAMEs productivities, with values of 109.68 and 11.53 mg/L • d, respectively, were obtained at higher $NaHCO_3$ concentrations in *N. oculata* cultures, since the supplement of inorganic carbon ($NaHCO_3$) allowed for a better assimilation of the nitrogen source.

Another way of adding inorganic carbon to the medium is enriching the air supply with CO_2.Gordillo et al. (2001) observed a dual effect between the addition of CO_2 (10 000 ppm) and the limitation of nitrates in cultures of *Ulva rigida*, without the addition of CO_2 the highest proportion of neutral lipids was obtained when there was a limitation of nitrates. Notwithstanding, the addition of CO_2 coupled to the limitation of nitrates promoted the highest accumulation of neutral lipids, obtaining up to 40% of lipid extract. A similar study performed on *Dunaliella viridis* cultures revealed that the limitation of nitrates coupled to the addition of 1% (v/v) CO_2 in the air supply promoted the highest accumulation of neutral lipids; however, there was no significant difference in the proportion of neutral lipids obtained through treatments with an abundance of nitrates with and without the addition of CO_2, and when there was limitation of nitrates without CO_2 enrichment. This points to the fact that the addition of CO_2 in the nitrogen limiting phase directs metabolism towards the accumulation of neutral lipids and reduces the production of polar lipids (Gordillo et al., 1998).

Furthermore, the addition of CO_2 in small concentrations (2-5%, v/v) allows the assimilation of NO_3^- and also increases the specific growth rate up to 3 times compared to a medium without CO_2(Chiu et al., 2009). However, certain CO_2 concentrations inhibit culture growth due to a decrease in pH, since the availability of inorganic carbon in CO_3^{2} form cannot be assimilated by the majority of microalgal species (Chai et al., 2012). This demands the establishment ofspecificCO_2supply mechanisms for each microalgal species.

Due to the fact that the accumulation of lipids occurs in conditions in which cell growth diminishes, it is necessary to combine different CO_2 limiting and supplementing strategies fitted to the culture regime to allow for the highest production of biomass with a high lipid content.

Table 1. Microalgal genomic sequences, available and in sequencing process (Tirichine y Bowler, 2011)

Class	Microalgalspecies	Strain	Project	Genome (Mb)
Prasinophyta	*Ostreococcustauri*	OTH95	Genome	13
	Ostreococcuslucimarinus	CCE9901	Genome	13
	Ostreococcussp.	RCC809	Genome	12
	Micromonaspusilla	CCMP1545	Genome	22
	Micromonaspusilla	RCC249	Genome	21
	Bathycoccusprasinos	BBAN7	Genome	18
Chlorophyta	*Chlamydomonasreinhartii*	CC-503	Genome	121
	Chlamydomonasincerta		EST	ND
	Volvox carteri		Genome	140
	Dunaliella salina	UTEX2908	Genome	130
	Chlorellavulgaris	CCAP19/18	Genome	46
	Coccomyxasp.	NC64A	Genome	49
	Mesostigmaviride	C-169	EST	
	Nephroselmisolivacea		EST	
Rhodophyta	*Cyanidioschyzonmerolae*	10D	Genome	140
	Galdieriasulphuraria		Genome	130
	Porphyrayezoensis		EST	46
	Chondruscrispus		Genome	49
Glaucophyta	*Cyanophoraparadoxa*		EST	
	Glaucocystisnostochinearum		EST	
Heterokon-tophyta	*Thalassiosirapseudonana*	CCMP1335	Genome	32
	Phaeodactylumtricornutum	CCP1055/1	Genome	27
	Fragilariopsiscylindrus	CCMP1102	Genome	81
	Pseudo-nitzschiamultiseries	CLN-47	Genome	
	Amphorasp	CCMP2378		
	Atheyasp.	CCMP212		
	Fragilariopsiskerguelensis	Ec32		214
	Ectocarpussiliculosus	CCMP1984	Genome	
	Aureococcusanophagefferens		Genome	
Haptophyta	*Emilianiahuxleyi*	CCMP1516	Genome	5
	Emilianiahuxleyi	RCC1217	Genome	5
	Emilianiahuxleyi	CCMP71	EST	
	Phaeocystisantartica			
	Phaeocystisglobosa			
	Pavlovalutheri		EST	
	Isochrysisgalbana	CCMP1323	EST	
Cryptophyta	*Guillardia theta*	CCMP2712	Genome	
	Goniomonas sp.	ATCC	EST	
	Goniomonas sp.	50108	EST	
Chlorarachniophytes	*Bigelowiellanatans*	CCMP2755		

EST: Expressed sequence tag.

In general, productivity and lipid content are inversely correlated, and stress conditions such as deprivation of nitrogen or phosphate, which limit cell growth, also increase lipid content. Nitrogen is the most commonly reported nutritional-limiting factor that triggers total lipid accumulation, mainly TAG in green microalgae (Hu et al., 2008; Pruvost et al., 2009, 2011). The underlying principle is that where there is insufficient N for protein synthesis as required for growth, excess carbon from photosynthesis is channeled into storage molecules, such as TAGs or starch, and protein content may be reduced. Some oleaginous microalgae seem to have the capacity for synthesizing *de novo* lipids when grown under nitrogen-deficiency conditions, channeling the excess of carbon and energy into storage lipids, mainly TAGs (Shifrin & Chisholm, 1981; Rodolfi et al., 2009).

A two-stage process has been suggested for N-limitation: 1) cells are first grown under nutrient-sufficient conditions for biomass accumulation followed by 2) nutrient deprivation for lipid synthesis. However, it is also possible to allow cells to deplete their natural nitrogen source rather than transfer them to a medium completely lacking N. At low nitrogen levels, *Scenedesmus obliquus* and *Chlorella vulgaris* contained high percentage of total lipids (45% of the biomass); mainly neutral lipids (Piorreck et al., 1984). In addition, higher lipid content values were reported under nitrogen-deficient conditions in *Chlorella emersonii* (63%, dw) and *C. minutissima* (56%, dw).

Metabolic Engineering of Lipid Production Pathways

With the advent of genome sequences and molecular tools for algae, there is the possibility that metabolic engineering may provide important and significant improvements for algal biodiesel production; for instance, by increasing yields of TAGs, or engineering pathways for novel biofuel molecules.

During the past decade, significant advances have been made in microalgae genomic databases. As such, the genomic sequences for some species of microalgae are available (Table 1). Certain metabolic pathways have been inferred from homology based searches inplant genes involved in lipid metabolism (Guschina and Harwood, 2006). Generally, TAG biosynthesis is composed of three main stages: acetyl-CoA carboxylation, elongation of the acyl group chain, and assembly of TAGs.

Based on the metabolic pathways of TAGs synthesis, models of overexpression of key enzymes for each stage of the process have been proposed (Courchesne et al., 2009).

The first checkpoint heavily linked to TAGs synthesis is the conversion of acetil-coenzymeA (CoA) into malonyl-CoA. This reaction is catalyzed by the acetyl-CoA carboxylase (ACC) enzyme. With the aim of increasing lipid production, ACC has been overexpressed in *Brassica napus* (rapeseed), *Solanum tuberosum* (potato), *Cyclotella cryptica* (diatom), and *Naviculas aprophila* (diatom).

The main result obtained in plants was a 5-fold increase in TAGs concentration in *S. tuberosum* with 0.058 mg TAG/mg wet weight (Klaus et al., 2004), whereas in the other organisms no positive effect was observed in the lipid content (Dunahay et al., 1995; Roesler et al., 1997). This suggests that overexpression of only ACC is not enough to increase lipids synthesis in diatoms; probably because the ACC-induced carboxylation is not a critical step in the metabolic pathway, or it could be ascribed also to the generation of another bottleneck due to the increase in ACC activity (Sheehan et al., 1998).

The reaction product of CoA carboxylation, CoA malonyl, is the main carbon donor for lipid synthesis, and participates in posterior reactions of fatty acids condensation. The first condensation reaction is accomplished by the carrier enzyme 3-ketoacyl-acyl-synthase III (KAS III). The KAS III enzyme generates a four-carbon product, which is the precursor of 6-18 carbon fatty acids (Hu et al., 2008). For this, increase in lipid productivity in *Nicotianatabacum* (tobacco), *Arabidopsis thaliana,* and *B. napus* has been attempted through overexpression of KAS III. However, the obtained results revealed a diminution in lipid content (Dehesh et al., 2001). This diminution can be attributed to KAS III belonging to the fatty acids synthetase (FAS) complex, where enzymatic activity depends on the interaction among its subunits and requires simultaneous manipulation of diverse subunits of the FAS complex.

On the other hand, manipulation of enzymes involved in the TAGs ensemble has yielded promising results. Overexpression of the enzyme glycerol 3-phosphate dehydrogenase (G3PDH) in *B. napus* allowed for a 40% increase of lipid content (Vigeolas et al., 2007). Likewise, overexpression of diacylglycerol acyltransferase(DAGAT), which catalyzes the last reaction of diacylglycerol to TAG, showed a higher lipid content in *Arabidopsis* plants, with a 10 to 70% increase (Jako et al., 2001). The success of the DAGAT overexpression strategy is attributed to the fact that the used substrate, diacylglycerol, can be used for the synthesis of TAG or for phospholipids formation. This suggests that DAGAT is a limiting step in lipids synthesis. However, there are no reports on DAGAT modification in microalgae.

Another used strategy to increase lipids production is by blocking metabolic pathways that promote the synthesis of energetic compounds (like carbohydrates) whose substrate is shared by the lipids pathway.

In this regard, silencing of the genes encoding for the enzymes ADP-glucose pyrophosphatase and isoamylase in mutants of the microalga *Chlamydomonas reinhartii* has been documented. These modifications exerted an effect during a nitrogen-limiting culture, reducing the starch of its cellular structure, which enabled an increase in fatty acids, 90% of which were TAGs (Wang et al., 2009).

Despite the efforts to increase lipid production through genetic modifications, the biotechnological processes based on transgenic microalgae are still under development due to the scarce information available from both the corresponding works and molecular databases, which has hindered the exploitation of the metabolism and lipid accumulation from oleaginous species. Hence, manipulation of culture conditions is the best documented strategy, and is the basis for bioprocesses at the commercial level.

Yusuf Chisti (2007) proposed different molecular level engineering that can be used to potentially:

1. Increase photosynthetic efficiency to enable increased biomass yield with light;
2. Enhance biomass growth rate;
3. Increase oil content in biomass;
4. Improve temperature tolerance to reduce the expense of cooling;
5. Eliminate the light saturation phenomenon
6. Reduce photo inhibition; and
7. Reduce susceptibility to photooxidation that damages cells.

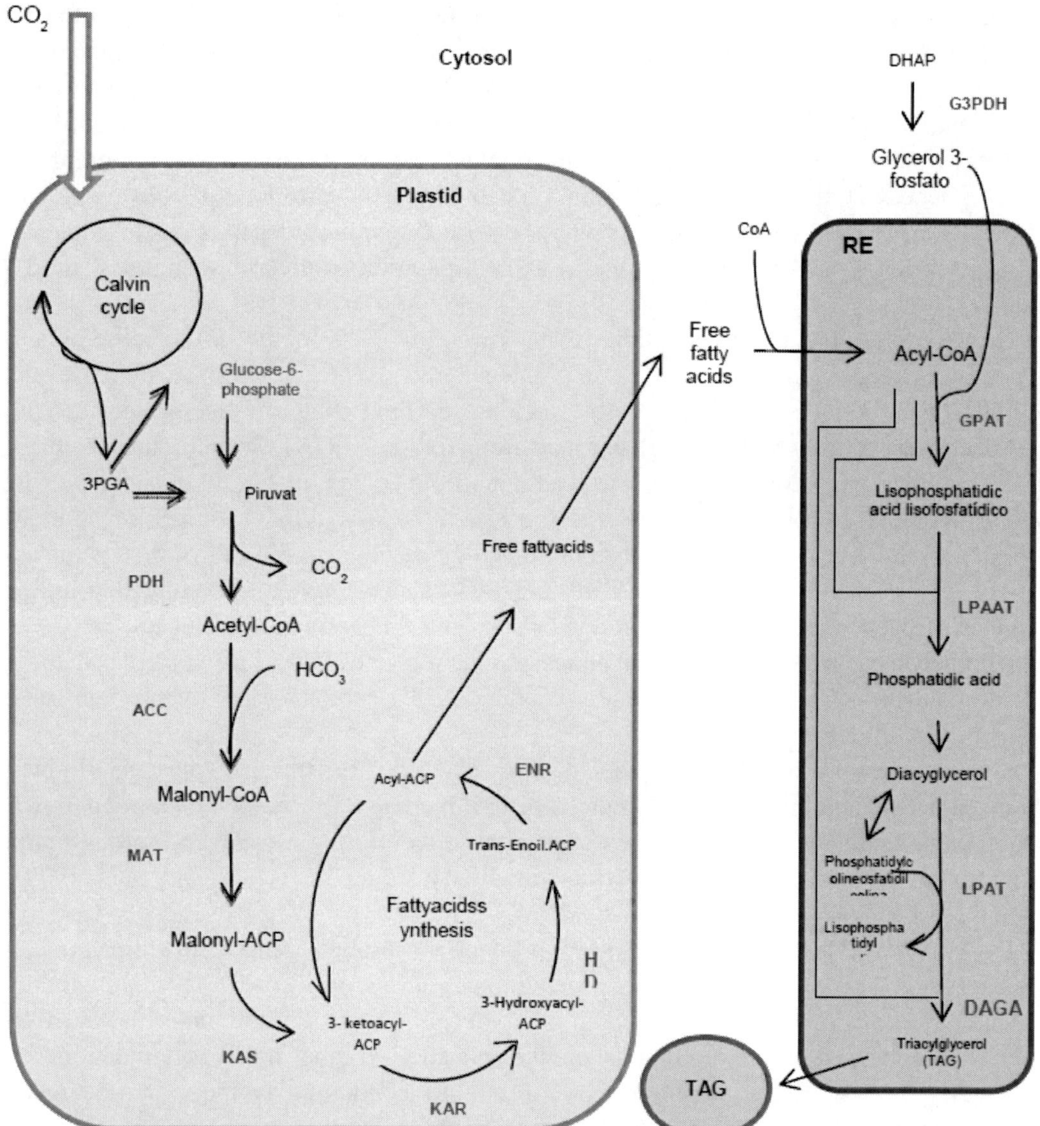

Figure 1. General scheme of lipids biosynthesis (Radakovits et al., 2010).
In blue, the main enzymes for lipids synthesis; **3PGA**, 3-phosphoglyceric acid; **ACP**, carrier protein of acyl groups; **PDH**, pyruvate dehydrogenase complex; **ACC**, acetyl CoA carboxylase; **MAT**, malonyl-CoA:ACP trans-acylass; **KAS**, 3-ketoacyl-ACP synthase; **KAR**, 3-ketoacyl-ACP reductase; **HD**, 3-didroxyacyl ACP dehydratase; **ENR**, enoyl-ACP reductase; **RE**, endoplasmic reticulum; **G3PDH**, glycerol 3-phosphate dehydrogenase; **GPAT**, glycerol 3-phosphateacyltransferase; **LPAAT**, phosphatidic acid acyltransferase; **LPAT**, lisophosphatidylcoline acyltransferase; **DAGAT**, diacylglycerol acyltransferase.

The use of genetically modified microalgae must take into account the worldwide biological safety standards as well as those of each country, particularly in open large scale systems.

Chromatographic Analysis of Fatty Acids

Analysis of FAMEs is an important tool to characterize fats and oils. The most common analytical methods is gas chromatography.

Owing to their low volatility, fatty acids from oils cannot be analyzed directly in a gas chromatographer; it is necessary to convert them into low molecular weight compounds that are more volatile, such as methyl esters. At the laboratory level, transesterification is achieved with a large excess of methanol (100 times higher than the required stoichiometric amount) and using either alkaline catalysts, such as KOH and NaOH, or acid ones, such as HCl, H_2SO_4 or BF_3.

FAMEs are separated into columns with polar phases such as polyethylenglycol (PEG) or cyanopropylsiloxane. Elution velocity of each methyl ester depends mainly on its volatility, which depends on the length of the carbon chain and the number and position of the molecule's unsaturations. The PEG polar columns resolve FAMEs by the number of carbons followed by the degree of unsaturation, whereas cyanosilicone columns are highly polar and resolve better isomers cis and trans, and other stereoisomers. In general, FAMEs with less number of carbons elute faster, and those with double bonds elute slower than saturated FAMEs (Figure 2).

Peaks can be identified based on the retention time, but must be confirmed through other analyses, such as thin-layer chromatography, fractionation column, or mass spectrometry.

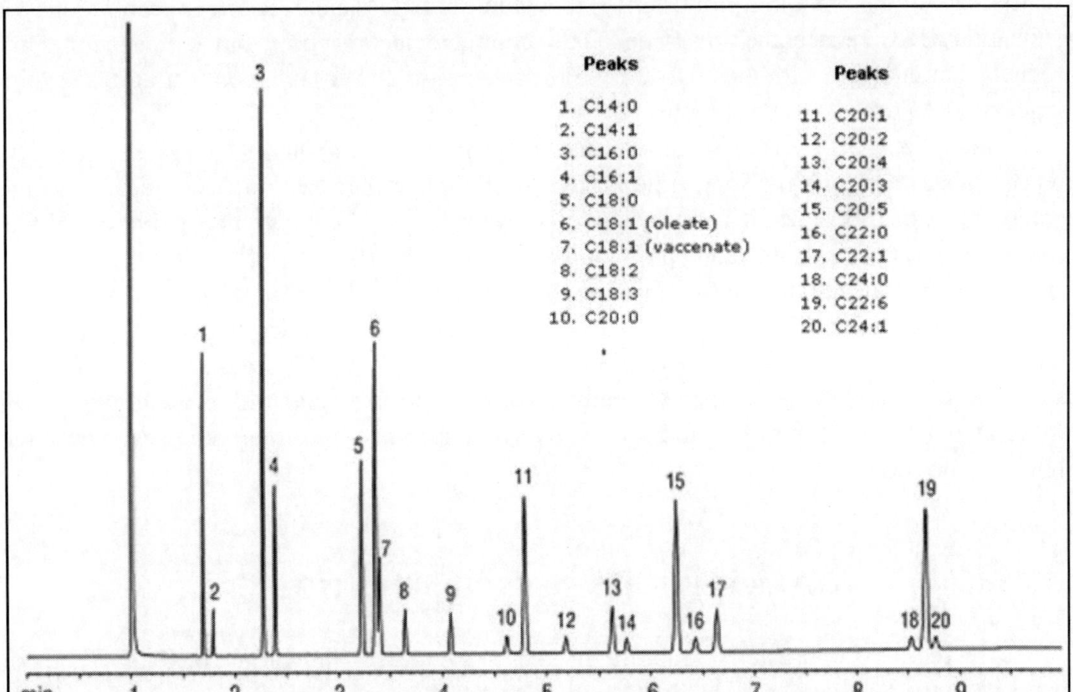

Figure 2. FAMEs standard in PEG FAMEWAX column (30m, ID 0.32mm, 0.25 um).

ACID TRANSESTERIFICATION OF FAMES

Acylglycerides and free fatty acids transesterify when heated with a large excess of anhydrous methanol in the presence of an acid catalyst (Figure 3).

$$RCOOR' + CH_3OH \underset{}{\overset{H^+}{\rightleftharpoons}} RCOOCH_3 + R'OH$$

$$RCOOH + CH_3OH \underset{}{\overset{H^+}{\rightleftharpoons}} RCOOCH_3 + H_2O$$

Figure 3. Transesterification reaction of fatty acids.

The presence of water prevents completion of the reaction; therefore, reagents must be water-free. It is possible to generate water-free conditions by adding anhydrous sodium sulfate to the reaction medium, but a more simple practice consists of operating with anhydrous reagents and materials. The reaction can be performed in a sealed tube, which allows for higher temperatures and less reaction time.

According to Christie (1993), the reagent most cited for the preparation of methyl esters is anhydrous HCl (5%) in methanol. Reaction conditions are usually: heating with reflux for 2 h or at 50°C overnight. However, other authors use smaller catalyst concentrations with different temperatures and incubation times. The appropriate condition depends on the specific conditions required to esterify the studied fatty acids. High concentrations of methanolic HCL reduce the time needed to complete the reaction, but can generate by-products that interfere with the analysis. On the other side, lower HCL concentrations require longer reaction times.

Methanolic HCl is the best general use reagent for esterification (Christie, 1993). Its greatest disadvantage is the long time needed to complete the reaction (as compared to the alkaline reaction). Besides, it is not adequate to esterify fatty acids with sensitive functional groups, such as cyclopropane and epoxy rings.

A 1 to 2% concentrated sulfuric acid solution in methanol has almost the same properties as the 5% methanolic HCl. The same reaction times are recommended for H_2SO_4- and HCl-methanol (Christie, 1993). Boron trifluoride (BF_3) can also be used, because it is a strong acid catalyst to transesterify most types of lipids. However, $BF_3$is a short life reagent, even when refrigerated, and can give rise to undesirable by-products when becoming old or when used at high concentrations.

ALKALINE TRANSESTERIFICATION

Acylglycerides transesterify rapidly in the presence of anhydrous methanol and an alkaline catalyst (Figure 4). At 50°C and with 100% excess methanol, triacylglycerols transesterify completely in 10 min. The disadvantage is that the free fatty acids do not esterify and the reaction must be anhydrous to prevent hydrolysis of lipids and undesirable saponification that can hinder the separation of phases. The alkaline catalysts are 0.5 to 2 M sodium or potassium methoxide. Although NaOH or KOH can also be used at similar concentrations, but maintaining strict water-free conditions.

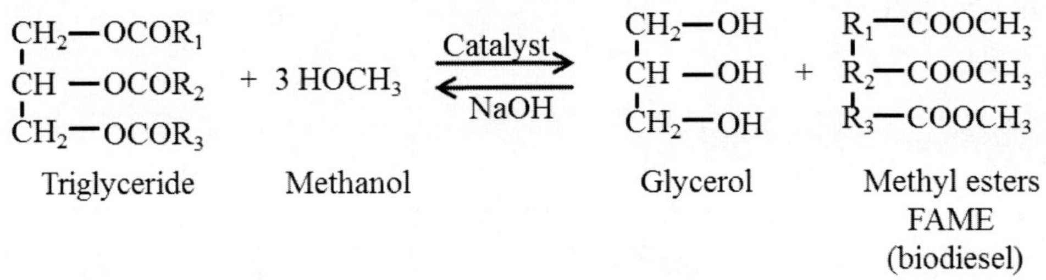

Figure 4. Transesterification reaction with an alkaline catalyst.

EFFECT OF NITROGEN LIMITATION AND CO2 SUPPLY ON THE FATTY ACID PROFILES OF MICROALGAL OILS FROM DUNALIELLATERTIOLECTA AND NANNOCHLOROPSISOCULATA

General Objective

To determine the fatty acids profiles of microalgal oils from *Dunaliellatertiolecta* and *Nannochloropsisoculata* under nitrogen limitation and CO_2 supply.

Specific Objectives

To establish the methylation technique for microalgal oils.

Standardize the gas chromatography method for the qualitative analysis of microalgal oils.

Methodology

Culture Systems

The lipid samples were obtained from two microalgal strains, *Nannochloropsisoculata* and *Dunaliellatertiolecta*, cultured under different nutrient conditions and culture systems. The systems from which the samples were obtained and their culture conditions are described in the following.

1-Lphotoreactors

N. oculata and *D. Tertiolecta* were cultured in F/2 medium in 1-L glass photobioreactors, with a luminous intensity of 100 μmol photon/m^2• s, 12:12 photoperiods, temperature controlled at 25°C and aeration at 2 vvm. In a first experiment, sudden nitrogen limitation was induced in both strains. Nitrates concentration was maintained at 250 mg/Land when reaching the stationary stage, the culture medium was replaced with a nitrates-free medium and cellular lipids were extracted at 24, 48, and 72 h of limitation. In another experiment, the simultaneous effect of NO_3concentrations and CO_2supply was tested in *D. tertiolecta*. The systems were enriched with CO_2 and NO_3, both variables at three levels: 0, 2, 4% (v/v) and

150, 200.250 mg/L, respectively. Cellular lipids were extracted daily between 5 and 10 days of culture.

200-L Raceways

N. oculata was cultured in 200-L semi-continuous raceway systems under nursery conditions. Harvesting cycles were performed when nitrates concentrations reached approximately 50 mg/L, removing half of the operation volume. All lipid samples were obtained 1 h before each harvest. In one experiment, the removed medium was substituted by: 1) fresh medium supplemented with nitrates adjusting the concentration to 250 mg/Land 2) fresh medium. Afterwards, the addition of CO_2at a flow of 2L/min during 30 min per day was evaluated, comparing with a culture without CO_2 addition. At each harvest, the nitrates-enriched medium was replaced by fresh medium, then adjusting the nitrogen concentration to 250 mg/L.

Lipids Extraction

The biomass of a 30-mL sample was separated by centrifugation at 6000 rpm for 15 min, and dried in an oven at 60°C for 12h. Hexane (4 mL) was added and incubated for 12 h at 4°C. Then, the biomass was washed with 3 mL hexane two times. The extract was transferred to a vial at constant weight for the gravimetric determination of total lipids.

Lipids Methylation

The lipids extract was transesterified with 750 μL of 0.5 methanolic HCl in closed tubes at 80°C for 3h. FAMEs were dissolved again in 1 mL hexane, filtered through a 2-μm membrane to a vial for its analysis by gas chromatography.

Chromatographic Analysis

Samples were analyzed in the Perkin Palmer Clarus 500 gas chromatographer. The AT-WAX capillary column (30 m long, 0.25 mm internal diameter. and 0.2 μm width of the stationary stage film) was used. Injector temperature was of 230°C. A flame ionization detector (FID) was used at a temperature of 250°C.The flame was kept on with an H_2 flow of45mL/min and an air flow of 450mL/min. Injection volume was 1 μL with a 1:1 split. Oven temperature was kept at 140°C for 5min, afterwards the temperature was increased 8°C/min until 240°C, and was kept for 15 min at 240°C. Helium was the carrier gas, at a pressure of 15.19 psi, a flow of 0.96 mL/min, and a linear velocity of 26.5 cm/s.

Results

Establishing the Gas Chromatography Analysis

The method by Morales et al. (2013) was followed with modifications, adjusting some parameters such as the oven temperature ramp and the split. The best response of the detector and best peaks separation, in a reasonable time, were obtained with the following method: 2 μL of injection volume, 1:1 split. Injector temperature at 230°C and of the detector at 250°C.Initial oven temperature 140°C for 5 min, increasing 8°C/min until 240°C, and maintained at 240°C for 15 min (Figure 5).

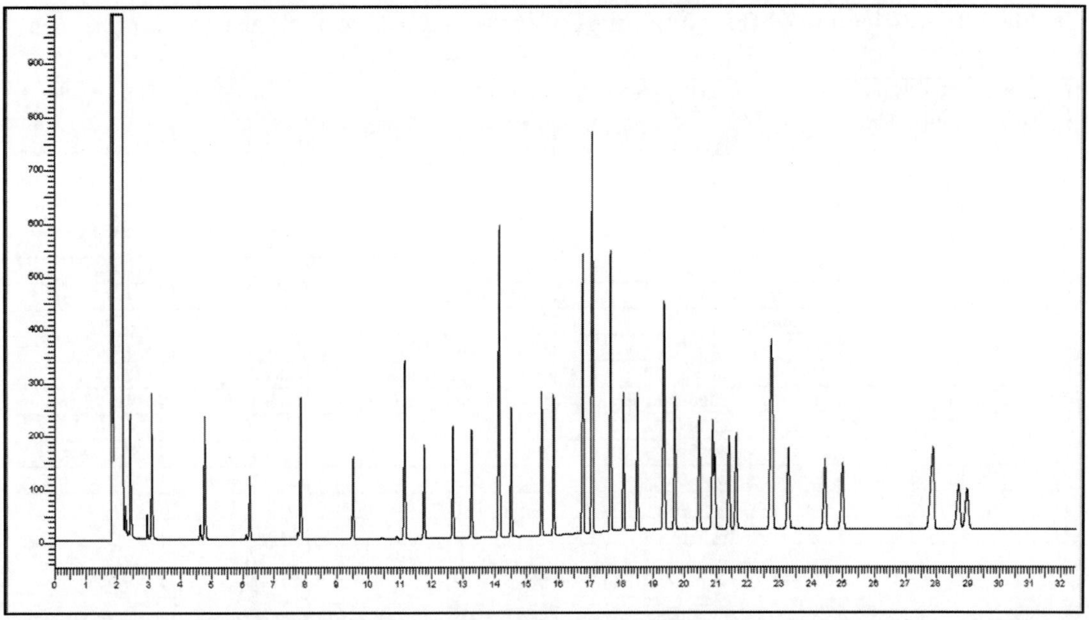

Figure 5. Supelco 37 (47885-U) standard analyzed in the AT-WAX column. Oven temperature program: 140°C for 5min, 8°C/min until 240°C, 240°C for 15 min.

Identification of the Peaks of the Standard

For the qualitative analysis of the fatty acid, the Supelco37 FAME mix (Sigma Aldrich 47885-U) was used. The elution order and peaks identification were achieved in a gas chromatographer coupled to a mass spectrometer, with a capillary AT-WAX column of 30 m length, 0.25 mm internal diameter, and 0.2 μm film width of the stationary phase. Likewise, we compared our results regarding the elution order with those reported in catalogs and publications by manufacturers of polyethylenglycol columns of similar dimensions.

Thirty-three of the thirty-seven components of the standard were found and identified with the used method. The non-separable compounds were:

1. *Butyric acid (C4:0); due to its low boiling point, it elutes practically at the same velocity as the solvent (dichlormethane).*
2. Elaidic acid (C18:1n9t); it is a trans isomer of the oleic acid and is not separable with the used method.
3. Linolelaidic acid (C18:2n6t); it is a trans isomer of the linoleic acid and is not separable with the used method.
4. Behenic acid (C22:0); it elutes immediately after the eicosapentaenoic acid (C20:5n3)in PEG columns; however, with the used method it was not possible to separate both fatty acids.

Separation of the aforementioned compounds is not relevant for the analysis of oils for biodiesel. The elution order and retention time of the compounds of the standard (Table 2) were obtained.

Table 2. Identification of the compounds of the standard through gas chromatography

Elution order	Retention time (min)	Fatty acid (methylester)	Common name	Formula	CAS Number
1	2.410	Hexanoic acid	Caproic acid	C6:0	10670-7
2	3.094	Octanoic acid	Caprilic acid	C8:0	111-11-5
3	4.778	Decanoid acid	Capric acid	C10:0	110-42-9
4	6.225	Undecanoic acid	Undecanoic acid	C11:0	1731-86-8
5	7.864	Dodecanoic acid	Lauric acid	C12:0	111-82-0
6	9.525	Tridecanoic acid	Tridecanoic acid	C13:0	1731-88-0
7	11.157	Tetradecanoic acid	Miristic acid	C14:0	124-10-7
8	11.765	Cis-9- tetradecenoic acid	Miristoleic acid	C14:1	56219-06-8
9	12.680	Pentadecanoic acid	Pentadecanoic acid	C15:0	7132-64-1
10	13.266	Cis-10-pentadecenoic acid	Cis-10-pentadecenoic acid	C15:1	90176-52-6
11	14.153	Hexadecanoic acid	Palmitic acid	C16:0	112-39-0
12	14.524	Cis-9-hexadecenoic acid	Palmitoleic acid	C16:1	1120-25-8
13	15.479	Heptadeconoic acid	Heptadeconoic acid	C17:0	1731-92-6
14	15.859	Cis-10-heptadecenoic acid	Cis-10-heptadecenoic acid	C17:1	90176-62-6
15	16.780	Octadecanoic acid	Stearic acid	C18:0	112-61-8
16	17.079	Cis-9-octadecenoic acid	Oleic acid	C18:1n9c	112-62-9
17	17.657	Cis,cis-9,12-octadecadienoic acid	Linoleic acid	C18:2n6c	112-63-0
18	18.061	Cis-6,9,12-octadecatrienoic acid	γ-Linolenic acid	C18:3n6	16326-32-2
19	18.506	Cis-9,12,15-octadecatrienoic acid	Linolenic acid	C18:3n3	301-00-8
20	19.363	Eicosanoic acid	Arachidic acid	C20:0	120-28-1
21	19.693	Cis-11-eicosenoic acid	Cis-11-eicosenoic acid	C20:1	2390-09-02.
22	20.470	Cis-11,14-eicosadienoic acid	Cis-11,14-eicosadienoic acid	C20:2	2463-02-7.
23	20.903	Cis-8,11,14-eicosatrienoic acid	Cis-8,11,14-eicosatrienoic acid	C20:3n6	21061-10-9
24	20.966	Heneicosanoic acid	Heneicosanoic acid	C21:0	6064-90-0
25	21.417	Cis-11,14,17-eicosatrienoic acid	Cis-11,14,17-eicosatrienoic acid	C20:3n3	55682-88-7
26	21.646	Cis-5,8,11,14-eicoatetraenoic acid	Araquidonic acid	C20:4n6	2566-89-4
27	22.746	Cis-5,8,11,14,17-eicosapentaenoic acid	Cis-5,8,11,14,17-eicosapentaenoic acid	C20:5n3	2734-47-6
28	23.274	Cis-13-docosenoic acid	Euricic acid	C22:1n9	1120-34-9
29	24.436	Cis-13,16-docosadienoic acid	Cis-13,16-docosadienoic acid	C22:2	61012-47-3
30	25.004	Tricosanoic acid	Tricosanoic acid	C23:0	2433-97-8
31	27.892	Tetracosanoic acid	Lignoceric acid	C24:0	2442-49-1
32	28.701	Cis-4,7,10,13,16,19-docosahexanoic acid	Cis-4,7,10,13,16,19-docosahexanoic acid	C20:6n3	2566-90-7
33	28.960	Cis-15-tetracosenoic acid	Nervonic acid	C24:1	2733-88-2

Effect of Sudden NO_3 Limitation on Fatty Acids Composition

In *D. tertiolecta*, larger times of sudden limitation diminished the proportion of saturated fatty acids C16:0 and, even more, C18:0, increasing the proportion of unsaturated fatty acids, mainly linoleic acid C18:3 (Figure 6). Chen et al. (2011) have also reported the increase of C18:3 in *D. tertiolecta* with nitrogen limitation.

Despite that the nitrates limitation time increased the production of cellular lipids in both strains, the composition of fatty acids did not change importantly between 24, 48, and 72 h of limitation in *N. oculata* (Figure 7). Pruvost et al. (2012) reported a decrement of C16:1 and C20:5, although at larger limitation times.

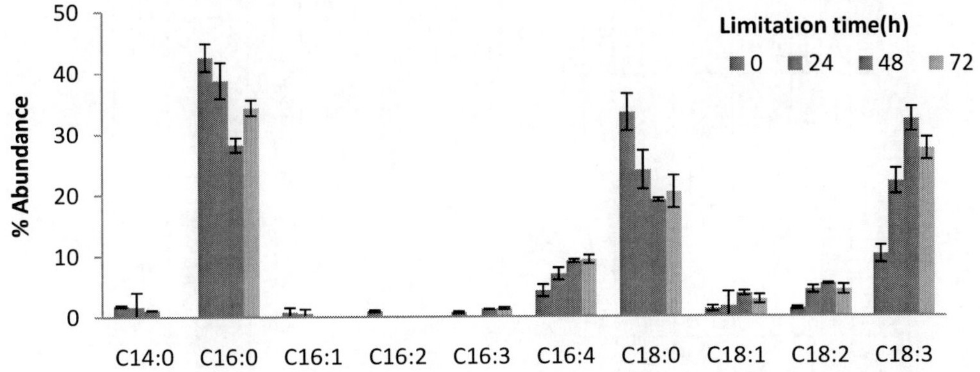

Figure 6. Variation in fatty acids composition along NO_3limitation time in *D. tertiolecta*.

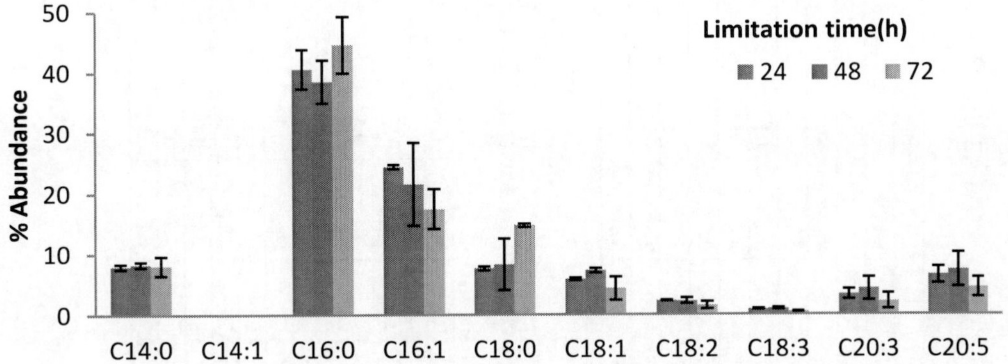

Figure 7. Variation in fatty acids composition along NO_3 limitation time in *N. oculata*.

Effect of CO_2 and no_3 Concentration on the Composition of D. Tertiolecta Fatty Acids

No important changes were found in fatty acids composition with respect to time, between days 5 and 10 of culture, in any of the analyzed levels. Variation in CO2 concentration in the culture did neither affect importantly the accumulation of specific fatty acids. These results agree with those reported by Tsuzuki et al. (1990): fatty acids composition of *D. tertiolecta* did not change importantly between a 5% CO_2-enriched culture and one without CO_2 enrichment. Changes in fatty acids composition were neither observed with changes in nitrogen concentration.

It was found that the main fatty acids are, in average: C16:0(42.6%), C18:0 (32.7%), C18:3 (10.3%), and C16:4 (4.3%), without important differences between the analyzed NO_3 and CO_2 levels (Table 3). Worthwhile mentioning is the large accumulation of stearic acid C18:0, which was the second most abundant fatty acid after C16:0. This finding is in contrast with the different fatty acid profiles reported for *D. tertiolecta* (Table 4), in which the main fatty acids are C16:0, C16:4, and C18:3, but C18:0 always appears at very low levels (below 2%).

Table 3. Fatty acids composition of *D. tertiolecta* with respect to CO_2 and NO_3 concentration

NO_3 (mg/L)	**CO_2 (%v/v)**	**% Abundance**										
		C14:0	C16:0	C16:1	C14:0	C16:3	C16:4	C14:0	C18:1	C18:2	C14:0	Total
250	4	0.2	250	4	0.2	250	4	0.2	250	4	0.2	250
	2	0.8	44.1	2	0.8	0.7	2	0.8	0.8	2	0.8	96.3
	0	1.8	45.2	0	1.8	0.6	0	1.8	0.8	0	1.8	97.7
200	4	1.0	200	4	1.0	200	4	1.0	200	4	1.0	200
	2	1.0	41.0	2	1.0	0.8	2	1.0	1.6	2	1.0	94.6
	0	1.6	41.4	0	1.6	0.5	0	1.6	1.2	0	1.6	96.7
150	4	2.4	150	4	2.4	150	4	2.4	150	4	2.4	150
	2	1.1	41.5	2	1.1	0.8	2	1.1	1.4	2	1.1	96.3
	0	1.9	41.2	0	1.9	0.9	0	1.9	1.7	0	1.9	96.9

Table 4. Reported fatty acid profiles for *D. tertiolecta*

N limit.	**CO_2Enrich.**	**C14:0**	**C14:1**	**C16:0**	**C16:1**	**C16:2**	**C16:3**	**C16:4**	**C18:0**	**C18:1**	**C18:2**	**C18:3**	**Reference**
No	0%	0.8		**22.9**	6.3	2.5	2.4	**16.1**		2.8	**10.8**	**35.2**	Tsuzuki et al. 1990
No	5%	1.4		**24.2**	2.3	3.4	2.8	**17.3**		5.0	**12.4**	**30.9**	
No	4%			**28.1**		2.8	1.4		0.6	19.3	**14.7**	**33.2**	Chen et al. 2011
Yes	4%			**26.4**		2.3	1.3		0.6	16.8	**13.1**	**39.6**	
No	0*	0.8	0.4	**14.8**	1.7	0.6		**17.1**	2.3	2.4	**3.6**	**51.4**	Evans et al. 1982
No	0*	0.3	0.1	**10.3**	5.2	2.0	4.7	**23.9**	0.3	2.2	**5.2**	**41.9**	Zhukova et al. 1995
No	0*	0.2		**14.7**	2.9	0.7	4.2		0.4	2.3	**4.8**	**43.5**	Volkman et al. 1989
	0*	0.5		**17.7**	6.0	3.0	1.2	**10.6**		4.9	**12.4**	**30.2**	Gouveia& Oliveira, 2009
	0*	1.3	0.1	**13.4**	2.5	1.2	3.4	**24.4**	1.0	3.4		**32.3**	Ackman et al. 1968
No	2-4%	1.8	0.0	**42.6**	0.7	0.9	0.7	4.2	**33.5**	1.2	1.3	**10.2**	This work
Yes	0%	0.9	0.0	**33.8**	0.2	0.0	0.9	8.5	**21.2**	2.7	4.6	**27.3**	This work

* without aeration.

Table 5. Fatty acid profiles reported for *N. oculata*

N limitation	CO_2 Enrich.	C14:0	C14:1	C16:0	C16:1	C18:0	C18:1	C18:2	C18:3	C20:4	C20:5	
No	pH*	7.0		26.0	29.0						28.0	Pruvost et al. 2012
Yes	pH*			35.0	21.0		21.0	5.5	5.5			
No	2%	2.4	0.0	14.5	15.7	1.8	10.1	3.6	2.5	6.0	21.5	Roncarati et al. 2004
No	2%	8.4	0.1	26.1	27.9	1.0	13.0	4.2	0.2	2.8	12.5	
No	0%			58.0		5.0	19.0	8.0	10.0			Converti et al. 2009
No	0%	3.4		24.2	26.2	0.8	0.8	2.5		3.8	27.8	Hodgson et al. 1991
No	pH*	7.0		30.0	25.0	2.0	5.0		9.0		15.0	Legrand et al. 2012
No	0**	3.9	0.2	20.5	25.2	1.8	3.6	0.5	2.9	5.6	29.7	Zhukova et al. 1995
No	0.5%	4.8		20.3	20.1	0.3	9.0	1.8	1.0	3.9	25.9	Dunstan et al. 1993
	0**	5.4		26.1	22.8	1.2	6.0	2.3	0.7	4.6	24.1	Renaud & Parry, 1994
No	0**	7.6		36.9	6.7	30.5	3.0	1.2	1.7		2.8	This work (RW)
Yes	0%	8.1		41.2	21.1	10.2	5.8	2.0	0.8	3.2	6.1	This work (PBR)

*CO_2 enrichment to adjust the culture's pH.**Without aeration.

Table 6. Proportion of SFAs, MUFAs, and PUFAs in the analyzed systems

	%SFAs	%MUFAs	%PUFAs
D. tertiolecta with limitation	55.9	2.9	41.2
D. tertiolecta without limitation	67.8	1.7	15.0
N. oculata with limitation	60.4	27.3	12.3
N. oculata semi-continuous RW (NO_3 adjustment)	77.2	14.5	8.3
N. oculata semi-continuous RW (without NO_3 adjustment)	81.8	12.3	5.9
N. oculata semi-continuous RW (CO_2+)	83.3	9.1	7.6
N. oculata semi-continuous RW (CO_2-)	83.2	8.9	7.8

Due to its composition, the oil from *D. tertiolecta* is adequate for its conversion to biodiesel as it has a high content of saturated fatty acids (85%) despite its low amount of MUFAs (4%) (Table 4). However, its high PUFAs content could be inconvenient when nitrogen limitation exists. When the culture is under nitrogen limitation, PUFAs can represent 34% of the fatty acids of *D. tertiolecta*. A solution to this problem could be to mix the *D. tertiolecta* oil obtained under nitrogen limitation with another oil having a higher proportion of saturated fatty acids, or by pretreatment with catalytic hydrogenation to produce a more stable biodiesel.

Fatty Acids Composition of N. oculata Cultured in Semi-continuous Raceways

In the 200-L raceway culture, the main fatty acids were: C16:0 (38.4%), C18:0 (28.5%), C16:1 (8.8%), C14:0 (3.6%).No important differences were found in the composition between the semi-continuous culture adjusted with NO_3at 250mg/L and the non-adjusted culture (Figure 8). This is because, although NO_3 concentration was not the same in both cultures, nitrogen was never a limiting factor for growth in both reactors.

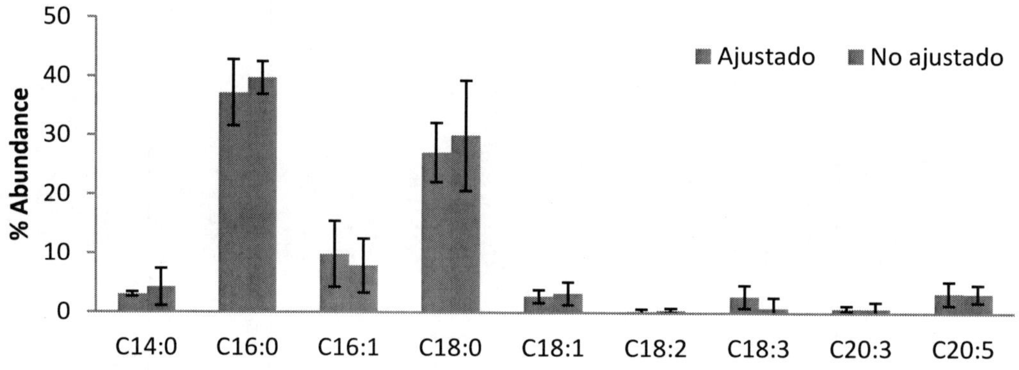

Figure 8. Fatty acids composition of the semi-continuous culture with NO_3 adjustment at 250mg/L with fresh medium at each harvest.

Effect of CO_2 Addition on the Composition of Fatty Acids of N. oculata Cultured in Semi-continuous Raceways

Although CO_2 (2%) addition can increase the proportion of long chain fatty acids, such as C18:0, C18:1, C18:3, and C20:0 (Roncararti et al., 2004);in this experiment, CO_2addition had no effect on the fatty acids composition of *N. oculata* (Figure 9).This is explained by the fact that the supplied CO_2concentration was very low (around 0.01%). These results are consistent with the evolution of the cultures, which were very similar in terms of biomass and lipids production, without any effect of the CO_2 supply.

The homogeneity in the fatty acids composition stands out during *N oculata* culture in raceways, the culture lasted 88 days. Since this is a system open to the environment there is no control on the pH or the temperature of the medium, which undergo important changes along a single day, as well as from one month to another. The proportion of the main fatty acids, C16:0, C18:0, and C18:1, was homogeneous in all experimental harvests. This finding could provide certainty that the fatty acids composition of *N. oculata* cultured in a raceway does not vary importantly despite the environmental changes, which is advantageous for the production of lipids at the industrial scale.

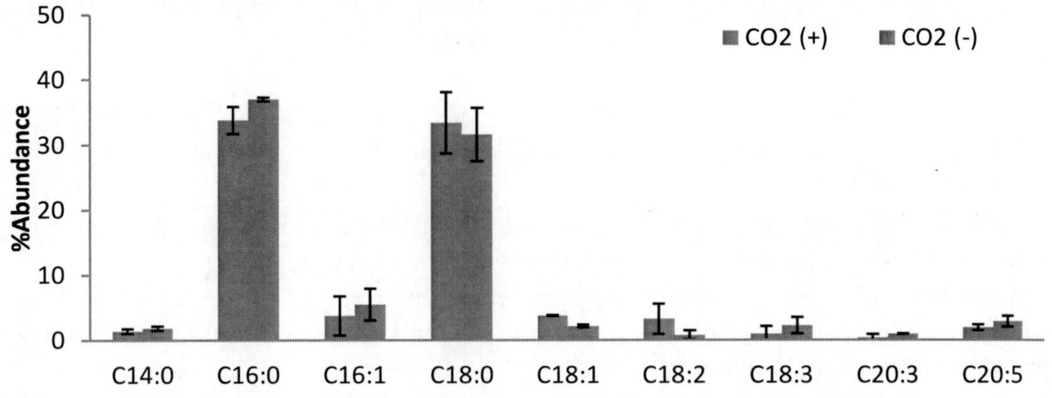

Figure 9. Fatty acids composition of the semi-continuous culture with CO_2 addition (2L/min for 30min daily) and the culture without CO_2 addition.

In contrast, the fatty acids proportion of *N. oculata*was different when cultured in the raceway as compared with the culture in the photobioreactor (PBR), in which a much lower proportion of C18:0 and a higher proportion of unsaturated fatty acids was obtained (Figure 10). This is a logical behavior since these are very different systems in terms of illumination, stirring, aeration, and temperature.

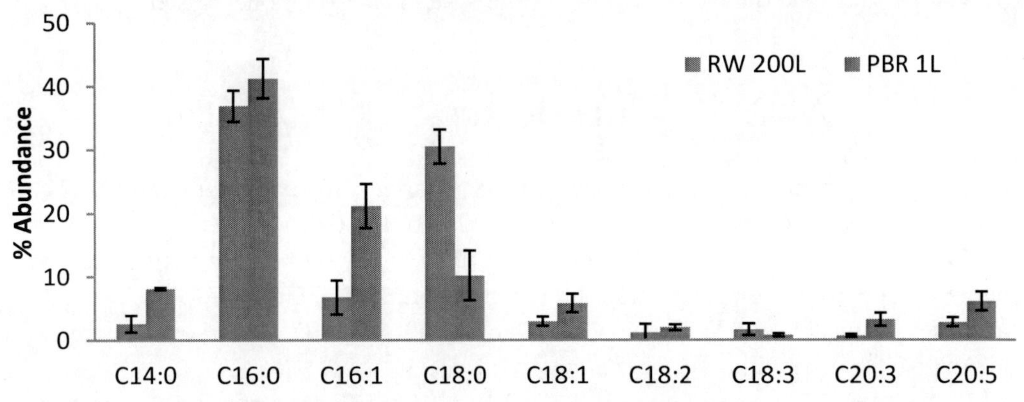

Figure 10. Fatty acids composition of *N. oculata* cultured in the 1-L photobioreactor (PBR) and in a 200-L RW.

Regarding the quality of the biodiesel, the raceway culture of *N. oculata* yields a large amount (81%) of saturated lipids, and a small amount of MUFAs (11%) and PUFAs (7%), which is convenient for the production of a biodiesel with a higher number of cetane and good flow properties at the same time. The photobioreactor culture contains fewer amounts of saturated lipids (60%) at the expense of a higher amount of MUFAs (27%) (Table 6).

The main fatty acids of *N. oculata* are C16:0, C16:1, C18:1, and C20:5 (Table 5), which agrees with the profiles found in this work. However, just like in *D. tertiolecta*, in the performed experiments, the stearic acid C18:0 was again among the main fatty acids of the culture.

Conclusion

In the photobioreactor culture, *D. tertiolecta* accumulated mainly C16:0 (42%), C18:0 (32%), and C18:3 (10%), whereas *N. oculata* accumulated C16:0 (41%), followed by C16:1 (21%) and C18:0 (10%).

Combining the variation in the $NO_3$and CO_2concentrations did not modify importantly the fatty acids composition of *D. tertiolecta* in any of the analyzed levels.

With sudden nitrogen limitation, the proportion of C18:3 in *D. Tertiolecta* increased from 10 to 30% of the fatty acids. In *N. oculata* no important change occurred in the fatty acids composition up to 72 h of limitation.

In the semicontinuous raceway culture, *N. oculata* presentedpredominantlyC16:0 (37%), followed by C16:1and C18:0.

TheC16:1 and C18:0 proportion was very different between the photobioreactor and the raceway cultures. C16:1 proportion was of 21% in the photobioreactor and of 6.7% in the raceway, whereas the proportion of C18:0 was of 10.2% in the photobioreactor and of 30.5% in the raceway.

Addition of 2 L/min during 30 min/day of CO_2did not affect the fatty acids composition of the culture.

The raceway culture presented a very homogeneous composition; high in saturated lipids C16:0 and C18:0, despite the lack of control of the system as it is an open system.

References

Ackman, R.G., Tocher, C.S., McLachlan J. (1968) Marine Phytoplankter Fatty Acids, *The Journal of Fishery Research Board of Canada.* 25: 1603-1620.

Allen, C.A., Watts, K.C., Ackman, R.G., Pegg, M.J. (1999).Predicting the viscosity of biodiesel fuels from their fatty acid ester composition. *Fuel.* 78: 1319–1326.

Brown, M.R., Jeffrey, S.W., Volkman, J.K., Dunstan, G.A. (1997) Nutritional properties of microalgae for mariculture. *Aquaculture,* 151:315–31.

Chen, M., Tang, H., Ma, H., Holland, T., Salley, S. (2011) Effect of nutrients on growth and lipid accumulation in the green algae *Dunaliellatertiolecta. Bioresource Technology* 102: 1649–1655.

Chisti, Y. (2007) Biodiesel from microalgae. *Biotechnology Advances* 25, 294-306.

Christie W.W. (1993) *Advances in Lipid Methodology*. Oily Press, Dundee; 69-111.

Converti, A., Casazza, A.A., Ortiz, E., Perego, P., Del Borghi, M. (2009) Effect of temperature and nitrogen concentration on the growth and lipid content of *Nannochloropsisoculata* and *Chlorella vulgaris* for biodiesel production. *Chemical Engineering and Processing,* 48: 1146–115.

Day, J.G., Edwards, A.P., Rodgers, G.A. (1991) Development of an industrial-scale process for the heterotrophic production of a micro-algal mollusc feed. *Bioresour.Technol.* 38.

Dorado, M.P. (2008) *Biofuels Refining and Performance;* McGraw Hill Professional; 107-148.

Dunn, R. O. (2002) Effect of oxidation under accelerated conditions on fuel properties of methyl soyate (biodiesel). *Journal of America Oil Chemists Society,* 79, 915–920.

Dunstan, G.A., Volkman, J.K., Barrett, S.M., Garland, C.D. (1993) Changes in the lipid composition and maximisation of the polyunsaturated fatty acid content of three microalgae growing in mass culture. *Journal of Applied Phycology,* 5:71–83.

Eisberg N. (2006) Harvesting energy. *Chem. Ind.,*17:24–25.

Evans, R.W., Kates, M., Ginzburg, Z. (1982)Lipid composition of halotolerant algae, *Dunaliellaparvalerche* and *Dunaliellatertiolecta.*BiochimicaetBiophysicaActa – *Lipids and Lipid Metabolism,* 712: 186–195.

Gouveia, L., & Oliveira, A.C. (1994) Microalgae as a raw material for biofuels production. *Journal of Applied Phycology,* 6: 347-356.

Guzman H.M., Valido, A.J., Duarte, L.C., Presmanes, K.F. (2010) Estimate by means of flow cytometry of variation in composition of fatty acids from *Tetraselmissuecica* in response toculture conditions. *Aquaculture International,* 18:189–99.

Halim, R., Gladman, B., Danquah, M.K., Webley, P.A. (2011) Oil extraction from microalgae for biodiesel production. *Bioresource Technology,* 102:178–85.

Huang, G.H., Chen, F., Wei, D., Zhang, X.W., Chen, G. (2010) Biodiesel production by microalgalbiotechnology. *Applied Energy,* 87:38–46.

Hodgson, P.A., Henderson, R.J., Sargent, J.R., Leftley, J.W. (1991)Patterns of variation in the lipid class and fatty acid composition of *Nannochloropsisoculata* (Eustigmatophyceae) during batch culture. *Journal of Applied Phycology, 3*: 169-181.

Hu, Q., Sommerfeld, M., Jarvis , E., Ghirardi, M., Posewitz, M., Seibert, M., Darzins, A. 2008. Microalgaltriacylglycerols as feedstocks for biofuel production: perspectives and advances. *Plant J.* 54, 621–639.

Imahara, H., Minami, E., Saka, S. (2006). Thermodynamic study on cloud point of biodiesel with its fatty acid composition. *Fuel,* 85: 1666–1670.

Knothe, G. (2005) Dependence of biodiesel fuel properties on the structure of fatty acid alkyl esters. *Fuel Processing Technol.,* 86: 1059–1070.

Knothe, G., Matheaus, A.C., Ryan, T.W. (2003)Cetane numbers of branched and straight-chain fatty esters determined in an ignition quality tester. *Fuel,* 82: 971–975.

Lv, X., Zou. L., Sun, B., Wang, J., Sun, M.Y. (2010)Variations in lipid yields and compositions of marine microalgae during cell growth and respiration, and within intracellularstructures. *Journal of Experimental Marine Biology and Ecology,* 391:73–83.

Mabee, W.E., Gregg, D.J., Saddler JN.(2005) Assessing the emerging biorefinery sector in Canada. *Applied BiochemBiotechnol,* 121–124:765–78.

Matheus, R., Schuchardta, U., Serchelia, R. (1998) Transesterification of Vegetable Oils: a Review. *J. Braz. Chem. Soc.,* 9.

Meng, X., Yang, J., Xu, X., Zhang, L., Nie, Q., Xian, M. (2009)Biodiesel production from oleaginous microorganisms. Renew. Energy, 34: 1–5.

Mittelbach, M., Remschmidt, C. *Biodiesel: The Comprehensive Handbook;* Martin Mittelbach: Graz, Austria, 2004.

Muradyan, E.A., Klyachko-Gurvich, G.L., Tsoglin, L.N., Sergeyenko, T.V., Pronina, N.A., (2004) Changes in lipid metabolism during adaptation of the *Dunaliellasalina* photosynthetic apparatus to high CO2 concentration. *Russian Journal of Plant Physiology,* 51: 53–62.

Ortega, J., López, A., Garcia, H.S. Hill, C.G. (2004) Lipase-mediated Acidolysis of Fully Hydrogenated Soybean Oil with Conjugated Linoleic Acid. *Journal of Food Science,* 69.

Ota, M., Kato, Y., Watanabe, H., Watanabe, M., Sato, Y., Smith, R.(2009) Fatty acid production from a highly CO2 tolerant alga, *Chlorocuccumlittorale*, in the presence of inorganiccarbon and nitrate. *Bioresource Technology,* 100:37–42.

Piorreck, M., Baasch, K.-H. , Pohl. P. 1984. Biomass production, total protein, chlorophylls, lipids and fatty acids of freshwater green and blue-green algae under different nitrogen regimes. *Phytochemistry,* 23 (2): 207–216.

Pruvost, J., Van Vooren, G., Cogne, G., Legrand, J., (2009). Investigation of biomass And lipids Production with Neochlorisoleoabundans in photobioreactor. *Bioresour. Technol.* 100, 5988–5995.

Pruvost, J., Van Vooren, G., Le Grand, F., Legrand, J., Cuiné, S., Peltier, G. (2012) Investigation of fatty acids accumulation in *Nannochloropsisoculata* for biodiesel application.Bioresource Technology, 124: 421–432.

Pruvost J, Van Vooren G, Le Gouic B, Couzinet-Mossion A, Legrand J. (2011). Systematic investigation of Biomass and lipid productivity by microalgaeinphotobioreactors for biodiesel application. *Bioresour Tech.* 102: 150-158.

Ramadan, M.F., Asker, M.H.S., Ibrahim, Z.K. (2008) Functional bioactive compounds and biological activities of *Spirulinaplatensis* lipids. *Czechoslovak Journal of Food Sciences,* 26:211–22.

Rao, A.R., Dayananda, C., Sarada, R., Shamala, T.R., Ravishankar, G.A. (2007) Effect of salinity on growth of green alga *Botryococcusbraunii* and its constituents. *Bioresource Technology,* 98:560–4.

Renaud, S.M. & Parry, D. L. (1994) Microalgae for use in tropical aquaculture II: Effect of salinity on growth, gross chemical composition and fatty acid composition of three species of marine microalgae. *Journal of Applied Phycology* 6: 347-356.

Rodolfi, L., Zittelli, G.C., Bassi, N., Padovani, G., Biondi, N., Bonini, G., Tredici, M.R.(2009). Microalgae for oil: strain selection, induction of lipid synthesis and outdoor mass cultivation in a low-cost photobioreactor. *Biotechnology and Bioengineering,* 102: 100–112.

Roncaraticarati, A., Meluzzi, A., Ciarri, S., Tallarico, N., Paolom, E. (2004) Fatty Acid Composition of Different Microalgae Strains According to the Culture Phase and the Carbon Dioxide Concentration. *Journal of the World Aquaculture Society,* 35.

Shifrin NS, Chisholm SW. 1981. Phytoplankton lipids: Interspecific differences and effects of nitrate, silicate and light-dark cycles. *J Phycol* 17: 374–384.

Soto, I., Campillo, P.J., Ortega, J., Rodríguez, M.T., Lercker, G., García, H. (2004) Cholesterol oxidation in traditional Mexican dried and deep-fried food products. *Journal of Food Composition and Analysis;* 21: 489– 495.

Teixeira, J.A., Dragone, G.M., Fernandes, B.D., Vicente, A.A. (2010) Third generation biofuels from microalgae. Current Research, *Technology and Education Topics in Applied Microbiology and Microbial Biotechnology.* Mendez-Vilas A (ed.), Formatex, 1355-1366.

Toledo, C., Morales, M., Eberto, N., Revah, S.(2013) Carbon dioxide fixation and lipid storage by *Scenedesmusobtusiusculus*. *BioresourceTechnology,* 130: 652–658.

Tsuzuki, M., Ohnuma, E., Sato, N., Takaku, T., & Kawaguchi, A. (1990) Effects of CO2 Concentration during Growth on Fatty Acid Composition in Microalgae. *Plant Physiol.* 93, 851-856.

Van Gerpen, B., Shanks, R., Pruszko.(2004) Biodiesel Production Technology. *National Renewable Energy Laboratory.*

Volkman, J.K., Jeffrey, S.W., Nichols, P.D., Rogers, G. I., Garland, C.D. (1989) Fatty acid and lipid composition of 10 species of microalgae used in mariculture. *J. Exp. Mur. Bioi. Ecol.,* 128: 219-240.

Williams, J.P., Khan, M. (1993).Microwave-mediated methanolysis of lipids and activation of thin-layer chromatographic plates. *Lipids,* 28: 953-955.

Williams, J.P., Khan, M.U., Wong, D. (1995). A simple technique for the analysis of positional distribution of fatty acids on di- and triacylglycerols using lipase and phopholipase A2. *J Lipid Res* 36:1407–12.

Zhukova, N.V., &Aizdaicher, N.A. (1995) Fatty acid composition of 15 species of marine microalgae. *Phytochemistry,* 39: 351-356.

In: Microalgae and Other Phototrophic Bacteria
Editor: Luis G. Torres Bustillos
ISBN: 978-1-63482-078-3

Chapter 5

MICROALGAE IN LIVESTOCK NUTRITION

L. Flórez-Miranda*, A. J. Martínez-Roldán. and R. O. Cañizares-Villanueva

Laboratorio de Biotecnología de Microalgas. Departamento de Biotecnología y Bioingeniería, Centro de Investigación y de Estudios Avanzados del Instituto Politécnico Nacional., Col. San Pedro Zacatenco, Delegación Gustavo A. Madero, México

ABSTRACT

The microalgae are a group of prokaryotic or eukaryotic photosynthetic organisms that contain pigments (chlorophyll and carotenoids); they are microscopic and can be unicellular or pluricellular. The microalgae are known to be the autotrophic microorganisms with the highest growth rates; they usually require minerals, water, light and CO_2 (Vonshak, 1990; Blanco et al., 2007; Mata et al., 2010; Koller et al., 2012). For many years, the microalgae have been studied to evaluate their capacity to produce diverse compounds, for example, dietary supplements, lipids, enzymes, biomass, polymers, toxins and pigments (Hintz et al., 1966; Soeder, 1980; Becker, 1994; Plaza et al., 2009).

Research concerning the use of microalgae as a food source began in Japan in the 1950s with the production of *Chlorella* for human nutrition; subsequently, processes for the production of other species of microalgae were developed (Soeder, 1980; Thomas et al., 1983; Beneman, 2013). Under optimum culture conditions (i.e., light, temperature, nutrient concentrations), biomass with a high protein content can be obtained and utilized to feed animals, fishes, mollusks and marine crustaceans (Bonotto, 1988, Beneman, 1992; Coutteau, 1996). Currently, interest in the use of microalgae in the formulation of animal diets has grown because of the presence of bioactive compounds in the microalgal biomass (Papadopoulus, 2002; Pulz & Gross, 2004; Beacker, 2007; Gouveia et al., 2008; Peiretti & Meineri, 2008; Kalogeropoulus, 2010).

* Corresponding author: lflores@cinvestav.mx.

1.1. Livestock Nutrition

Nutritional requirements for animals vary according to species and age, and whether they are to be used for breeding. To ensure complete nutrition, it is necessary to formulate specific diets (Cheeke, 1991; George et al., 2001; Thornton, 2010). Nutrients are defined as chemical substances necessary for proper cellular maintenance, growth and organic function. All animals absorb nutrients from their diet. Commodities such as milk, meat, and eggs are supplied for human consumption from animal production systems (Church & Pond, 1974). There are five types of nutrients that are necessary in the diet: (1) energy sources, (2) proteins, (3) vitamins, (4) minerals and (5) water (Gillespie & Flanders, 2009).

The principal energy source in food is carbohydrates, which can be simple (sugars and starch) or complex (fiber); carbohydrates are most commonly found in grains, cereals and fodders (Church & Pond, 1974). Different types of livestock can use different sources of carbohydrates; ruminants (cows, sheep and goats) can use high amounts of fiber because of their complex digestive apparatus, while in pigs, cereals are preferred because they are more easily digestible. (Overton et al., 1998; Gillespie & Flanders, 2009). Fats and oils also provide an energetic contribution 2.25 times higher than that provided by carbohydrates. Fats and oils are also very digestible and provide a vehicle for fat-soluble vitamins. Fats can be obtained from plant (i.e., soy oil) or animal (fish oil) sources. Cereals used in formulations as a protein source also supply fat to the feed, which can account for 1% to 10%, of the total lipid content (Church & Pond, 1974; Doreau & Chilliard, 1997).

Proteins are very important for maintenance of bodily structure in animals; each different species needs a particular protein supplement for proper nutrition (Cheeke, 1991; Schroeder, 2013). Non-ruminants must have all the essential amino acids supplied in the diet, while ruminants can obtain some of them from the microorganisms present in the rumen, because a portion of the ruminal flora is digested and the resultant peptides and amino acids are absorbed in the small intestine (Minson, 2012). Protein sources can be of plant or animal origin; plant sources include soymeal, linseed, cotton and Lucerne, while animal sources include fishmeal, dry whey, casein and lactoalbumin or milk powder (Morales et al., 1994; FAO, 2004).

Vitamins are feed additives that influence the proper function of diverse metabolic pathways and impact reproduction, the immune response and embryonic development (Galyean et al., 1999; Mahan et al., 2007). In animal feed formulations, vitamins A, E, K, D and C and the B-complex vitamins are very important. Ruminants require only the addition of fat-soluble vitamins (A, D, and E) because the rest (the B-complex vitamins and vitamin K) are obtained from fermentation in the rumen. Some minerals, such as Fe, F, Ca, K, Na, I and Se are also important in animal nutrition to maintain proper metabolic function; minerals occur naturally in cereals, fodder and other dietary components. (Miller & Kornegay, 1983). Depending on the species, geography and breeding type, mineral deficiencies sometimes occur, requiring specific dietary supplementation. (Gillespie & Flanders, 2009).

During the last two decades, animal production systems have faced two dilemmas - satisfaction of the growing demand for food and quality control of food products (i.e., meeting quality standards and consumer demands) (Thornton, 2010). Specific animal feed products vary worldwide, depending on cultural preferences in some regions (1). Because the

USA is the principal consumer and producer of animal feed, recommendations from the United States Food and Drug Administration (FDA) and United States Department of Agriculture (USDA) are used by many countries. These two government agencies are responsible for animal nutrition in the USA (USITC, 2000).

Sources of Novel Nutrients for Livestock

Animal production is one of the principal food sources for human nutrition - approximately 40% of the protein consumed worldwide originates from animals. Commensurate with increasing global population, demand for food is also growing rapidly, necessitating the conversion of animal production to an intensive practice. The FAO estimates meat production at 300 million of tons per year and milk production at 700 million of tons per year, along with an increase of 30% for egg production by 2020 (FAO, 2011). This is disturbing, considering that from 2 to 15 kg of plant material is necessary to produce 1 kg of animal product; this causes a large drain on land, energy, water, and other resources to satisfy the demand for animal products (Harmansen, 2003; Van der Spielgel et al., 2013).

To maintain nutritional balance and food quality, different nutrient sources were studied in order to discover new nutrient sources and substitutes for conventional additives used in animal nutrition. An example is the use of the biomass from microorganisms such as yeast, fungi, bacteria or microalgae as single-cell proteins (SCPs) (Becker, 2007). Although consumption of microorganisms is not novel, specific technologies and detailed studies are necessary to guarantee the quality and safety of SCP.

In general, the production of SCP is carried out by fermentation processes (submerged or solid state) employing diverse substrates (even agro-industrial residues). After fermentation, the biomass is harvested, washed, and dried, and in some cases, further steps such as cellular disruption and metabolite extraction are added to the process. Feed production from microorganisms relies on: (1) organism-dependent factors such as growth rate, biomass productivity, and the presence of toxins, etc., and (2) operational variables such as the growth substrate employed, contamination risk, and industrial scale, etc. (Anupama & Ravindra, 2000). Additionally, the amount of nucleic acids is an important factor that limits the use of the different SCP sources, because these molecules are metabolized to uric acid, which can accumulate and cause several diseases. Bacteria, yeasts and filamentous fungi have a high nucleic acid content which necessitates removal of nucleic acids when these organisms are used as food. For microalgae, it is not necessary to remove the nucleic acids, so the entire protein content (30-60%) of the biomass can be utilized (Becker, 2007; Nasseri et al., 2011), and lipids, carbohydrates and vitamins are also provided (Hemaiswarya et al., 2011; Tang & Suter, 2011).

2. Algae in Livestock Nutrition

Microalgae have been used as food since ancient times, in México in the XVI century and in Chad by the Kanembu tribe, which collected and consumed *Spirulina* (*Arthrospira*) biomass. Technological development of microalgae production began in the 1960's with two objectives: (1) the production of food for humans and animals and (2) the production of energy sources. Because of the progress of research on the production process, operational conditions for the autotrophic production of microalgae, such as *Spirulina*, *Chlorella*, *Haematococcus* and *Dunaniella* in raceways, are well established today (Harun et al., 2010; Beneman, 2013). Although the majority of the microalgae are autotrophic, some species have the capability to grow in the dark via consumption of an organic substrate, (this metabolism is called heterotrophic), and some organisms exhibit a combination of autotrophic and heterotrophic metabolism that is referred to as mixotrophic (Mata et al., 2010; Bumbak et al., 2011).

Heterotrophic growth of some microalgae is used to obtain cultures with high cellular density, which is useful in the production of certain high value metabolites such as fatty acids (FA), polyunsaturated fatty acids (PUFA), pigments, etc. (Kobayashi et al., 1992; Shi et al., 1999; Sun et al., 2008). The heterotrophic cultivation of microalgae is carried out using stirred tanks very similar those employed for the production of yeast and bacteria. The most common organic substrates are glucose, acetate, glycerol, etc. (Bumbak et al., 2011).

2.1. Chemical Composition of Microalgae

The chemical composition of the microalgal biomass its consequent nutritional value depends on the species and the culture conditions (Mata et al., 2010). In Table 1 the specific content of protein, carbohydrates and lipids of produced by some species of microalgae for animal-nutritional supplements are presented.

One of the principal appeals of microalgal biomass is its protein content (Soeder, 1980; Spolaore et al., 2006), because it is higher than that in cereals and leguminous plants. This allows the partial or total replacement of conventional protein sources in poultry and aquacultural feed (Becker, 2004; Harum et al., 2010; Shields & Lupatsch, 2012). However, the production cost of microalgal biomass is higher with respect to other protein sources (Table 1), which is an obstacle for the utilization of many species. Thus it is necessary to optimize the production process with the objective of cost reduction so the incorporation of the microalgal biomass in animal feed becomes profitable.

The composition of the biomass is influenced by the culture conditions, allowing for the production of many compounds with biological activity (Shi et al., 1999; Hu, 2004; Shen et al., 2010); Table 2 shows functional compounds found in different species of microalgae with utility as dietary supplements for animals.

2.2. Microalgal Compounds with Biological Activity

A functional food is one that has a specific function in an organism or prevents disease, in addition to its nutritional input. In the last decade, interest in this kind of food has grown considerably (Andonov et al., 2011). Many foods from animal sources, such as milk and its derivatives, eggs, meat, etc., can be converted to functional foods by enriching them with compounds such as FA, PUFA, vitamins, minerals, antioxidants, etc. and can be obtained from microalgal biomass.

Table 1. Chemical composition and mean price of different species of microalgae (% of DW) and ingredients employed in livestock diets

Source	Protein (%)[a]	Lipids (%)[a]	Carbohydrates (%)[a]	Energetic Contribution (MJ/Kg)[a]	Cost 2014 (US $/kg)
Fishmeal	63.0	11.0	NR	20.1	1.82[b]
Poultry meal	58.0	11.3	NR	19.1	NR
Corn gluten	62.0	5.0	18.5	21.3	0.216 (Maíz)[b]
Soy	44.0	2.2	39.0	18.2	0.542[b]
Wheat flour	12.2	2.9	69.0	16.8	0.334 (Trigo)[b]
Spirulina sp.	58.0	11.6	10.8	20.1	20[c]
Chlorella sp.	52.0	7.5	24.3	19.3	44[c]
Dunaliella salina	57[d]	6[d]	32[d]		1-221.5[e]
Haematococcus pluvialis					50-800[e]

[a] Shields & Lupatsh, 2012; [b] www.indexmundi.com; [c] www.oilgae.com; [d] Becker, 2007; [e]www.alibaba.com.
NR Not reported

Table 2. Functional ingredients from microalgae with applications in the feed-industry

Functional ingredient	Nutritional application
Carotenoids	
b-carotene	Antioxidant, colorant, feed additive (provitamin A)
Astaxanthin	Antioxidant, immunomodulation and cancer prevention activity, pigment in aquaculture
Cantaxanthin	Antioxidant, pigment in poultry
Lutein	Antioxidant, prevents macular degeneration, pigment in poultry
Fatty Acids	
EPA (eicosapentaenoic acid)	Omega 3, cerebral development in children, cardioprotector
DHA (docosahexaenoic acid)	Omega 3, cerebral development in children, cardioprotector
Vitamins	
Tocopherol (Vitamin E)	Antioxidant

[a] Ben-Amozt & Avron, 1983; [b] Raja et al., 2007; [c] Ceron et al., 2007; [d] Del Campo et al.,2004; [f] Gouveia et al., 1996; [g] Blanco et al., 2007; [h] Sánchez et al., 2008; [i] García-Cañedo et al,.2011; [j] Wu et al., 2009; [k] Wen et al., 2002; [l] Atalah et al., 2007;[m] Khozin-Goldberg & Cohen, 2006; [p] Yokochi et al., 1998;

2.2.1. Polyunsaturated fatty acids

PUFA of the ω-3 type such as docosahexaenoic acid (DHA) and eicosapentaenoic acid (EPA) are recognized as hormones and neurotransmitter precursors and help to regulate the immune response (Rodriguez-Cruz et al., 2005; Legardé, 2008); these compounds also diminish the risk of diabetes, arterial hypertension, arthritis, cancer and cardiovascular disease (Simopoulus, 1991; Simopoulos, 1999). The ω-3 FA differs from the ω-6 in the location of the first double bond adjacent to the terminal methyl group. These two families of FA are important and considered as essential FA because animals (including humans) cannot synthetize them and must get them from the diet.

To maintain metabolic equilibrium in the organism a ω-6:ω-3 ratio of 2:1 is recommended; however, an imbalance may be caused because ω-6 FAs are more common and can be obtained from foodstuffs such as corn oil, sunflower oil, soy, eggs, meat and leguminous plants. A ω-6:ω-3 ratio of 17:1 in occidental diets is estimated, and many metabolic and nervous diseases are attributed to the imbalance (Simopoulus, 1999; Moloney, 2007). The most convenient way to compensate for this deficiency is the enrichment of eggs, milk and meat with ω-3 FAs.

The employment of marine species of microalgae has been emphasized for the production of PUFA. De Swaaf et al. (2003) achieved 114.6 mg l^{-1} d^{-1} of DHA using *Crypthecodinium cohnii*; while Wen et al. (2002), reported a productivity of 49.7 mg l^{-1} d^{-1} for EPA in a heterotrophic feed batch culture of the diatom *Nitzschia laevis*. García *et al.* (2004), obtained 43.13 mg l^{-1} d^{-1} of EPA with *Phaeodactylum tricornutum*. Actually, it is possible to find commercial products such as DHASCO (Market Biosciences Co.) - a supplement rich in DHA obtained from *Crypthecodinium conhii* (Yap & Chen, 2002).

2.2.2. Carotenoids

The carotenoids are a large group of pigments that are characterized by a structure containing 40 carbon atoms with many conjugated double bonds (Breithaupt, 2007); they have an important role in photosynthesis and the photoprotection of photosynthetic organisms as part of the light-harvesting complex. In animals and humans they act as antioxidants - blue light filters protecting the skin and eyes from strong irradiation - and can be used as food colorants. For all of the above reasons, the market for carotenoids is rapidly growing (Margalith, 1999; Cardozo et al., 2007; Fernández- Sevilla et al., 2010).

Animals cannot synthesize carotenoids *de novo*; they are present in the body because they are absorbed from the diet, and can be accumulated or transformed into other compounds in the tissues (Hudon, 1994). The carotenoids are divided into two groups according to the occurrence of oxygen in the molecule; carotenoids containing O2 are called xanthophylls while those that do not contain oxygen are called carotenes (Breithaupt, 2007; Del Campo et al., 2007). Because xanthophylls are fat-soluble, they can accumulate in meat, eggs, and the skin of fishes and in the shells of certain crustaceans; in fowl, they can accumulate in the skin, liver, teguments and feathers, and in the egg yolk (Breithaupt, 2007; Hammershøj et al., 2010; Englmaierová et al., 2013).

There are more than 600 different carotenoids but only few have commercial applications. β-carotene produced by *Dunaliella salina* (Hejazi et al., 2004; Raja et al., 2007), astaxanthin produced by *Haematococcus pluvialis* (Lorenz & Cysewski, 2000; Orosa et al., 2005; Kang et al., 2010), and lutein, zeaxantin and cantaxantin produced by different species of green algae (Sánchez et al., 2008; Harun et al., 2010; Chu et al., 2011) are important carotenoids with commercial value, produced by microalgae.

3. Potential of Microalgal Biomass for Feed

Research concerning the application of microalgal biomass in animal feed shows that the addition of low quantities to diets for chickens, pigs and cattle causes metabolic benefits to the animals, such as improvement of the immune system, and reproductive development, weight control and an increase in the food conversion rate (g weight gained/g food consumed) (Harel & Clayton, 2004).

Since microalgae are basic to the aquatic environment, one of their major applications is in aquaculture (Benemann, 1992; Coutteau, 1996). Factors that influence the important role of microalgae in aquaculture are: 1) all the developmental stages of bivalves use microalgal biomass as food; 2) gastropod mollusks and urchins need a diet enriched in benthic diatoms; 3) crustaceans such as shrimp consume microalgae during their larval stages and 4) the larvae of many fishes feed on zooplankton, which grow by consuming microalgae (George et al., 2001; Hemaiswarya et al., 2010; Shields & Lupatsch, 2012).

After aquaculture, the commercial sector second in importance is poultry production; as early as 1952 there are reports about supplementation of chicken diets with 20% microalgal biomass, and a positive effect was observed on growth and food conversion efficiency (Becker, 2004; Lum et al., 2013).

As for poultry, it is possible to employ microalgae in livestock production to supplement the diets of pigs, sheep and cattle; many studies have been carried out on the improvement of the nutritional characteristics of meat and milk from various animals (Frankil, 1999; Sardi et al., 2006; Peiretti & Meineri, 2008). Today, the dual use of microalgae in biorefinery applications and animal feed has been proposed (Oswald et al., 1957; Subhadra et al., 2011; Lum et al., 2013) which entails the possibility of the complete utilization of the biomass. First, lipids for biodiesel production are extracted and subsequently, the biomass (rich in protein, pigments and other compounds) is used to feed animals (Li *et al*, 2008; Lundquist et al., 2010; Stephens et al., 2010).

In a study carried out with defatted biomass in *Staurosira* sp. (the lipids were extracted to produce biodiesel), Austic e*t al.* (2013) demonstrated that there is no delay in the growth of fowl when their diet is enriched with 7.5% microalgal biomass, along with some amino acids (Met, Lys, Ile, Thr, Trp and Val). Something similar was observed when biomass of *Staurosira* sp. was provided to laying hens, regarding weight gain, feed consumption and egg production (Leng et al., 2014).

3.1. Cattle

The supplementation of livestock diets using microalgal biomass has provided different benefits depending on the type of animal studied; this is because microalgal biomass has a protein content from 7 to 70% of the dry weight (Becker, 2004a; Bai et al., 2012; Holman & Malau-Aduli, 2013) along with PUFA, phytosterols, and some vitamins (Cañizares-Villanueva et al., 1995; Marshall et al., 2002; Becker, 2004a) that are useful in the formulation of diets for farm animals.

Meat has a low content of long chain ω-3 FA which should necessarily be added to the diet (Ashes et al., 1992); however in ruminants, a portion of the PUFA accumulated are hydrogenated by the bacteria present in the rumen (Doreau & Chilliard, 1997; Cooper et al., 2004); which necessitates the inclusion of large PUFA in amounts to ensure that a portion is included in the muscle. It has been observed that the employment of microalgal PUFA decreases hydrogenation compared with other fat sources such as fish or linseed oils (Givens et al., 2006).

Meat from pigs fed with a diet supplemented with microalgal biomass has better color (Saeid et al., 2013) and the flavor is not modified (Sardi et al., 2006); also, the amount of cholesterol is lower (Yang et al., 2006) making it healthier to consume. Another advantage of the use of microalgal biomass in the formulation of pig diets is that this biomass contains some micronutrients (i.e., Iodine), making the use of inorganic sources unnecessary and increasing its availability. The addition of this micronutrient is necessary to ensure the proper function of different glandules in pigs (Bańoch et al., 2013).

In rabbits, the addition of microalgal biomass (*Chlorella*), causes a reduction in the amount of fat compared with commercial diets (Marounek et al., 2009)

In lamb, the use of microalgal biomass causes an off flavor, diminishing acceptance of the meat by consumers, so microalgae are not recommended for use as supplements in the diet of sheep (Elmore et al., 2005; Nute et al., 2007). The alteration in the taste is caused by a high amount of PUFA accumulated in the muscle, which produces aldehydes when oxidized and causes the bad taste (Elmore et al., 1997; Vasta & Priolo, 2006; Van Durme et al., 2012)

In chickens, the use of 50 g of marine microalgal biomass in the diet causes an increase in the amount of essential FA, because the FA added to the feed are absorbed and accumulated in muscle, increasing the nutritional value of the meat (Mooney et al., 1998;) The flavor and odor of the meat are not modified, ensuring its acceptance by the consumer (Mooney et al., 1998; Ribeiro et al., 2013).

It is necessary to enrich milk with nutraceuticals such as PUFA and monounsaturated FA (Lock & Bauman, 2004); but this is not a simple process, because hydrogenation of PUFA in the rumen of the cow is prevalent and lowers milk quality (Franklin et al., 1999; Donovan et al., 2000; Whitlock et al., 2006). Recently, there has been interest in finding a way to enrich milk with conjugated linoleic acids (CLA), because milk consumption has demonstrated health benefits and anticancer activity (Ip et al., 1994; Belury, 2002a & 2002b). CLA are isomers of linoleic acid (C18:2); the double bonds are in different locations (Donovan et al., 2000). Two of the most common CLA are rumenic and vaccenic acid produced by the ruminal flora (Kramer et al., 1998; Lock & Bauman, 2004).

The addition of *Schizochytrium* biomass (which is rich in PUFA) to dairy cow diets favors the accumulation of unsaturated FA and also modifies the bacterial community in the rumen; this reduces the production of FA with an impair number of carbons produced by the ruminal flora (Boeckaert et al., 2007; AbuGhazaleh, 2009). The accumulation of mono and polyunsaturated FA in milk is desirable because of the health benefits of these compounds (Ip et al., 1994; Belury, 2002). For dairy sheep, the addition of DHA Gold™ (a product rich in *Schizochytrium marine* PUFA) increased the amount of essential FA in the milk, especially araquidonic (C20:4), eicosapentaenoic (C20:5) and docosahexaenoic (C22:6) acids; furthermore, the amount of CLA increased and the total fat decreased, ensuring a healthy product (Reynolds et al., 2006; Toral et al., 2010; Bichi et al., 2013).

3.2. Aquaculture

Macro- and microalgae (phytoplankton) are the basis of trophic chains in the aquatic environment (Muller-Feuga, 2000; Muller-Feuga, 2004; Guedes & Malcata, 2012). In the aquaculture industry, microalgae are very useful for the assembly of artificial trophic chains or for the production of certain filtering organisms and fish larvae (Benemann, 1992). The genera most used in aquaculture are: *Chlorella*, *Tetraselmis*, *Isochrysis*, *Pavlova*, *Pheodactylum*, *Chaetoceros*, *Nannochloropsis*, *Skeletonema* y *Thalassiosira*. These are sometimes administered individually, but the use of mixtures is preferable to promote balanced nutrition (Spolaore et al., 2006; Hemaiswarya et al., 2011; Priyadarshani et al., 2012). These genera have all the characteristics necessary for use in aquaculture: (1) appropriate size with respect to consumers, (2) high nutritional value, (3) high digestibility, and (4) ease of cultivation (Conceição et al., 2010; Guedes & Malcata, 2012).

Filtering mollusks (the most important groups are oysters, clams, scallops and mussels) consume microalgal biomass during their entire life span because they are herbivorous (Muller-Feuga, 2000; Muller-Feuga, 2004). Herbivorous adult fish are rare, but the larval stages use algae as feed because the digestive apparatus is not completely developed and complex diets cannot be digested (Conceição et al., 2010).

3.2.1. Live feed

The development of the technology for the breeding of mollusks has been ongoing for at least 50 years and different feed sources have been proposed (bacteria, yeast, frizzed or lyophilized microalgae biomass, etc.); however, none has been as useful as fresh microalgal biomass (Muller-Feuga, 2000; Muller-Feuga, 2004). Actually, in aquaculture there are two strategies to supply microalgal biomass, transparent and green ponds (Moriarty, 1997); in the former the feed is produced in independent containers and added only when necessary. In the latter, the microalgae are produced in the fish-production ponds without controlling the cellular concentration (Moriarty, 1997; Neori, 2011).

In aquaculture of filtering bivalves, it is very important to consider the nutritional requirements of the animals produced because of differences due to the physiological stage. For the larvae, the most important factors are the nutrimental and bacteriological qualities of the feed, in the latter stages feed quantity is transient because consumption increases 100-fold. Finally, in the reproductive phase both quality and quantity are important even though the number of organisms changes considerably (Muller-Feuga, 2000; Guedes & Malcata, 2012).

In fish, microalgae are important as feed in the first larval stage, but in later phases the majority of species is carnivorous and feed on zooplankton (Chakraborty et al., 2007; Hemaiswarya et al., 2011). Microalgae are also important because they are the food of cladocerans and copepods, principally of the genera *Moina*, *Daphnia*, *Euterpina*, *Tigriopus*, *Brachionus and Artemia*; these microcrustaceans are the food for the adult stages of carnivorous fish (Muller-Feuga, 2004; Becker, 2004b; Conceição et al., 2010; Hemaiswarya et al., 2011).

3.2.2. Astaxanthin

Of the different products used in aquaculture, astaxanthin has a high commercial value because it is important in the production of salmonids (salmons, trout, etc.), shrimp, etc. The diet of these animals must be supplemented with astaxanthin, a natural red pigment that is produced by yeast (*Xanthophyllomyces dendrorhous*) and microalgae (*Haematococcus pluvialis*, *Chlorella zofingiensis*, *Chlorococcum* sp., etc.) (Yuan et al., 2011; Barbachano-Torres et al., 2014).

The huge demand for salmonids has generated the growth of fish farms to produce them, which in turn has caused an increase in the demand for natural astaxanthin in the last few years (Lorenz & Cysewki, 2000). The necessity for astaxanthin in salmonid cultures is not only because the pigment causes the characteristic roseate color (Plaza et al., 2009; Yuan et al., 2011); but also because it was demonstrated that its consumption promotes proper development (Sommer et al., 1991; Christiansen, & Torrissen, 1995). Natural astaxanthin is more easily assimilated than the synthesized product. The maximum concentration of this pigment is found in *Haematococcus* (Choubert & Heinrich, 1993; Lorenz & Cysewki, 2000; Li et al., 2011; Zhang et al., 2014). *Haematococcus* biomass is used in the formulation of salmonid diets or directly added to salmonid cultures; approximately 20 ppm of astaxanthin is needed to ensure bright pigmentation and development (Christiansen et al., 1995; Bazyar-Lakeh et al., 2010).

In shrimp production, *Haematococcus* can be directly added to the diet at a concentration of at least 50 ppm (Choubert & Heinrich, 1993; Menasveta et al., 1993; Chien et al., 2003; Tume et al., 2009). Deficiency of astaxanthin in shrimp causes "blue color syndrome" that results in a pale yellow color when the shrimp are cooked instead of the characteristic roseate color; this considerably diminishes the commercial value of the shrimp because of diminished consumer acceptance (Menasveta et al., 1993).

Other organisms, such as lobsters and crayfish also require supplementation with astaxanthin-rich microalgal biomass, but at concentrations above 400 ppm (Barcalay et al., 2006; Daly et al., 2013).

3.2.3. Ornamental fishes

Ornamental fishes are another quickly growing commercial sector in aquaculture; actually, hundreds of the 4000 species are cultivated in aquaria and fishbowls. The most popular species are *Symphysodon aequifasciatus*, *Poecilia reticulata*, *Xiphophorus helleri*, *Poecilia sphenops*, *Poecilia latipinna Carassius auratus Paracheirodon innesi*, *Brachydanio rerio*, *Pelvicachromis pulcher*, *Pterophyllum scalare*, *Pseudotropheus lombardoi*, etc. (Ezhil et al., 2013; Kouba et al., 2013; Velasco-Santamaría & Corredor-Santamaría, 2011; Sales & Janssens, 2004; Pannevis & Earle, 1994) All of these species are popular because of their bright coloration. These colors are caused by pigmentation of the skin by carotenoids obtained in the diet, the most important of which are β-carotene, lutein, astaxanthin, tunaxanthin, etc. (Sales & Janssens, 2003; Bjerkeng, 2008). Even if the fishes are unable to produce a particular pigment, they can transform a pigment obtained in the diet to obtain the one required (for example, zeaxanthin → astaxanthin; β-carotene → lutein) (Hata & Hata, 1993).

The use of microalgal biomass as live feed for ornamental fish production is very effective for enhancing the color of fishes produced in hatcheries (Gouveia et al., 2003; Velasco-Santamaría, & Corredor-Santamaría, 2011). The genera most employed are *Chlorella*, *Spirulina*, *Haematococcus*, etc. (Gouveia et al., 2003; Alagappan et al., 2004; James et al., 2006; Bjerkeng, 2008; Dernekbasi et al., 2010; Güroy et al., 2012). However, *Haematococcus* generum is difficult for certain species, such as *Cyprinus carpio* and *Carassius auratus* to digest. (Gouveia et al., 2003). However, it should be considered that the type and amount of microalgae added depends on which fish is being cultivated because pigmentation is genus dependent (Bjerkeng, 2008).

4. POULTRY

The employment of microalgae in poultry feed has two principal objectives, (1) enrichment of the meat and eggs with functional ingredients (Ginzberg et al., 2000; Andonov et al., 2009; Kalogeropoulus et al., 2010), and (2) to confer desired pigmentation on the products (Gouveia et al., 1996; Fredriksson et al., 2006). Foods produced by the poultry industry are very important because this protein source is easily accessible by the population (Wint & Robinson, 2007) because of its low cost compared with other livestock products such as beef, seafood, etc.

Many studies have focused on obtaining "designer food" via the addition of PUFA, vitamins, selenium, folic acid, etc. to animal diets on the farm; this increases the amount of bioactive compounds in the end products (Andonov et al., 2011; Walker et al., 2012).

Coloration is an important characteristic in poultry products because is related to the nutritional quality of the product and the health of the animals (Breithaup, 2008; Liu et al., 2008). Xanthophylls such as lutein, zeaxanthin, astaxanthin and cantaxanthin, are the carotenoids most employed as feed additives for poultry; lutein and zeaxanthin are the most important commercially because they have beneficial effects on human health (Wang et al., 2006; Selvaraj, 2010; Christaki et al., 2013).

The microalgae, specifically the Chlorophyceae, can produce and accumulate different carotenoids such as lutein, by genera such as *Muriellopsis,* reaching a specific content ranging from 0.4% to 0.6% of the dry weight (DW) (Del Campo et al., 2000; Blanco et al., 2007), or *Scenesdemus,* which produces 0.55% of the DW (Sánchez et al., 2008a). These data show the great potential of microalgae as a pigment source for poultry. Because the pigments are fat-soluble, their absorption depends on the amount of lipids in the diet (Hudon, 1994; Koutsos et al., 2003; Burri & Clifford, 2004; Wu et al., 2009). A direct relationship between the pigmentation of the egg yolk and the long-chain lipid content in the feed has been reported regarding the supplementation of laying hens' diets with carotenoids (Nitzan, 1999; Fredriksson, 2006).

Imborr & Waldenstedt (2000) observed that the addition of dry biomass of *Haematococcus pluvialis* (1.8 g to 8.9 g per kg of feed) in the diet of chickens caused a decrease in the number of pathogens (*Clostridium perfringens* and *Campylobacter jejuni*) in the feces and an increase the amount of astaxanthin in the muscle, improving the meat's appearance without modifications to the growth. Similar results have been observed in eggs, causing nutritional enhancement of the egg yolk pigmentation.

Ginzberg et al. (2000) employed lyophilized biomass of *Porphyridium* sp. to supplement laying hens' diets to evaluate the effect of the biomass on the cholesterol content in the blood and egg yolk. The results show that the addition of 5% to 10% of *Porphyridium* biomass reduces the cholesterol concentration in both blood and egg yolk. Also they observed changes in the FA composition of the egg yolk, specifically an increased content of linoleic and arachidonic acids.

The ω-3 FA are important in human nutrition (Legardé, 2008; Lukiw & Bazan, 2008; Calder, 2009; Chapkin et al., 2009) - marine microalgae produce principally EPA (20:5 w3) and DHA (22:6w3) which are accumulated in lipid globules in the cytosol (Liu & Lin, 2001; Přibyl et al., 2013). Many studies have demonstrated that the addition of biomass rich in PUFA in poultry causes an increase in the content of ω-3 FA in the meat and eggs. Nitsan *et al.* (1999), studied the effect of the supplementation with a flour rich in *Nannochloropsis* sp. biomass with high amount of EPA (27.7% of the total lipids) of a commercial diet for laying hens; the EPA was converted to DHA and accumulated in the lipids in the egg yolk.

The addition of 1% *Nannochloropsis* increased the DHA content in the egg yolk, but did not change the concentration in muscle; this is attributed to the fact that in laying hens the fat from the feed is preferentially deposited in the egg yolk with only the excess accumulated in muscle, neccesitating the addition of large amounts if the objective is to enrich the meat.

In research carried out by Fredriksson *et al.* (2006) employing laying hens fed with 20% *Nannochloropsis oculata* biomass, the content of the carotenoids luteína/zeaxantina increased from 8 ppm to 22 ppm with cantaxanthin (7.7 ppm) and β-carotene (1.3 ppm) found in the eggs, which improved the egg yolk color.

Gouveia *et al.* (1996) reported on the enrichment of laying hens with dry biomass of *Chlorella vulgaris*, as a carotenoid source, using the same quantity present in the commercial diet (Lutein 34% and cantaxanthin 66%). The eggs produced by hens fed with *C. vulgaris* biomass had a more intense pigmentation on the Roche scale, and the amount with respect to the total carotenoids was increased, reaching values of 86% and 14% for lutein and cantaxanthin, respectively.

CONCLUSION

The first report of the employment of microalgal biomass for human consumption dates to 400 years ago in the Mexico's Valley. In 1950s the possibility of incorporating microalgal biomass into livestock diets was studied, sparked by the high nutritional value of the biomass and the scarcity of grains generated by both World Wars. Actually, the use of microalgae as feed is important not because of scarcity but because population increases have enhanced the demand for high quality livestock products. Also, interest in functional foods has promoted the use of microalgae biomass as a source of various bioactive molecules.

In the last two decades, many studies have demonstrated that the addition of microalgae to livestock diets not only enhances weight gain but also causes enrichment of livestock products with biologically active compounds such as PUFA, antioxidants, carotenoids, etc. These compounds confer beneficial nutritional characteristics on the end-products (milk, eggs, meat, etc.).

Today, the most important commercial sector in which microalgae are employed is aquaculture. in which they are used as feed for species cultivated in hatcheries. Application can be as live feed, in the formulation of diets or as a source of specific metabolites. Due the effect of the consumption of microalgae on fertility, fecundity and survival of the different stages of aquatic species, its economic importance in aquaculture has increased.

Recently, the biorefinery concept has been adopted, which proposes exploitation of microalgal biomass as an energy source, producing high amounts of defatted-biomass with high amounts of proteins; the biomass is proposed for use as livestock feed principally for ruminants, due their capacity to digest high quantities of cellulose.

All the benefits of microalgae in animal feed analyzed in this chapter ensure the persistence of microalgal biotechnology as a productive economic sector that responds to the requirements of energetics, nutraceutics, food and feed.

REFERENCES

Abu-Ghazaleh, A. A., Potu, R. B. & Ibrahim, S. (2009). The effect of substituting fish oil in dairy cow diets with docosahexaenoic acid-micro algae on milk composition and fatty acids profile. *Journal of Dairy Science*, *92*(12), 6156 – 6159.

Alagappan, M., Vijula, K. & Sinha, A. (2004). Utilization of *Spirulina* algae as a source of carotenoid pigment for blue gouramis (*Trichogaster trichopterus* Pallas). *Journal of Aquaculture and Aquatic Sciences*, *10*, 1 – 11.

Andonov, A., Šefer, D., Marković, R., Šobajić, S. & Cilev, G. (2011). Influence of ω-3 fatty acids supplemented to laying hens diet on laying performance. *Macedonian Journal of Animal Science*, *1*, 139 – 145.

Anupama & Ravindra, P. (2013). Value-added food: Single cell protein. *Biotechnology Advances*, *18*, 459 – 479.

Ashes, J. R., Siebert, B. D., Gulati, S. K., Cuthbertson, A. Z. & Scott, T. W. (1992). Incorporation of *n*-3 fatty acids of fish oil into tissue and serum lipids of ruminants. *Lipids*, *27*(8), 629 – 631.

Atalah, E., Hernández-Cruz, C. M., Izquierdo, M. S., Rosenlund, G., Caballero, M. J., Valencia, A. & Robaina, L. (2007). Two microalgae *Crypthecodinium cohnii* and *Phaeodactylum tricornutum* as alternative source of essential fatty acids in starter feeds for seabream (*Sparus aurata*). *Aquaculture*, *270*, 178 – 185.

Austic, R. E., Mustafa, A., Jung, B., Gatrell, S. & Lei, X. G. (2013). Potential and limitation of a new defatted diatom microalgal biomass in replacing soybean meal and corn in diets for broiler chickens. *Journal of Agricultural and Food Chemistry*, *61*, 7341 – 7348.

Bai, A., Stündl, L., Bársony, P., Fehér, M., Jobbágy, P., Herpergel, Z. & Vaszkó, G. (2012). Algae production on pig sludge. *Agronomy for Sustainable Development*, *32*(3), 611 – 618.

Baňoch, T., Svoboda, M., Kuta, J., Saláková, A. & Fajt, Z. (2013). The effect of iodine from iodine-enriched alga *Chlorella* spp. on the pork iodine content and meat quality in finisher pigs. *Acta Veterinaria Brno*, *81*(4), 339 – 346.

Barbachano-Torres, A., Castelblanco-Matiz, L. M., Ramos-Valdivia, A. C., Cerda-García-Rojas, C. M., Salgado, L. M., Flores-Ortiz, C. M. & Ponce-Noyola, T. (2014). Analysis of proteomic changes in colored mutants of *Xanthophyllomyces dendrorhous* (*Phaffia rhodozyma*). *Archives of Microbiology*, *196*(6), 411 – 421.

Barclay, M. C., Irvin, S. J., Williams, K. C. & Smith, D. M. (2006). Comparison of diets for the tropical spiny lobster *Panulirus ornatus*: astaxanthin-supplemented feeds and mussel flesh. *Aquaculture Nutrition*, *12*(2), 117-125.

Bazyar-Lakeh, A. A., Ahmadi, M. R., Safi, S., Ytrestøyl, T. & Bjerkeng, B. (2010). Growth performance, mortality and carotenoid pigmentation of fry offspring as affected by dietary supplementation of astaxanthin to female rainbow trout (*Oncorhynchus mykiss*) broodstock. *Journal of Applied Ichthyology*, *26*(1), 35 – 39.

Becker, E. W. (1994). Microalgae: Biotechnology and Microbiology. Cambridge: Cambridge University Press.

Becker, E. W. (2004a). 18 Microalgae in Human and Animal Nutrition. In A. Rrichmond (Ed.) Handbook of microalgal culture: biotechnology and applied phycology (pp. 312 – 351). Oxford, UK: Publishing Ltd.

Becker, E. W. (2004b). Microalgae for Aquaculture. In A. Rrichmond (Ed.) Handbook of microalgal culture: biotechnology and applied phycology (pp. 380 – 391). Oxford, UK: Publishing Ltd.

Becker, E. W. (2007). Micro-algae as a source of protein. *Biotechnology Advances*, *25*, 207 – 210.

Belury, M. A. (2002b). Inhibition of carcinogenesis by conjugated linoleic acid: potential mechanisms of action. *The Journal of Nutrition*, *132*(10), 2995 – 2998.

Belury, M. A. (2002). Dietary conjugated linoleic acid in health: physiological effects and mechanisms of action. *Annual Review of Nutrition*, *22*, 505 – 531.

Ben-Amozt, A. & Avron, M. (1983). On the factors which determine massive β-carotene accumulation in the halotolerant alga *Dunaliella bardawil*. *Plant Physiology*, *72*, 593 – 597.

Beneman, J. (2013). Microalgae for biofuels and animal feeds. *Energies*, *6*, 5869 – 5886. doi:10.3390/en6115869

Benemann, J. R. (1992). Microalgae aquaculture feeds. *Journal of Applied Phycology*, *4*(3), 233 – 245.

Bichi, E., Hervás, G., Toral, P. G., Loor, J. J. & Frutos, P. (2013). Milk fat depression induced by dietary marine algae in dairy ewes: Persistency of milk fatty acid composition and animal performance responses. *Journal of Dairy Science*, *96*(1), 524 – 532.

Bjerkeng, B. (2008). Carotenoids in aquaculture: fish and crustaceans. In G. Britton, S. Liaaen-Jensen & H. Pfander (Eds.) Carotenoids (Volume 4: Natural Functions, pp. 237 – 254). Basel, Switzerland: Birkhäuser Basel.

Blanco, A. M., Moreno, J., Del Campo, J. A., Rivas, J. & Guerrero, M. G. (2007). Outdoor cultivation of lutein-rich cells of *Muriellopsis* sp. in open ponds. *Applied Microbiology and Biotechnology*, *73*, 1259 – 1266.

Boeckaert, C., Fievez, V., Van Hecke, D., Verstraete, W. & Boon, N. (2007). Changes in rumen biohydrogenation intermediates and ciliate protozoa diversity after algae supplementation to dairy cattle. *European Journal of Lipid Science and Technology*, *109*(8), 767 – 777.

Bonotto, S. (1988). Food and chemicals from microalgae. *Progress in Oceanography*, *21*, 207 – 215.

Breithaupt, D. E. (2007). Modern application of xanthophylls in animal feeding - a review. *Trends in Food Science & Technology*, *18*, 501 – 506.

Breithaupt, D. E. (2008). Xanthophylls in poultry feeding. In: Britton, G., Liaaen-Jensen, S. & Pfander, H. (Eds.) *Carotenoids*, *Vol. 4: Natural Functions*, (pp. 255 - 263). Basel, Switzerland: Birkhäuser.

Bumbak, F., Cook, S., Zachleder, V., Hauser, S. & Kovar, K. (2011). Best practices in heterotrophic high-cell-density microalgal processes: achievements, potential and possible limitations. *Applied Microbiology and Biotechnology*, *93*, 31 – 46.

Burri, B. J. & Clifford, A. J. (2004). Carotenoid and retinoid metabolism: insights from isotope studies. *Archives of Biochemistry and Biophysics*, *430*, 110 – 119.

Calder, P. C. (2009). Polyunsaturated fatty acids and inflammatory processes: New twists in an old tale. *Biochimie*, *91*, 791 – 795.

Cañizares-Villanueva, R. O., Dominguez, A. R., Cruz, M. S. & Rios-Leal, E. (1995). Chemical composition of cyanobacteria grown in diluted, aerated swine wastewater. *Bioresource Technology*, *51*(2), 111 – 116.

Cerón , M. C., García-Malea, M. C., Rivas, J., Alcien, F. G., Fernández, J. M., Del Rio, E., Guerrero, M. G. & Molina, E. (2007). Antioxidant activity of *Haematococcus pluvialis* cells grown in continuous culture as a function of their carotenoid and fatty acid content. *Applied Microbiology and Biotechnology*, *74*, 1112 – 1119.

Ceron , M. C., García-Malea, M. C., Rivas, J., Alcien, F. G., Fernandez, J. M., Del Rio, E., Guerrero, M. G. & Molina, E. (2007). Antioxidant activity of *Haematococcus pluvialis* cells grown in continuous culture as a function of their carotenoid and fatty acid content. *Applied Microbiology and Biotechnology*, *74*, 1112 – 1119.

Chakraborty, R. D., Chakraborty, K. & Radhakrishnan, E. V. (2007). Variation in fatty acid composition of *Artemia salina* nauplii enriched with microalgae and Baker's yeast for use in larviculture. *Journal of Agricultural and Food Chemistry*, *55*(10), 4043 – 4051.

Chapkin, R. S., Kim, W., Lupton, J. R. & McMurray, D. N. (2009). Dietary docosahexaenoic and eicosapentaenoic acid: Emerging mediators of inflammation. *Prostaglandins, Leukotrienes and Essential Fatty Acids*, *81*, 187 – 191.

Cheeke, P. R. (1991). *Applied Animal Nutrition : feeds and feeding*.Englewood, New Jersey, USA: Prentice Hall.

Chien, Y. H., Pan, C. H. & Hunter, B. (2003). The resistance to physical stresses by *Penaeus monodon* juveniles fed diets supplemented with astaxanthin. *Aquaculture*, *216*(1), 177 – 191.

Choubert, G. & Heinrich, O. (1993). Carotenoid pigments of the green alga *Haematococcus pluvialis*: assay on rainbow trout, *Oncorhynchus mykiss*, pigmentation in comparison with synthetic astaxanthin and canthaxanthin. *Aquaculture*, *112*(2), 217 – 226.

Christaki, E., Bonos, E., Giannenas, I. & Florou-Paneri, P. (2013). Functional properties of carotenoids originating from algae. *Journal of the Science of Food and Agriculture*, *93*, 5 – 11.

Christiansen, R., Lie, O. & Torrissen, O. J. (1995). Growth and survival of Atlantic salmon, *Salmo salar* L., fed different dietary levels of astaxanthin. First-feeding fry. *Aquaculture Nutrition*, *1*(3), 189 – 198.

Chu, F. L., Pirastru, L., Popovic, R. & Sleno, L. (2011). Carotenogenesis up-regulation in *Scenedesmus* sp. using a targeted metabolomics approach by liquid chromatography−high-resolution mass spectrometry. *Journal of Agricultural and Food Chemistry*, *59*, 3004 – 3013.

Church, D. C. & Pond, W. G. (1974). *Basic Animal Nutrition and Feeding*. Corvallis, Oregon: Oregon State Univ.

Conceição, L. E., Yúfera, M., Makridis, P., Morais, S. & Dinis, M. T. (2010). Live feeds for early stages of fish rearing. *Aquaculture Research*, *41*(5), 613 – 640.

Cooper, S. L., Sinclair, L. A., Wilkinson, R. G., Hallett, K. G., Enser, M. & Wood, J. D. (2004). Manipulation of the n-3 polyunsaturated fatty acid content of muscle and adipose tissue in lambs. *Journal of Animal Science*, *82*(5), 1461 – 1470.

Coutteau, P. (1996). Microalgae. In: Lavens, P. & Sorgeloos, P. (Eds.) *Manual on the production and use of live food for aquaculture: FAO Fisheries Technical Paper. No. 361*. (pp. 7 – 47)Rome, Italy : FAO.

Daly, B., Swingle, J. S. & Eckert, G. L. (2013). Dietary astaxanthin supplementation for hatchery-cultured red king crab, *Paralithodes camtschaticus*, juveniles. *Aquaculture Nutrition*, *19*(3), 312 – 320.

De Swaaf, M. E., Sijtsma, L. & Pronk, J. T. (2003). *High-cell-density fed-batch cultivation of the docosahexaenoic acid producing marine alga Crypthecodinium cohnii. Biotechnology and Bioengineering*, *81*, 666 – 672.

Del Campo, J. A., Moreno, J., Rodríguez, H., Vargas, M. A., Rivas, J. & Guerrero, M. G. (2000). Carotenoid content of chlorophycean microalgae: factors determining lutein accumulation in *Muriellopsis* sp. (Chlorophyta). *Journal of Biotechnology*, *76*, 51 – 59.

Del Campo, J. A., García-González, M. & Guerrero, M. G. (2007). Outdoor cultivation of microalgae for carotenoid production: current state and perspectives. *Applied Microbiology and Biotechnology*, *74*, 1163 – 1174.

Del Campo, J. A., Rodriguez, H., Moreno, J., Vargas, M. Á., Rivas, J. & Guerrero, M. G. (2004). Accumulation of astaxanthin and lutein in *Chlorella zofingiensis* (Chlorophyta). *Applied Microbiology and Biotechnology*, *64*, 848 – 854.

Dernekbasi, S., Unal, H., Karayucel, I. & Aral, O. (2010). Effect of dietary supplementation of different rates of spirulina (*Spirulina platensis*) on growth and feed conversion in Guppy (*Poecilia reticulata* Peters, 1860). *Journal of Animal and Veterinary Advances*, *9*(9), 1395 – 1399.

Donovan, D. C., Schingoethe, D. J., Baer, R. J., Ryali, J., Hippen, A. R. & Franklin, S. T. (2000). Influence of dietary fish oil on conjugated linoleic acid and other fatty acids in milk fat from lactating dairy cows. *Journal of Dairy Science*, *83*(11), 2620 – 2628.

Donovan, D. C., Schingoethe, D. J., Baer, R. J., Ryali, J., Hippen, A. R. & Franklin, S. T. (2000). Influence of dietary fish oil on conjugated linoleic acid and other fatty acids in milk fat from lactating dairy cows. *Journal of Dairy Science*, *83*(11), 2620 – 2628.

Doreau, M. & Chilliard, Y. (1997). Digestion and metabolisms of dietary fat in animals farm. *British Journal of Nutrition*, *78*, S15 – S16.

Doreau, M. & Chilliard, Y. (1997). Effects of ruminal or postruminal fish oil supplementation on intake and digestion in dairy cows. *Reproduction Nutrition Development*, *37*(1), 113 – 124.

Elmore, J. S., Cooper, S. L., Enser, M., Mottram, D. S., Sinclair, L. A., Wilkinson, R. G. & Wood, J. D. (2005). Dietary manipulation of fatty acid composition in lamb meat and its effect on the volatile aroma compounds of grilled lamb. *Meat Science*, *69*(2), 233 – 242.

Elmore, J. S., Mottram, D. S., Enser, M. & Wood, J. D. (1997). Novel thiazoles and 3-thiazolines in cooked beef aroma. *Journal of Agricultural and Food Chemistry*, *45*(9), 3603 – 3607.

Englmaierová, M., Skřivan, M. & Bubancová, I. (2013). A comparison of lutein, spray-dried *Chlorella*, and synthetic carotenoids effects on yolk colour, oxidative stability, and reproductive performance of laying hens. *Czech Journal of Animal Science*, *58*, 412 – 419.

Ezhil, J. & Narayanan, M. (2013). Enhancement of pigmentation in blue morph, *Pseudotropheus lombardoi* through feeding different carotenoid sources. *World Journal of Fish and Marine Sciences*, *5*(6), 655 – 659.

Fernández-Sevilla, J. M., Acién-Fernández, F. G. & Molina-Grima, E. (2010). Biotechnological production of lutein and its applications. *Applied Microbiology and Biotechnology*, *86*, 27 – 40.

Food and Agriculture Organization of the United Nations (FAO) (2004). *Protein Sources for the Animal Feed Industry*. Roma, Italy: FAO.

Food and Agriculture Organization of the United Nations (FAO) (2011). *Protein Global livestock Production Systems*. Roma, Italy: FAO.

Franklin, S. T., Martin, K. R., Baer, R. J., Schingoethe, D. J. & Hippen, A. R. (1999). Dietary marine algae (*Schizochytrium* sp.) increases concentrations of conjugated linoleic, docosahexaenoic and transvaccenic acids in milk of dairy cows. *Journal of Nutrition*, *129*, 2048 – 2054.

Franklin, S. T., Martin, K. R., Baer, R. J., Schingoethe, D. J., Hippen, A. R. (1999). Dietary marine algae (*Schizochytrium* sp.) increases concentrations of conjugated linoleic, docosahexaenoic and transvaccenic acids in milk of dairy cows. *Journal of Nutrition*, *129*(11), 2048 – 2052

Fredriksson, S., Elwinger, K. & Pickova, J. (2006). Fatty acid and carotenoid composition of egg yolk as an effect of microalgae addition to feed formula for laying hens. *Food Chemistry*, *99*, 530 – 537.

Galyean, M. L., Perino, L. J. & Duff, G. C. (1999). Interaction of cattle health/immunity and nutrition. *Journal of Animal Science*, *77*, 1120 – 1134.

García-Cañedo, J. C., Flores-Ortiz, C. M. & Cañizares-Villanueva, R. O. (2011). Carotenoid profile and content of the green microalgae *Scenedesmus incrassatulus*. *Journal of the American Dietetic Association*, *111*, A94.

George, S. B., Lawrence, J. M., Lawrence, A. L., Smiley, J. & Plank, L. (2001). Carotenoids in the adult diet enhance egg and juvenile production in the sea urchin *Lytechinus variegates*. *Aquaculture*, *199*, 353 – 369.

Gillespie, J. R. & Flanders, F. B. (2009). *Modern Livestock and Poultry Production: Texas sciences series* (8 th. Edition). New York, USA: Cengage Learning.

Ginzberg, A., Cohen, M., Sod-Moriah, U. A., Shany, S., Rosenshtrauch, A. & Arad, S. (2000). Chickens fed with biomass of the red microalga *Porphyridium* sp. have reduced blood cholesterol level and modified fatty acid composition in egg yolk. *Journal of Applied Phycology*, *12*, 325 – 330.

Givens, D. I., Kliem, K. E. & Gibbs, R. A. (2006). The role of meat as a source of *n* 3 polyunsaturated fatty acids in the human diet. *Meat Science*, *74*(1), 209 – 218.

Gouveia, L., Batista, A. P., Sousa, I., Raymundo, A. & Bandarra, N. M. (2008). Microalge in novel food products. In: Konstantinos, N. & Papadopoulos (Eds) *Food Chemistry Research Developments*. (pp. 75 – 111) New York, USA: Nova Science Publishers, Inc.

Gouveia, L., Rema, P., Pereira, O. & Empis, J. (2003). Colouring ornamental fish (*Cyprinus carpio* and *Carassius auratus*) with microalgal biomass. *Aquaculture Nutrition*, *9*(2), 123 – 129.

Gouveia, L., Veloso, V., Reis, A., Fernandes, H., Novais, J. & Empis, J. (1996). *Chlorella vulgaris* used to Colour Egg Yolk. *Journal of the Science of food and Agriculture*, *70*, 167 – 172.

Guedes, A. C. & Malcata, F. X. (2012). Nutritional value and uses of microalgae in aquaculture. *Aquaculture*, *10*(1516), 59 – 78.

Güroy, B., Şahin, İ., Mantoğlu, S. & Kayalı, S. (2012). *Spirulina* as a natural carotenoid source on growth, pigmentation and reproductive performance of yellow tail cichlid *Pseudotropheus acei*. *Aquaculture International*, *20*(5), 869 – 878.

Hammershøj, M., Kidmoseb, U. & Steenfeldtc, S. (2010). Deposition of carotenoids in egg yolk by short-term supplement of coloured carrot *(Daucus carota*) varieties as forage material for egg-laying hens. *Journal of the Science of Food and Agriculture*, *90*, 1163 – 1171.

Harel, M. & Clayton, D. (2004). Feed formulation for terrestrial and aquatic animals. US Patent 20070082008 (WO/2004/080196).

Harun, R., Singh, M., Forde, M. G. & Danquah, M. K. (2010). Bioprocess engineering of microalgae to produce a variety of consumer products. *Renewable and Sustainable Energy Reviews*, *14*, 1037 – 1047.

Hata, M., Hata, M., 1973. Carotenoid pigments in goldfish. V. Conversion of zeaxanthin to astaxanthin. *Bulletin of the Japanese Society of Scientific Fisheries. 38*, 339 – 343.

Hejazi, M. A., Kleinegris, D. & Wijffels, R. H. (2004). Mechanism of extraction of □-Carotene from microalga *Dunaliellea salina* in two-phase bioreactors. *Biotechnology and Bioengineering*, *88*, 593 – 600.

Hemaiswarya, S., Raja, R., Kumar, R. R., Ganesan, V. & Anbazhagan, C. (2011). Microalgae: a sustainable feed source for aquaculture. *World Journal of Microbiology and Biotechnology*, *27*, 1737 – 1746.

Hermansen, J. E. (2003). Organic livestock production systems and appropriate development in relation to public expectations. *Livestock Production Science*, *80*, 3 – 15.

Hintz, H. F. & Heitman, H. (1967). Sewage-grown algae as a protein supplement for swine. *Animal Production*, *9*, 135 – 140. doi:10.1017/S0003356100038393.

Hintz, H. F., Heitman, H. Jr., Weir, W. C., Toreix, D. T. & Meyer, J. H. (1966). Nutritive value of algae grown on sewage. *Journal of Animal Science*, *25*, 675 – 681.

Holman, B. W. B. & Malau-Aduli, A. E. O. (2013). *Spirulina* as a livestock supplement and animal feed. *Journal of Animal Physiology and Animal Nutrition*, *97*(4), 615 – 623.

Hu, Q. (2004). Environmental effects on cell composition. In: Richmond, A. (Ed.). *Handbook of Microalgal Culture: Biotechnology and Applied Phycology*. (83 - 93). Oxford, England: Blackwell Science.

Hudon, J. (1994). Biotechnological applications of research on animal pigmentation. *Biotechnology Advances*, *12*, 49 – 69.

Inborr, J. & Waldenstedt, L. (2000). Effect of astaxanthin-rich algal meal (*H. pluvialis*) on growth performance, caecal *Campylobacter* and *Clostridium* counts and tissue astaxanthin concentration of broiler chickens. *Proceedings of the Australian Poultry Science Symposium*, *12*, 174 – 177.

Ip, C., Singh, M., Thompson, H. J. & Scimeca, J. A. (1994). Conjugated linoleic acid suppresses mammary carcinogenesis and proliferative activity of the mammary gland in the rat. *Cancer Research*, *54*(5), 1212 – 1215.

James, R., Sampath, K., Thangarathinam, R. & Vasudevan, I. (2006). Effect of dietary spirulina level on growth, fertility, coloration and leucocyte count in red swordtail, *Xiphophorus helleri*. *The Israeli Journal of Aquaculture - Bamidgeh*, *58*(2), 97 – 104.

Kalogeropoulos, N., Chiou, A., Gavala, E., Christea, M. & Andrikopoulos, N. K. (2010). Nutritional evaluation and bioactive microconstituents (carotenoids, tocopherols, sterols and squalene) of raw and roasted chicken fed on DHA-rich microalgae. *Food Research International*, *43*, 2006 – 2013.

Kang, C. D., Han, S. J., Choii, S. P. & Sim, S. J. (2010). Fed-batch culture of astaxanthin-rich *Haematococcus pluvialis* by exponential nutrient feeding and stepwise light supplementation. *Bioprocess and Biosystems Engineering*, *33*, 133 – 139.

Karina H. M. Cardozo, K., Guaratini, T., Barros, M. P., Falcão, V. R., Tonon, A. P., Lopes, N. P., Campos, S., Torres, M. A., Souza, A. O., Colepicolo, P. & Pinto, E. (2007). Metabolites from algae with economical impact. *Comparative Biochemistry and Physiology, Part C, 146*, 60 – 78.

Khozin-Goldberg, I. & Cohen, Z. (2006). The effect of phosphate starvation on the lipid and fatty acid composition of the fresh water eustigmatophyte *Monodus subterraneus. Phytochemistry, 67*, 696 – 701.

Khozin-Goldberg, I. & Cohen, Z. (2006). The effect of phosphate starvation on the lipid and fatty acid composition of the fresh water eustigmatophyte *Monodus subterraneus. Phytochemistry, 67*, 696 – 701.

Kobayashi, M., Kakizono, T., Yamaguchi, K., Nishio, N. & Nagai, S. (1992). Growth and astaxanthin formation of *Haematococcus pluvialis* in heterotrophic and mixotrophic conditions. *Journal of Fermentation and Bioengineering, 74*, 17 – 20.

Koller, M., Salerno, A., Tuffner, P., Koinnig, M., Böchzelt, H., Shober, S., Pieber, S., Schnitzer, H., Mittelbach, M. & Braunegg, G. (2012). Characteristics and potential of micro algal cultivation strategies: a review. *Journal of Cleaner Production, 37*, 377 – 388.

Kouba, A., Sales, J., Sergejevová, M., Kozák, P. & Masojídek, J. (2013). Colour intensity in angelfish (*Pterophyllum scalare*) as influenced by dietary microalgae addition. *Journal of Applied Ichthyology, 29*(1), 193 – 199.

Koutsos, E. A., Clifford, A. J., Calvert, C. C. & Klasing, K. C. (2003). Maternal carotenoid status modifies the incorporation of dietary carotenoids into the immune tissues of growing chickens (*Gallus gallus domesticus). Journal of Nutrition, 133*, 1132–1138.

Kramer, J. K., Parodi, P. W., Jensen, R. G., Mossoba, M. M., Yurawecz, M. P. & Adlof, R. O. (1998). Rumenic acid: a proposed common name for the major conjugated linoleic acid isomer found in natural products. *Lipids, 33*(8), 835 – 835.

Legardé, M. (2008). Docosahexaenoic acid: Nutrient and precursor of bioactive lipids. *European Journal of Lipid Science and Technology, 110*, 673 – 678.

Leng, X., Hsu, K., Austic, R. E. & Lei, X. (2014). Effect of dietary defatted diatom biomass on egg production and quality of laying hens. *Journal of Animal Science and Biotechnology, 5*, 3 – 9.

Li, J., Zhu, D., Niu, J., Shen, S. & Wang, G. (2011). An economic assessment of astaxanthin production by large scale cultivation of *Haematococcus pluvialis. Biotechnology Advances, 29*(6), 568 – 574.

Li, Q., Du, W. & Liu, D. (2008). Prespectives of microbial oils for biodiesel production. *Applied Microbiology and Biotechnology, 80*, 749 – 756.

Liu, CP. & Lin, LP. (2001). Ultrastructural study and lipid formation of *Isochrysis* sp. CCMP1324. *Botanical Bulletin-Academia Sinica Taipei, 42*, 207 – 214.

Liu, G. D., Hou, G. Y., Wang, D. J., Lv, S. J., Zhang, X. Y., Sun, W. P. & Yang, Y. (2008). Skin pigmentation evaluation in broilers fed different levels of natural okra and synthetic pigments. *Journal of Applied Poultry Research, 17*, 498 – 504.

Lock, A. L. & Bauman, D. E. (2004). Modifying milk fat composition of dairy cows to enhance fatty acids beneficial to human health. *Lipids, 39*(12), 1197 – 1206.

Lorenz, R. T. & Cysewski, G. R. (2000). Commercial potential for *Haematococcus* microalgae as a natural source of astaxanthin. *Trends in Biotechnology, 18*(4), 160 – 167.

Lukiw, W. J. & Bazan, N. G. (2008). Docosahexaenoic acid and the aging brain. *The Journal of Nutrition*, 2510 – 2514.

Lum, K. K., Kim, J. & Lei, X. G. (2013). Dual potential of microalgae as a sustainable biofuel feedstock and animal feed. *Journal of Animal Science and Biotechnology*, *4*, 53 – 59.

Lundquist, T. J., Woertz, I. C., Quinn, N. W. T. & Benemann, J. R. (2010). *A realistic Technology and engineering assessment of algae biofuel production.* Berkeley, California: Energy Biosciences Intitute.

Mahan, D. C., Carter, S. D., Cline T. R., Hill, G. M., Kim, S. W., Miller, P. S., Nelssen, J. L., Stein, H. H. & Veum, T. L. (2007). Evaluating the effects of supplemental B vitamins in practical swine diets during the starter and grower-finisher periods--a regional study. *Journal of Animal Science*, *85*, 2190 – 2197.

Margalith, P. Z. (1999). Production of ketocarotenoids by microalgae. *Applied Microbiology and Biotechnology*, *51*, 431 – 438.

Marounek, M., Dokoupilová, A., Volek, Z. & Hoza, I. (2009). Quality of meat and selenium content in tissues of rabbits fed diets supplemented with sodium selenite, selenized yeast and selenized algae. *World Rabbit Science*, *17*(4), 207 – 212.

Marshall, J. A., Nichols, P. D. & Hallegraeff, G. M. (2002). Chemotaxonomic survey of sterols and fatty acids in six marine raphidophyte algae. *Journal of Applied Phycology*, *14*(4), 255 – 265.

Mata, T. M., Martins, A. A. & Caetano, N. S. (2010). Microalgae for biodiesel production and other applications: A review. *Renewable and Sustainable Energy Reviews*, *14*, 217 – 232.

Menasveta, P., Worawattanamateekul, W., Latscha, T. & Clark, J. S. (1993). Correction of black tiger prawn (*Penaeus monodon* fabricius) coloration by astaxanthin. *Aquacultural Engineering*, *12*(4), 203 – 213.

Miller, E. R & Kornegay, E. T. (1983). Mineral and vitamins nutrition of swine. *Journal of Animal Science*, *57*, 315 – 329.

Minson, D. J. (2012). *Forage in ruminants nutrition.* San Diego, California: Academic Press, Inc.

Moloney, A. P. (2007). Enrichment of omega-3 fatty acids and CLA in beef by diet modification. *Irish Veterinary Journal*, *60*, 180 – 185.

Mooney, J. W., Hirschler, E. M., Kennedy, A. K., Sams, A. R. & Van Elswyk, M. E. (1998). Lipid and flavour quality of stored breast meat from broilers fed marine algae. *Journal of the Science of Food and Agriculture*, *78*(1), 134 – 140.

Morales, A. E., Cardenete, G., De la Higuera, M. & Sanz, A. (1994). Effects of dietary protein source on growth, feed conversion and energy utilization in rainbow trout, *Oncorhynchus mykiss*. *Aquaculture*, *124*, 117 – 126.

Moriarty, D. J. (1997). The role of microorganisms in aquaculture ponds. *Aquaculture*, *151*(1), 333 – 349.

Muller-Feuga (2004). Microalgae for aquaculture: the current global situation and future trends. In A. Rrichmond (Ed.) Handbook of Microalgal Culture: Biotechnology and Applied Phycology (pp. 352 – 364). Oxford, UK: Publishing Ltd

Muller-Feuga, A. (2000). The role of microalgae in aquaculture: situation and trends. *Journal of Applied Phycology*, *12*(3-5), 527 – 534.

Nasseri, A. T., Rasoul-Amini, S., Morowvat, M. H. & Ghasemi, Y. (2011). Single cell protein: Production and process. *American Journal of Food Technology*, *6*, 103 – 116.

Neori, A. (2011). "Green water" microalgae: the leading sector in world aquaculture. *Journal of Applied Phycology*, *23*(1), 143 – 149.

Nitsan, Z., Mokady, S. & Sukenik, A. (1999). Enrichment of poultry products with ω3 fatty ácids by dietary supplementation with the alga *Nannochloropsis* and Mantur oil. *Journal of Agricultural and Food Chemistry*, *47*, 5127 – 5132.

Nute, G. R., Richardson, R. I., Wood, J. D., Hughes, S. I., Wilkinson, R. G., Cooper, S. L. & Sinclair, L. A. (2007). Effect of dietary oil source on the flavour and the colour and lipid stability of lamb meat. *Meat Science*, *77*(4), 547 – 555.

Orosa, M., Franqueira, D., Cid, A. & Abalde, J. (2005). Analysis and enhancement of astaxanthin accumulation in *Haematococcus pluvialis*. *Bioresource Technology*, *96*, 373 – 378.

Oswald, W. J., Gotaas, H. B., Golueke, C. G., Kelleb, W. R. (1957). Algae in waste treatment. *Sewage and Industrial Wastes*, *29*, 437 – 457.

Overton, T. R., Emmert, L. S. & Clark, J. H. (1998). Effects of source of carbohydrate and protein and rumen-protected methionine on performance of cows. *Journal of Dairy Sciences*, *81*, 221 – 228.

Pannevis, M. C. & Earle, K. E. (1994). Maintenance energy requirement of five popular species of ornamental fish. *Journal of Nutrition*, *124*, 2616 – 2618.

Papadopoulos, G., Goulas, C., Apostolaki, E. & Abril, R. (2002). Effects of dietary supplements of algae, containing polyunsaturated fatty acids, on milk yield and the composition of milk products in dairy ewes. *Journal of Dairy Research*, *69*, 357 – 365.

Peiretti , P. G. & Meineri, G. (2008). Effects of diets with increasing levels of *Spirulina platensis* on the performance and apparent digestibility in growing rabbits. *Livestock Science*, *118*, 173 – 177.

Plaza, M., Herrero, M., Cifuentes, A. & Ibanez, E. (2009). Innovative natural functional ingredients from microalgae. *Journal of Agricultural and Food Chemistry*, *57*(16), 7159 – 7170.

Přibyl, P., Cepák, V. & Zachleder, V. (2013). Production of lipids and formation and mobilization of lipid bodies in Chlorella vulgaris. *Journal of Applied Phycology*, *25*, 545 – 553.

Priyadarshani, I., Sahu, D. & Rath, B. (2012). Algae in aquaculture. *International Journal of Health Sciences and Research*, *2*(1), 108 – 114.

Pulz, O. & Gross, W. (2004). Valuable products from biotechnology of microalgae. *Applied Microbiology and Biotechnology*, *65*, 635 – 648.

Raja, A., Hemaiswarya, S. & Rengasamy, R. (2007). Exploitation of *Dunaliella* for β-carotene production. *Applied Microbiology and Biotechnology*, *74*, 517 – 523.

Reynolds, C. K., Cannon, V. L. & Loerch, S. C. (2006). Effects of forage source and supplementation with soybean and marine algal oil on milk fatty acid composition of ewes. *Animal Feed Science and Technology*, *131*(3), 333 – 357.

Ribeiro, T., Lordelo, M. M., Alves, S. P., Bessa, R. J. B., Costa, P., Lemos, J. P. C., Ferreira L.M. A., Fontes, C. M. G. A. & Prates, J. A. M. (2013). Direct supplementation of diet is the most efficient way of enriching broiler meat with n-3 long-chain polyunsaturated fatty acids. *British Poultry Science*, *54*(6), 753-765.

Ritzenthaler, K. L., McGuire, M. K., Falen, R., Shultz, T. D., Dasgupta, N. & McGuire, M. A. (2001). Estimation of conjugated linoleic acid intake by written dietary assessment methodologies underestimates actual intake evaluated by food duplicate methodology. *The Journal of Nutrition*, *131*(5), 1548 – 1554.

Rodríguez-Cruz, M., Tovar, A. R., Del Prado, M. & Torres, N. (2005). Mecanismos moleculares de acción de los ácidos grasos poliinsaturados y sus beneficios en la salud. *Revista de Investigación Clínica*, *57*, 457 – 472.

Saeid, A., Chojnacka, K., Korczyński, M., Korniewicz, D. & Dobrzański, Z. (2013). Effect on supplementation of *Spirulina maxima* enriched with Cu on production performance, metabolical and physiological parameters in fattening pigs. *Journal of Applied Phycology*, *25*(5), 1607 – 1617.

Sales, J. & Janssens, G. P. (2003). Nutrient requirements of ornamental fish. *Aquatic Living Resources*, *16*(06), 533 – 540.

Sánchez, J. F., Fernández, J. M., Acién, F. G., Rueda, A., Pérez-Parra, J. & Molina, E. (2008a). Influence of culture conditions on the productivity and lutein content of the new strain *Scenedesmus almeriensis*. *Process Biochemistry*, *43*, 398 – 405.

Sánchez, J. F., Fernández-Sevilla, J. M., Acién, F. G., Cerón, M. C., Pérez-Parra, J. & Molina-Grima, E. (2008b). Biomass and lutein productivity of *Scenedesmus almeriensis*: influence of irradiance, dilution rate and temperature. *Applied Microbiology and Biotechnology*, *79*, 719 – 729.

Sardi, L., Martelli, G., Lambertini, L., Parisini, P. & Mordenti, A. (2006). Effects of a dietary supplement of DHA-rich marine algae on Italian heavy pig production parameters. *Livestock Science*, *103*(1), 95 – 103.

Schroeder, J. W. Forage nutrition for ruminants: Quality Forage AS12150. April, 2013. Available from: http://www.ag.ndsu.edu/pubs/ansci/dairy/as1250.pdf.

Selvaraj, R. K., Shanmugasundaram, R. & Klasing, K. C. (2010). Effects of dietary lutein and PUFA on PPAR and RXR isomer expression in chickens during an inflammatory response. *Comparative Biochemistry and Physiology*, *Part A*, *157*, 198 – 203.

Shen, Y., Yuan, W., Pei, Z. & Mao, E. (2010). Heterotrophic culture of *Chlorella protothecoides* in various nitrogen sources for lipid production. *Applied Biochemistry and Biotechnology*, *160*, 1674 – 1684.

Shi, X., Liu, H., Zhang, X. & Chen, F. (1999). Production of biomass and lutein by *Chlorella protothecoides* at various glucose concentrations in heterotrophic cultures. *Process Biochemistry*, *34*, 341 – 347.

Shields, R. J. & Lupatsch, I. (2012). Algae for aquaculture and animal feeds. *Technikfolgenabschätzung – Theorie und Praxis*, *21*, 23 – 37.

Simopoulos, A. P. (1991). Omega-3 fatty acids in health and disease and in growth and development. *The American Journal of Clinical Nutrition*, *54*, 438 – 463.

Simopoulos, A. P. (1999). Essential fatty acids in health and chronic disease. *The American Journal of Clinical Nutrition*, *70*, 560S – 569S.

Soeder, C. J. (1980). Massive cultivation of microalgae: results and prospects. *Hydrobiologia* *72*, 197 – 209.

Sommer, T. R., Potts, W. T. & Morrissy, N. M. (1991). Utilization of microalgal astaxanthin by rainbow trout (*Oncorhynchus mykiss*). *Aquaculture*, *94*(1), 79 – 88.

Spolaore, P., Joannis-Cassan, C., Duran, E. & Isambert, A. (2006). Commercial applications of microalgae. *Journal of Bioscience and Bioengineering*, *101*(2), 87 – 96.

Stephens, E., Ross, I. L., Mussgnug, J. H., Wagner, L. D., Borowitzka, M. A., Posten, C., Kruse, O. & Hankamer, B. (2010). Future prospects of microalgal biofuel production systems. *Trends in Plant Science*, *15*, 554 – 564.

Subhadra, B. & Grinson-George (2011). Algal biorefinery-based industry: an approach to address fuel and food insecurity for a carbon-smart world. *Journal of the Science of Food and Agriculture*, *91*, 2 – 13.

Sun, N., Wang, Y., Li, Y. T., Huang, J. C., Chen, F., (2008). Sugar-based growth, astaxanthin accumulation and carotenogenic transcription of heterotrophic Chlorella zofingiensis (Chlorophyta). *Process Biochemistry 43*, 1288 – 1292.

Tang, G. & Suter, P. M. (2011). Vitamin A,nutrition, and health values of algae: Spirulina, Chlorella, and Dunaliella. *Journal of Pharmacy and Nutrition Sciences*, *1*, 111 – 118.

Thomas, W. H., Seibert, D. L. R., Alden, M., Eldridge, P. & Neori, A. (1983). Yields, photosynthetic efficiencies, and proximate chemical composition of dense cultures of marine microalgae. San Diego: Universidad de California, SERI/STR-231-1896.

Thornton, P. K. (2010). Livestock production: recent trends,future prospects. *Philosophical Transactions of the Royal Society B. 365*, 2853 – 2867.

Toral, P. G., Hervás, G., Gómez-Cortés, P., Frutos, P., Juárez, M. & De la Fuente, M. A. (2010). Milk fatty acid profile and dairy sheep performance in response to diet supplementation with sunflower oil plus incremental levels of marine algae. *Journal of Dairy Science*, *93*(4), 1655 – 1667.

Tume, R. K., Sikes, A. L., Tabrett, S. & Smith, D. M. (2009). Effect of background colour on the distribution of astaxanthin in black tiger prawn (*Penaeus monodon*): Effective method for improvement of cooked colour. *Aquaculture*, *296*(1), 129 – 135.

U.S International Trade Commission (2000). Industry & Trade Summary: Animal feeds. Washinton, DC: Diane Publishing Co.

Van der Spielgel, M., Noordam, M. Y. & Fels-Klerx, H. J. (2013). Safety of novel protein sources (Insects, microalgae, seaweed, duckweed, and rapeseed) and legislative aspects for their application in food and feed production. *Comprehensive Reviews in Food Science and Food Safety*, *12*, 662 – 678.

Van Durme, J., Goiris, K., De Winne, A., De Cooman, L. & Muylaert, K. (2013). Evaluation of the volatile composition and sensory properties of five species of microalgae. *Journal of Agricultural and Food Chemistry*, *61*(46), 10881 – 10890.

Vasta, V. & Priolo, A. (2006). Ruminant fat volatiles as affected by diet. A review. *Meat Science*, *73*(2), 218 – 228.

Velasco-Santamaría, Y. & Corredor-Santamaría, W. (2011). Nutritional requirements of freshwater ornamental fish: a review. *Revista MVZ Córdoba. 16*(2), 2458 – 2469.

Vonshak, A. (1990). Recent advances in microalgal biotechnology. *Biotechnology Advances*, *8*, 709 – 727.

Walker, L. A.,Wang, T., Xin, H. & Dolde, D. (2012). Supplementation of laying-hen feed with palm tocos and algae astaxanthin for egg yolk nutrient enrichment. *Journal of Agricultural and Food Chemistry*, *60*, 1989 – 1999.

Wang, M., Tsao, R., Zhang, S., Dong, Z., Yang, R., Gong, J. & Pei, Y. (2006). Antioxidant activity, mutagenicity/anti-mutagenicity, and clastogenicity/anti-clastogenicity of lutein from marigold flowers. *Food and Chemical Toxicology*, *44*, 1522 – 1529.

Wen, Z.Y., Jiang, Y. & Chen, F. (2002). High cell density culture of the diatom *Nitzschia leavis* for eicosapentaenoic acid production: fed-batch development. *Process Biochemistry*, *37*, 1447 – 1453.

Whitlock, L. A., Schingoethe, D. J., AbuGhazaleh, A. A., Hippen, A. R. & Kalscheur, K. F. (2006). Milk production and composition from cows fed small amounts of fish oil with extruded soybeans. *Journal of Dairy Science*, *89*(10), 3972 – 3980.

Wint, G. R. W. & Robinson, T. P. (2007). *Gridded livestock of the world.* Rome, Italy: FAO.

Wu, L., Huang, X., Shi, K. & Tan, R. (2009). Bioavailability comparison of free and esterified lutein for layer hens. *Brazilian Journal of Poultry Science*, *11*, 95 – 98.

Wu, Z. Y., Qu, C. B. & Shi, X. M. (2009). Biochemical system analysis of lutein production by heterotrophic *Chlorella pyrenoidosa* in a fermentor. *Food Technology and Biotechnology*, *47*, 450 – 455.

Wu, Z. Y., Qu, C. B. & Shi, X. M. (2009). Biochemical system analysis of lutein production by heterotrophic *Chlorella pyrenoidosa* in a fermentor. *Food Technology and Biotechnology*, *47*, 450 – 455.

Yang, Y. X., Kim, Y. J., Jin, Z., Lohakare, J. D., Kim, C. H., Ohh, S. H., Lee, S. H., Choi, J. Y. & Chae, B. J. (2006). Effects of dietary supplementation of astaxanthin on production performance, egg quality in layers and meat quality in finishing pigs. *Asian Australasian Journal of Animal Sciences*, *19*(7), 1019 – 1025.

Yap, C. Y. & Chen, F. (2002). Polyunsaturated fatty acids: Biological significance, biosynthesis, and production by microalgae and microalgae-like organisms. In: Chen, F. & Jiang, Y.(Eds.) *Algae and their Biotechnological Potential.* (pp. 1 – 32). Netherlands: Springer.

Yokochi, T., Honda, D., Higashihara,T. & Nakahara, T. (1998). Optimization of docosahexaenoic acid production by *Schizochytrium limacinum* SR21. *Applied Microbiology and Biotechnology*, *49*, 72 – 76.

Yuan, J. P., Peng, J., Yin, K. & Wang, J. H. (2011). Potential health-promoting effects of astaxanthin: A high-value carotenoid mostly from microalgae. *Molecular Nutrition & Food Research*, *55*(1), 150 – 165.

Zhang, W., Wang, J., Wang, J. & Liu, T. (2014). Attached cultivation of *Haematococcus pluvialis* for astaxanthin production. *Bioresource Technology*, *158*, 329 – 335.

In: Microalgae and Other Phototrophic Bacteria
Editor: Luis G. Torres Bustillos
ISBN: 978-1-63482-078-3

Chapter 6

HYDROGEN AND POLYHIDROXYBUTYRATE (PHB) PRODUCTION BY A PHOTOHETEROTROPHIC MIXED CULTURE

E. I. García-Peña[1*] and R. A. Gonzalez-Garcia[2]
[1]Bioprocesses Department, UPIBI-Instituto Politecnico Nacional, Mexico
[2]The University of Queensland, Brisbane, Australia

ABSTRACT

Biological Hydrogen (H_2) production has been mainly developed via Anaerobic Digestion (AD). Most of the studies demonstrate that the main drawback of this process is the low H_2 yields due to the generation of some organic acids as by-products. Those can be used as feedstock for other processes.

Phototrophic bacteria from the *Rhodobacter* and *Rhodopseudomonas* genus have been identified as potential H_2 producers since they have the metabolic machinery for its production when growing on organic acids. However, H_2 is produced by a complex enzymatic system which is not yet fully understood. Additionally, the use of mixed phototrophic cultures for the H_2 production has the potential to increase the yield when the final products of AD are used as substrate.

The aim of this project was to evaluate H_2 and PHB production by mixed phototrophic cultures when feeding with organic acids

1. INTRODUCTION

H_2 is an attractive source of energy due to its well-known advantages. H_2 is extensively used, not only as energy source but also for different processes as for example in food industry. Nowadays it is mainly produced through chemical process, which involves an

* Corresponding author: Bioprocesses Department, Unidad Profesional Interdisciplinaria de Biotecnología, Instituto Politecnico Nacional. P.O. Box 07340, Mexico City, Mexico. Ph.: (+52) 5557 296000 ext 56474; Fax: (+52) 5557 296000 ext 56305. Email: egarciap@ipn.mx.

extensive use of energy and high cost. That is the reason why during the last decades the study and the development of biological process for H_2 production have been a matter of concern. Three different ways to obtain H_2 by biological processes have been identified: biophotolysis, dark fermentation, photofermentation. The dark fermentation is one of the most studied technologies to due to the simplicity and robustness of the process, in addition because it is a relatively low price technology. Additionally, the dark fermentation allows the use of organic solid waste and wastewater as feedstock, allowing the simultaneous energy production and pollutants elimination. Some renewable feedstock sources such as complex wastewaters and agro-industrial wastes has been evaluated for H_2 production, for example food wastes (Nazlina et al., 2009), vinasses (Buitrón and Carvajal, 2010), water hyacinth, soybean oil extraction residues, mushroom wastes (Chuang et al., 2011), and wastewater from cheese processing (Yang et al., 2007). Data obtained in these studies shows that the H_2 production could be reduced or even inhibit due to some specific characteristics of the wastewater or organic wastes, such as deficient buffer capacity, nutrient imbalance, and presence of microbial populations that may consume H_2 or produce methane (Yang et al., 2007). Additionally, it is well known that in dark fermentation up to 20% of the substrates are stoichiometric converted to H_2, with a maximum conversion of 33% (Hallenbeck and Ghosh, 2009, Esquivel et al. 2014). The remaining carbon is accumulated as byproducts during the process. This means that at list two-thirds of the carbon in the substrate remains, creating a waste disposal problem.

As a result of the aforementioned problems and others, relatively low H_2 yields have been reported. In order to overcome these and to increase the H_2 production yield, some strategies have been evaluated: the use of genetically modified microorganisms (Chittibabu et al., 2004), modifications in the environmental conditions established in the reactor, and hybrid systems that integrate a dark fermentation process with methane or a photoheterotrophic processes (Argun y Kargi, 2011, Loss et al., 2013).

Regarding the modification of the operational conditions there were evaluated the pre-treatment of the inoculum and the application of 2-Bromoethanesulfonic acid (BES) pulses during fermentation to suppress proliferation of H_2 consumers (Abreu et al., 2011, Lee et al., 2009). Also the reduction of the H_2 partial pressure, which constitute a thermodynamic condition that favors the H_2 yield due to its influence on metabolic pathways (Mandal et al., 2006, Esquivel-Elizondo et al, 2013). The use of co-digestion processes has been also evaluate due to the positive synergistic effects of the mixed materials with complementary characteristics and the supply of missing nutrients by the co-substrate (adequate carbon/nitrogen (C/N) ratio, and the macro- and micro-nutrients concentration). A balanced C/N ratio allows enhancing the buffer capacity of the system. Additionally, the co-digestion process also reduces the possibility of inhibitory effects, which, in turn, increases H_2 production (Gomez-Romero et al., 2014).

Regarding the use of hybrid systems, there exists some interest in the use of a photehetorotrophic system feed with the residues of the dark fermentation process. In these system, the purple non-sulfur bacteria (PNSB) could be use because these microorganism naturally produce H_2 using the enzyme nitrogenase while growing on organic acids. Nitrogenases allow the excess electrons to be disposed-off and the redox cofactors are recycled when assimilating reduced carbon compounds under nitrogen limited conditions in the light. Thus, photofermentation could be used as a downstream process, after other fermentation processes where carbohydrates (including these in organic waste) are converted to organic acids. PNSB can overcome the thermodynamic barrier of converting organic acids into H_2 by using light energy. Moreover, PNSB offer an additional advantage since they are able to produce simultaneously Polyhydroxyalkanoates (PHA). The PHA are usually accumulated as polyhydroxybutyrate (PHB) and polydroxyvalerate (PHV) (Wen et al., 2010). These compounds constitute a storage biopolymers produced by some microorganisms at excess carbon source and nutrient limited conditions, and play a role as sink for carbon and reducing equivalents (Mohan and Reddy, 2013). The PHB production could be advantageous as both economically and environmentally, because of their production from renewable resources and their high biodegradability (Mohan and Reddy, 2013). Moreover, the chemical properties of the polymer are similar to those of petrochemical plastics like polypropylene (Marang et al., 2013).

The factors controlling both, the H_2 and the PHB production, by PNSB will be discussed in the next paragraphs.

A number of authors have been study the H_2 production by photosynthetic bacteria evaluating a wide range of substrates and some different microorganisms, either in pure or mixed culture. These studies showed that the main factors controlling the process are the substrate, the C/N ratio and the light intensity.

It is well known that PNSB have the capacity to ferment various organic acids as acetate, lactate, and butyrate with relatively high conversion efficiency (Barbosa et al., 2001, Fang et al., 2005, Ren et al., 2009, Tao et al., 2008). Also some studies have been developed with glucose and ethanol byproduct (Oh et al., 2004).

Recently, Yilmaz et al. 2010 studied the partitioning of electrons from different substrates (i. e. glucose, lactate, succinate, fumarate, and pyruvate). Authors found by using an electron equivalent analysis that 85% of the electrons from the substrates are distributed into four sinks: H_2, cell biomass, polyhydroxybutyrate (PHB), and soluble microbial products (SMP). This aspect was evaluated during growth and stationary phases of *Rhodobacter sphaeroides* cultures. Data showed that the highest total cumulative amount of electrons directed toward H_2 production (55%) was found using lactate as carbon and energy source. Biomass and PHB production were the main electron sinks during the growth phase. While during the stationary phase the biomass was almost neglected and the major amount of electrons was evolved as H_2 and SMP. Additionally, it was demonstrate that the higher percentages of electrons directed toward H_2 were found in cultures with succinate and lactate as substrate. The lowest percentage of electrons partitioned into H_2 was determined with glucose as carbon and energy source. The behavior of the electrons evolved as PHB production was inversely correlated, the higher was obtained with glucose and the lower with the acids as substrate.

Oh et al. (2004) studied a *Rhodopseudomonas palustris* P4 strain able to produce H_2 either from CO_2 by water-gas shift reaction or from various sugars by anaerobic fermentation. The use of this strain for H_2 production is interesting because it is fast, however its yield is relatively low due to the formation of various organic acids. To increase H_2 production yield from glucose, the use of acetate was evaluated. Experiments were performed in batch cultures using both light-grown and dark-grown cells. When the dark-grown P4 was challenged with light and acetate, H_2 was produced with the consumption of acetate after a lag period. H_2 production was inhibited when a nitrogen source, especially ammonium, was present. The fact that a single strain can perform both dark- and light-fermentation gives a great advantage in process development. Compared to a one-step dark-fermentation, the combined dark- and light-fermentation can increase the H_2 production yield on glucose by two-fold.

Some others works with *Rhodobacter capsulatus* JP91 showed that photofermentative H_2 production from glucose is feasible with a maximum reported hydrogen yield of 3 mol H_2/mol of glucose (Abo-Hashesh et al., 2011). In that study, a single glucose concentration at two different glutamate concentrations keeping light intensity constant was investigated, suggesting that a more thorough investigation of the various process parameters could lead to higher hydrogen yields. More recently, Ghosh et al., 2012 study the effect of light intensity, glucose and glutamate concentrations over the nitrogenase activity, and in turn H_2 production, in a culture of *Rhodobacter capsulatus*. By using surface methodology (Box–Behnken design) for H_2 optimization, the authors found that the main factor determining nitrogenase activity was the light intensity, under the initial hypothesis that higher nitrogenase activity is correlated with the higher H_2 production. When the H_2 yields were tested by using the same analysis, data shown that glucose and glutamate concentrations had much stronger effects than light intensity. This suggests that hydrogen yields may not be directly proportional to nitrogenase activity. This strongly suggests that factors other than the total amount of enzyme present are controlling activity and hydrogen production. These factors are probably related to flux restriction of either high-energy electrons or ATP at the level of metabolic supply to nitrogenase, as previously suggested by studies of the relationship between nitrogenase protein content, nitrogenase activity, and nitrogen supply (Yakunin et al.1999). The hydrogen yield predicted by the model, 5.29 mol H_2/mol glucose, is higher than the previously reported molar hydrogen yield of 3 mol H_2/mol glucose (Abo-Hashesh et al., 2011), and higher than the vast majority of previous studies which examined single-stage fermentation of glucose.

Light intensity is one of the most important factors in H_2 production due to its potential effects on nitrogenase enzyme regulation and flux through the electron transport chain (Jouanneau et al., 1985). Light intensity could potentially modulate end product distribution, as well as the duration of the lag phase and growth kinetics.

Some studies showed that the optimal light intensity for strain *Rhodopseudomonas palustris* RLD-53 was 4.4–7.32 W/m^2 (Ren et al., 2009). *Rhodopseudomonas palustris* CQK 01 increased the H_2 production with light intensity increased to a threshold of 7000 lux (7.32 W/m^2), and when the light intensity exceeds the threshold, H_2 production was inhibited. Increased light intensity resulted in an increase in the total volume and production rate of hydrogen, and light intensity at 15 W/m^2 highest hydrogen yield in combination system of dark- and photo-fermentation process was obtained (Nath and Das, 2009).

The initial ratio of carbon to nitrogen in the production medium also plays an important role. A low initial carbon to nitrogen ratio results in lower rates of fermentation and lower H_2 yields, whereas total fermentation times increase with a higher initial carbon to nitrogen ratio. In H_2 production by photofermentation, this effect is mainly due to the influence of the carbon to nitrogen ratio on nitrogenase synthesis with nitrogenase repression occurring when there is excessive fixed nitrogen in the medium. However, since hydrogen production by nitrogenase is mainly observed in actively growing cells, fixed nitrogen must be supplied to support cell growth. One way around this problem experimentally is to supply fixed nitrogen in the form of an amino acid, many of which have only minor effects on nitrogenase repression when supplied at moderate concentrations (Ghosh et al., 2012)

Photofermentative H_2 production has been also evaluated by using different types of wastewater from a variety of industries: olive mill effluent (Eroglu et al., 2004), sugar refinery wastewater (Yetis et al., 2000), tofu industry wastewater (Zhu et al., 1999), brewery wastewater (Seifert et al., 2010) and glycerol (Sabourin-Provost and Hallenbeck, 2009).

Additionally, studies to develop a two-stage system capable of converting the fermentation byproducts from a first stage to H_2 has been recently performed (Hallenbeck and Ghosh, 2009, Hallenbeck, 2011, Keskin et al., 2011). High H_2 yields (13.7 mol H_2/mol sucrose) have been obtained from beet molasses in two stage systems (Avcioglu et al., 2011; Ozgur et al., 2010a, 2010b) and (Ozgur et al., 2010b). This yield is similar to those results obtained with pure sucrose, 14.2 mol H_2/mol sucrose (Chen et al., 2008). H_2 yields ranged from 0.93 for a mixed culture (Ozkan et al., 2011) and 14.9 molH_2/mol substrate for *Rhodobacter capsulatus* JP91 (Keskin and Hallenbeck, 2012) for two stages. Differences in the H_2 production are a function of the microorganism and substrates used during photofermentation processes.

During this study the activation of the mixed phototrophic cultures obtained from Winogradsky columns was performed. The characterization and identification of the microbial population present in the mixed cultures was also developed by using molecular analysis, DGGE and pyrosequencing. Additionally, the growth kinetics and the potential production of H_2 and PHB were study with different substrates, acetate and butyrate. Further investigation is required to understand the factors controlling nitrogenase modification during photofermentation of organic acids and how this might affect overall hydrogen productivity. In this initial study a metabolite analysis was not performed. The logical extension of this research would be to carry out an identification and analysis of metabolic end products of the mixed cultures to obtain H_2 or PHB as final products.

2. Materials and Methods

2.1. Batch Cultures

Batch fermentations for growth and H_2 production were performed initially in serological bottles (80 mL operation volume) Figure 1. All the experiments were conducted at initial pH 6.8. Two different media were used for the experiments. 1) A sterile complete media (AT) which contains (g/L): KH_2PO4 1.0, $MgCl_2*6H_2O$ 0.5, $CaCl_2*2H_2O$ 0.1, NH_4Cl 1.0, $NaHCO_3$ 3, NaCl 1.0, Vitamin solution 1mL, SLA solution 1 mL, sodic ascorbate 0.5, yeast extract 0.5 and acetate or butyrate 1.0 as energy and carbon source. The vitamin solution and the SLA solution compositions are shown in Table 1, respectively. 2) A nitrogen limited modified mRCBV media containing (mg/L): $(NH_4)_2SO_4$ 100, $FeSO_4$ 12, $MgSO_4*7H_2O$ 200, $CaCl_2*2H_2O$ 75, KH_2PO4 49.5, Na_2HPO_4 $*H_2O$ 129.5, Disodic EDTA 20, yeast extract 100, vitamin and SLA solutions (1mL). Acetate was used as the only carbon source (2 g/L) for the H_2 production test.

After inoculation the bottles were sealed with butyl rubber septum and aluminum crimps. Bottles were flushed with Ar (99.9%) for a period of 5 min to ensure anaerobic conditions. The experiments were incubated at 30°C and 150 rpm in an incubator shaker. Constant illumination was supplied at 2000 lux. Samples of the headspace were periodically taken to evaluate H_2. Biomass, PHB and acetate concentration were also measured. The experimental systems were run by triplicates.

Table 1. Vitamin and SLA compositions

Vitamin solution		SLA solution	
Compound	mg in 100 mL	Compound	mg in 1000 mL
Biotine	10	$FeCl_2*4H_2O$	1800
Niacin	35	$CoCl_2*6H_2O$	250
Dichlore thiamine	30	$NiCl_2*6H_2O$	10
p-Aminobezoic acid	20	$CuCl_2*2H_2O$	10
Pyridoxolium hydrochloride	10	$MnCl_2*4H_2O$	70
Ca-panthothenate	10	ZnCl2	100
Vitamin B_{12}	5	H_3BO_3	500
		$Na_2MoO_4*2H_2O$	30

Photo-bioreactor Experimets

A reactor of 6 L of total volume (New Brunswick BioFlo 115 Fermentor) was used for the experiments using a working volume of 5 L (Figure 2). For the initial stage of operation the AT medium was used and then the RCBV was fed to the reactor to favor the H_2 production. The pH was kept and maintained at a value of 6.8 by automatic addition of NaOH or H_2SO_4 solution. The temperature was set and controlled by a water bath operated by using the control systems of the reactor. To assessing anaerobic conditions the system was flushed with Argon gas during 10 min once the system was inoculated. A continuous agitation of 150 rpm was used through the time culture. Constant illumination of 17.5 μmole/cm^2/s was obtained by a led as shown in Figure 2.

2.3. Analytical Methods

Hydrogen production were measured using a gas chromatograph (Gow Mac 550) equipped with a thermal conductivity detector and 18' x 1/8' x 0.085'' silica-gel column (Alltech). The column temperature was 30°C, the injector and detector temperatures were 75°C and 120°C respectively. Nitrogen was used as a carrier at flow of 30 ml/min.

Acetate was measured with a high performance liquid chromatography (HPLC) series 200 of Perkin Elmer equipped with a UV detector and a Prevail Organic Acid 5u (Alltech), 150 x 4.6mm column. The analysis conditions were: aqueous solution of KH_2PO4 with concentration of 25 mMol (pH =2.5) as mobile phase, wavelength of 210 nm, a flow rate of 1 ml/min and with an injection volume of 30 μl of sample.

Biomass concentration was quantified by measuring optical density at 600 nm and Bradford technique.

The PHB was separated from the culture and cells by centrifugation (3000 g for 30 min) and the resulting pellet was washed with acetone and ethanol separately to remove unwanted materials. The pellet with lysed cells was then washed with acetone and ethanol and dissolved in chloroform. Then a thermal extraction was developed.

2.4. Microbial Population Identification

2.4.1. DNA Extraction

Liquid samples (~1.5 mL) were taken for microbial analysis and further DNA extraction from the already adapted inoculum. DNA was extracted by using the CTAB protocol as described in Gomez-Romero et al., 2014. Cells from 1.5 mL of the broth were harvested in a tube by centrifugation at 14,000 x g for 5 min followed by decantation of the supernatant. The pellet was re-suspended in 0.5 mL of a pre-heated CTAB extraction buffer, the mixture was incubated during 1 hour at 65°C. Once the samples were at room temperature, a chloroform-octanol solution (24:1) was added and then centrifuged at 1,813 x g during 10 min. The aqueous supernatant was transferred into a sterile tube (1.5mL) and 30 μL of RNase solution (10 mg/mL) were added. Samples were incubated during 50 min at room temperature. The DNA pellet was obtained by adding one volume of iso-propanol and centrifugation at 2,414 x g for 10 min. The DNA pellet was re-suspended in TE buffer (1M Tris pH 8 and 0.5M EDTA) and incubated overnight at room temperature. Then, 1 mL phenol- chloroform solution (1:1) was added and the mixture was centrifuged at 1,813 x g during 10 min. Afterwards, the total nucleic acid was precipitated with NaCl-ethanol solution and centrifuged at 9,659 x g and 4°C for 1 min. The purified DNA was eluted with 40 μL of miliQ water and kept at -20°C before using it as template DNA for DGGE and pyrosequencing analysis.

2.4.2. DGGE Analysis

Before the DGGE analysis the 16S rRNA gene was amplified using the following universal bacterial primers: forward primer F-GC Clamp 5′CGCCCGCCGCGCGG CGGGCGGGGCGGGGGCACGGGGGGCCTACXXctcctacgg gaggcagcag 3″) and reverse primer R (5″TCCCGCAACGAGCGCAACCC 3″). An 800- bp PCR product was obtained corresponding to the nucleotides between positions 290-309 and 1045-1064 of the E. coli 16S rRNA gene. The 50 μL of PCR reaction mixture contained 1 μL of template DNA, 5 μL of 10×X buffer, 1 μL of 25 mM $MgCl_2$, 1 μL of dNTPs (5 μM each of dATP, dCTP, dGTP, and dTTP), 1 μL of 10 mM forward primer, 1 μL of 10 mM reverse primer, 45 μL of sterile water, and 1 unit of Taq polymerase. hePCR was conducted as follows: 94°C pre-denaturing for 5 min; 30 cycles of 95°C denaturing for 30 s, 56°C annealing for 30 s, and 72°C extension for 30 s. This process was followed by 72°C extension for 10 min. The PCR-amplified DNA products were separated by DGGE on 8% polyacrylamide gels with a linear gradient of 30–50% denaturant (100% denaturant was 40% [v/v] formamide plus 42% [w/v] urea) using the DCodeTM System (Bio-Rad, Hercules, CA, USA).

2.4.3. 454-Pyrosequencing

Bacterial tag-encoded FLX amplicon pyrosequencing (bTEFAP) was performed at the Research and Testing Laboratory (Lubbock, TX, USA). Bacterial primers 104F and 530R were used to amplify the V2 and V3 hyper-variable regions of the 16S rRNA gene spanned nucleotides 137-242 and 433-497 base pair, respectively (numbering based on the *E. coli* 16S rRNA gene). Amplicons were sequenced with the FLX-Titanium System Genome Sequencer. QIIME 1.7.0 pipeline was used to process raw sequences. Sequences with at least one of the following characteristics were omitted for the downstream analysis: shorter than 200 bps, quality score of 25 or below, any primer or barcode mismatches, more than 6 homopolymers.

From the sequences that passed the quality filtering, OTUs were picked based on 97% sequence similarity, using uClust algorithm as described in Esquivel-Elizondo 2013.

3. Results

3.1. Initial Characterization of Phototrophic Mixed Cultures

Most of the photoheterotropic systems for H_2 production have been studied using pure strains of different PNSB and under sterile conditions. However, for field applications and for the hybrid systems the use of mixed cultures over the pure strains could have some advantages, especially for biotechnology processes that use wastewater and organic solid waste. Mixed cultures could have the ability to adapt the microbial diversity during the handling of mixed substrates and the metabolic versatility to use different organic substrates, which could be produce in an upstream process, as a dark fermentation (Cai et al., 2012).

Three photo-heterotropic mixed cultures were isolated from Winogradsky columns. The mixed cultures arbitrary label as C2, C4 and C5 were initially characterized by a DGGE analysis. Figure 3 shows the band profile obtained for each of the cultures. This data showed less microbial diversity for the C2 culture compared with C4 and C5 cultures.

Table 2. Microbial characterization of the mixed photo-heterotrophic cultures by pyrosequencing analysis

	% of the total bacterial population		
Microorganisms	**C2**	**C4**	**C5**
Citrobacter	12.029	12.029	
Bacteroides	24.07	24.07	24.07
Rhodopseudomonas	29.01	29.01	29.01
Dysgonomonas		8.06	26.45
Others	34.89	26.83	20.47

The DGGE profiles were further corroborated by a pyrosequecing analysis, the result at genus level is shown in Table 2. All the cultures contain *Phodosepudonomas palustris* at slightly different percentages and constituted around 30% of the total microbial populations. Other microorganisms as *Citrobater* and *Bacteriodetes* were present in lower percentages. *Disgonomonas* was found in cultures C4 and C5. Recent pyrosequencing analysis and characterization of the culture C2 showed that at the maximum H_2 production time *Rhodopseudomonas palustri* constitutes the main population of the mixed culture (67%, data not shown).

Our data are similar to that recently reported by Loss et al., 2013. These authors demonstrated the effectiveness of the Winogradsky column as an effective methodology for the enrichment of environmental sample with PNS bacteria. During this study the authors use a DGGE analysis to determined that the mixed consortium was constitute of only 4 species. Some other studies showed similar results, a mixed phototrophic sludge from an anaerobic digestion system contains only three phototrophic H_2-producing species (Zhang et al., 2002). The enrichment of a mixed phototrophic sludge from the sediment of a local river for continuous hydrogen production from acetate and butyrate have been also reported, data obtain showed that 3 of 5 species were identified as phototrophs (Ying et al., 2008).

3.2. Microbial Growth with Acetate and Butyrate under Non Limiting Nitrogen Conditions

The mixed cultures were evaluated under two different conditions: growth media with acetate and ammonium as nitrogen source and nitrogen limited conditions. These conditions allowed to determine their capacity to growth with different substrates (acetate and butyrate) and to produce H_2 under nitrogen limited conditions.

Growth at mCRVB media with acetate and butyrate as carbon and energy source was evaluated as first attempt to evaluate the capacity of the different mixed cultures to use these substrates. As seen in Figure 4 all mixed cultures growth with acetate as substrate. Lower growth was determined in butyrate fed cultures.

Table 3 summarized the results of the final biomass production for each of the cultures, all cultures growth well with acetate. In general growth was lower with butyrate, the higher was determined with C4 and C5 and lower growth was observed with C2 culture.

Table 3. Growth with acetate and butyrate as carbon source

Culture	AT acetate	mRCVB acetate	mRCVB butyrate
C2	++++	++++	+
C4	++++	++++	++
C5	++++	++++	++

Once it was determined that all cultures were able to growth with acetate and butyrate. The complete profile/behavior of the growth was determined with acetate as substrate.

In Figure 5 the biomass production during 400 h of culture is presented.

Similar behavior was observed for the three mixed cultures, after a lag phase of approximately 80 h an exponential growth phase was determined. The maximum growth was reached at 200 h of culture and after that a stationary phase was established. No H_2 was determined under these conditions. The kinetic conditions maximum growth rate (umax) and biomass yield (Yx/s) are summarize in Table 4 for each of the photohetetotrophic mixed cultures. Cultures C2 and C4 showed higher umax values as compared with the one determined for C4. The biomass yield was similar for all cultures, but slightly higher for C5.

Table 4. Kinetic parameters under growth conditions

Culture	Umax (h-1)	Yx/s (g/g)
C2	0.0064	0.21
C4	0.0051	0.23
C5	0.0063	0.25

3.3. Microbial Growth and H_2 and PHB Production with Acetate under Limited Nitrogen Conditions

When a limited nitrogen medium was used the period of growth initiated at 60 h of culture, followed by a period of 200 h of H_2 production with an slightly increase in growth (data not shown). Table 5 shows the biomass yield, the maximum growth rate and the H_2 production as it was expected under limited nitrogen conditions the Yx/s and the υmax were lower than those obtained under growth conditions of culture. The highest H_2 production (75%) was determined for the C2 mixed culture and the lowest for the C5.

The PBH trend was inversely at the one for H_2 production, the higher PHB production was determined for the C4 and C5 photoheterotrophic mixed cultures. The highest H_2 production was associated with the low diversity microbial community.

Table 5. Kinetic parameters and H_2 and PHB production under limited nitrogen conditions

Culture	Initial acetate concentration (g/L)	Initial pH	Yx/s (g/g)	υ_{max} (h^{-1})	% H_2	PHB (g/L)
C2	2.0	6.8	0.012	0.0019	74.39	1.0
C4	2.0	6.8	0.08	0.0017	48.71	1.9
C5	2.0	6.8	0.08	0.0033	15.22	2.0

3.4. Microbial Growth and H_2 and PHB Production with Acetate under Limited Nitrogen Conditions

Since the C2 culture was the best H_2 producer, an experiment with this mixed microbial culture was performed in the photobioreactor. In Figure 6 two stages of the culture are shown. At the first stage growth was promoted, approximately 50% of the initial acetate concentration was consumed during the 100 h of culture, this consumption was associated with an increment of the biomass.

The H_2 production was associated with the biomass growth. After the initial consumption of acetate a stationary phase was observed, no acetate consumption was determined during approximately 100 h of culture, and then the H_2 production achieved its higher production rate until it reached a maximum of 43%. This H_2 production (43%) was lower compare with the maximum H_2 percentage (v/v in gas phase) obtained in the serological bottles.

The H_2 production behavior was most likely driven by nitrogen metabolism. That is, when the cells grew presumably used all the available fixed nitrogen (ammonium) for growth. During this time nitrogenase was inhibited and no H_2 could be produced. However, as soon as the fixed nitrogen source was consumed, cellular growth almost stopped, provided that reduced carbon was still available and at the same time the repressive effect on nitrogenase was removed and H_2 started to be produced.

Conclusion

During this work the effectiveness of the Winogradsky column as a methodology for the enrichment of environmental sample with PNS bacteria was demonstrated. Bacterial 16s rRNA gene analysis showed that the mixed cultures microbial population is composed for 3 to 4 species among then *Rhodopseudomonas palustri* (around 30%). At the higher H_2 production time this specie constitutes the main microbial population (67%). Additionally it was demonstrated that all cultures were able to grow on acetate and butyrate.

When growing under non-limiting condition, all cultures showed similar values for specific growth rate (0.006 h-1) and biomass yield (0.23 g/g). Production of H_2 occurred when a completely ammonia limiting condition was tested. The mixed photo-heterotrophic culture C2 showed the highest H_2 production (70%) on the gas phase and the highest H_2 production yield. The PBH trend was inversely at the one for H_2 production, the higher PHB production was determined for the C4 and C5 photoheterotrophic mixed cultures.

References

Abo-Hashesh M., Ghosh D., Tourigny A., Taous A., Hallenbeck P.C. 2011. Single stage photofermentative hydrogen production from glucose: An attractive alternative to two stage photofermentation or co-culture approaches. *Int. J. Hydrogen Eng.* 36, 13889-13895.

Abreu AA, Alves JI, Pereira MA, Sousa DZ, Alves MM. 2011. Strategies to suppress hydrogen-consuming microorganisms affect macro and micro scale structure and microbiology of granular sludge. *Biotechnol. Bioeng.* 108, 1766-1775.

Argun H., Kargi F. (2011) Bio-hydrogen production by different operational modes of dark and photo-fermentation: An overview. *Int J Hydrogen Energy,* 36, 7443-7459.

Aparecida Loss R., Fontes M.L., Reginatto V., Vasconcellos Antônio R. 2013. Biohydrogen production by a mixed photoheterotrophic culture obtained from a Winogradsky column prepared from the sediment of a southern Brazilian lagoon. *Renewable Energy* 50, 648-654.

Avcioglu S.G., Ozgur E., Eroglu I., Yucel M., Gunduz U., 2011. Biohydrogen production in an outdoor panel photobioreactor on dark fermentation effluent of molasses. *Int. J. Hydrogen Energy* 36, 11360-11368.

Barbosa, M.J., Rocha, J.M.S., Tramper, J., Wijffels, R.H., 2001. Acetate as a carbon source for hydrogen production by photosynthetic bacteria. *J. Biotechnol.* 85, 25–33.

Buitrón G, Carvajal C., 2010. Biohydrogen production from Tequila vinasses in an anaerobic sequencing batch reactor: Effect of initial substrate concentration, temperature and hydraulic retention time. *Bioresour. Technol.* 101, 9071-9077.

Cai J, Wang G, Pan G. 2012. Hydrogen production from butyrate by a marine mixed phototrophic bacterial consort. *Internat. J. Hydrogen Energy* 37, 4057-4067.

Chittibabu G, Nath K, Das D. 2006. Feasibility studies on fermentative hydrogen production by recombinant Escherichia coli BL-21. *Process Biochem.* 41, 682-688.

Chen C.Y., Yang M.H., Yeh K.L., Liu C.H., Chang J.S., 2008. Biohydrogen production using sequential two-stage dark and photo fermentation processes. *Int. J. Hydrogen Energy* 33, 4755-4762.

Chuang YS, Lay CH, Sen B, Chen CC, Gopalakrishnan K, Wu JH, Lin CS, Lin CY. 2011. Biohydrogen and biomethane from water hyacinth (Eichhorniacrassipes) fermentation: Effects of substrate concentration and incubation temperature. *Int. J. Hydrogen Energy* 36, 14195-14203.

Eroglu E., Gunduz U., Yucel M., Turker L., Eroglu I., 2004. Photobiological hydrogen production by using olive mill wastewater as a sole substrate source. *Int. J. Hydrogen Energy* 29, 163-171.

Esquivel-Elizondo S.V. 2013. *MsC degree Thesis.* Producción biológica de hidrógeno bajo diferentes condiciones de presión parcial de este gas. Maestría en Ciencias en Bioprocesos. UPIBI-IPN. Junio 2013.

Esquivel-Elizondo S., Chairez I., Salgado E., Aranda J. S., Baquerizo G., Garcia-Peña E. I. 2014. Controlled continuous bio-Hydrogen production using different biogas release strategies. *Appl Biochem Biotechnol,* in press doi 10.1007/s12010-014-0961-8.

Fang H.H.P., Zhu H.G., Zhang T., 2006. Phototrophic hydrogen production from glucose by pure and co-cultures of *Clostridium butyricum* and *Rhodobacter sphaeroides. Int. J. Hydrogen Energy* 31, 2223-2230.

Ghosh D., Flore Sobro I., Hallenbeck P.C. 2012. Optimization of the hydrogen yield from single-stage photofermentation of glucose by *Rhodobacter capsulatus* JP91 using response surface methodology. *Bioresource Technology* 123, 199–206.

Gomez-Romero J., Gonzalez-Garcia A., Chairez I., Torres L. , García-Peña E.I. 2014. Selective adaptation of an anaerobic microbial community: biohydrogen production by codigestion of cheese whey and vegetables fruit waste. *Int. J. Hydrogen Energy in press* http://dx.doi.org/10.1016/j.ijhydene.201406.2050.

Hallenbeck P.C., 2011. Microbial paths to renewable hydrogen production. *Biofuels* 2, 285-302.

Hallenbeck PC, Ghosh D. 2009. Advances in fermentative biohydrogen production: the way forward? *Trends Biotechnol.* 27, 287-297.

Jouanneau Y., Wong B., Vignais P.M., 1985. Stimulation by light of nitrogenase synthesis in cells of Rhodopseudomonas capsulatus growing in N-limited continuous cultures. *Biochimica et Biophysica Acta* 808, 149-155.

Keskin T., Abo-Hashesh M., Hallenbeck P.C., 2011. Photofermentative hydrogen production from wastes. *Bioresour. Technol.* 102, 8557-8568.

Keskin T., Hallenbeck P.C. 2012. Hydrogen production from sugar industry wastes using single-stage photofermentation. *Bioresource Technology* 112, 131–136.

Lee HS, Krajmalnik-Brown R, Zhang H, Rittmann BE. 2009. An Electron-Flow Model Can Predict Complex Redox Reactions in Mixed-Culture Fermentative BioH_2: Microbial Ecology Evidence. *Biotechnol. Bioeng.* 104, 687-697.

Loss R.A., Fontes M.L., Reginatto V., Vasconcellos R.A. 2013. Biohydrogen production by a mixed photoheterotrophic culture obtained from a Winogradsky column prepared from the sediment of a southern Brazilian lagoon. Renew. *Energy,* 50, 648-654.

Mandal B., Nath K., Das D. 2006. Improvement of biohydrogen production under decreased partial pressure of H2 by *Enterobacter cloacae*. *Biotechnology Letters,* 28, 831-835.

Marang L., Jiang Y., van Loosdrecht M.C.M., Kleerebezem R. 2013. Butyrate as preferred substrate for polyhydroxybutyrate production. *Biores. Technol.,* 142, 232-239.

Mohan S.V., Reddy M.V. 2013. Optimization of critical factors to enhance polyhydroxyalkanoates (PHA) synthesis by mixed culture using Taguchi design of experimental methodology. *Biores. Technol.,* 128 409–416.

Nath K., Das D. 2009. Effect of light intensity and initial pH during hydrogen production by an integrated dark and photo-fermentation process. *Int. J. Hydrogen Eng.* 34, 7497-7501.

Nazlina HMY, Aini ARN, Ismail F, Yosof MZM, Hassan MA. 2009. Effect of different temperature, initial pH and substrate compositions on biohydrogen production from food waste in batch fermentation. *Asian J. Biotechonol.,* 1, 42-50.

Oh Y.K., Seol E.H., Kim M.S., Park S., 2004. Photoproduction of hydrogen from acetate by a chemoheterotrophic bacterium *Rhodopseudomonas palustris* P4. *Int. J. Hydrogen Eng.* 29, 1115–1121.

Ozgur E., Afsar N., De Vrije T., Yucel M., Gunduz U., Claassen P.A.M., Eroglu I., 2010a. Potential use of thermophilic dark fermentation effluents in photofermentative hydrogen production by Rhodobacter capsulatus. *J. Clean Prod.* 18, 23-28.

Ozgur E., Mars A.E., Peksel B., Louwerse A., Yucel M., Gunduz U., Claassen P.A.M., Eroglu I., 2010b. Biohydrogen production from beet molasses by sequential dark and photofermentation. *Int. J. Hydrogen Energy* 35, 511-517.

Ozkan L., Erguder T.H., Demirer G.N., 2011. Effects of pretreatment methods on solubilization of beet-pulp and bio-hydrogen production yield. *Int. J. Hydrogen Energy* 36, 382-389.

Ren N.Q., Liu B.F., Ding J., Xie G.J., 2009. Hydrogen production with *Rhodopseudomonas faecalis* RLD-53 isolated from freshwater pond sludge. *Bioresource Technol.* 100, 484–487.

Ren, N.Q., Liu, B.F., Ding, J., Guo, W.Q., Cao, G.L., Xie, G.J., 2008. The effect of butyrate concentration on photo-hydrogen production from acetate by *Rhodopseudomonas faecalis* RLD-53. *Int. J. Hydrogen Eng.* 33, 5981–5985.

Sabourin-Provost G., Hallenbeck P.C., 2009. High yield conversion of a crude glycerol fraction from biodiesel production to hydrogen by photofermentation. *Bioresour. Technol.* 100, 3513-3517.

Seifert K., Waligorska M., Laniecki M., 2010. Hydrogen generation in photobiological process from dairy wastewater. *Int. J. Hydrogen Energy* 35, 9624-9629.

Tao Y., He Y., Wu Y., Liu F., Li X., Zong W., Zhou Z., 2008. Characteristics of a new photosynthetic bacterial strain for hydrogen production and its application in wastewater treatment. *Int. J. Hydrogen Energy* 33, 963-973.

Wen Q., Chen Z., Tian T., Chen W. 2010. Effects of phosphorus and nitrogen limitation on PHA production in activated sludge. *J. Environ. Scienc.* 22,1602-1607.

Yakunin A.F., Laurinavichene T.V., Tsygankov A.A., Hallenbeck P.C., 1999. The presence of ADP-ribosylated Fe protein of nitrogenase in *Rhodobacter capsulatus* is correlated with cellular nitrogen status. *J. Bacteriol.* 181, 1994-2000.

Yang P, Zhang R, McGarvey JA, Benemann JR. 2007. Biohydrogen production from cheese processing wastewater by anaerobic fermentation using mixed microbial communities. *Int. J. Hydrogen Energy* 32, 4761-4771.

Yetis M., Gunduz U., Eroglu I., Yucel M., Turker L., 2000. Photoproduction of hydrogen from sugar refinery wastewater by *Rhodobacter sphaeroides* O.U.001. *Int. J. Hydrogen Energy* 25, 1035-1041.

Yilmaz L.S., Kontur W.S., Sanders A.P., Sohmen U., Donohue Timothy J., Noguera Daniel R. 2010. Electron partitioning during light- and nutrient-powered hydrogen production by *Rhodobacter sphaeroides*. *Bioenerg. Res.,* 3,55-66.

Ying LR, Zhang T, Fang HHP. 2008. Characteristics of a phototrophic sludge producing hydrogen from acetate and butyrate. *Int. J. Hydrogen Energy* 33, 2147- 2155.

Zhang T, Liu H, Fang HHP. 2002. Microbial analysis of a phototrophic sludge producing hydrogen from acidified wastewater. *Biotechnology Letters* 24, 1833-1837.

Zhu H., Suzuki T., Tsygankov A.A., Asada Y., Miyake J., 1999. Hydrogen production from tofu wastewater by *Rhodobacter sphaeroides* immobilized in agar gels. *Int. J. Hydrogen Energy* 24, 305–310.

In: Microalgae and Other Phototrophic Bacteria
Editor: Luis G. Torres Bustillos
ISBN: 978-1-63482-078-3

Chapter 7

SOFTWARE SENSOR FOR MICROALGAE CULTURE: THEORETICAL DEVELOPMENT, NUMERICAL SOLUTIONS AND REAL TIME IMPLEMENTATIONS

N. Martínez*[1], *A. Poznyak*[2] *and I. Chairez*[1]
[1]Bioprocesses Department, UPIBI-IPN, Mexico
[2]Automatic Control Department, CINVESTAV-IPN, Mexico

Abstract

This chapter describes a fixed-time convergent step-by-step high order sliding mode observer for a certain type of microalgae culture system. The observer was developed using a finite-time hierarchical structure based on a modified super twisting algorithm. The modification included nonlinear gains of the output error that were used to prove uniform convergence of the estimation error. An energetic function similar to a Lyapunov one was used for proving the convergence between the observer and the bioreactor variables. A non-smooth analysis was proposed to prove the fixed-time convergence of the observer states to the microalgae culture variables. The observer was tested to solve the state estimation problem of an aerobic reaction governed by the Droop model described by the time evolution of biomass, substrate and nitrogen quota. The former variable was used as the output information because it is feasible to measure it on-line by regular sensors. Numerical simulations showed the superior behavior of this observer compared to the one having linear output error injection terms (high-gain type) and one having an output injection obtaining first order sliding mode structure.

PACS 05.45-a, 52.35.Mw, 96.50.Fm
Keywords: Microalgae culture, software sensor, super-twisting algorithm, step-by-step observer
AMS Subject Classification: 53D, 37C, 65P

*Corresponding author: E-mail address: ichairezo@gmail.com

1. Introduction

Microalgae are sunlight-driven cell biological factories that convert carbon dioxide to potential biofuels, foods, feeds and high-value bioactivities. In addition, they can be used in bioremediation applications and as nitrogen fixing biofertilizers [1]. Microalgae are prokaryotic or eukaryotic photosynthetic microorganisms that can grow rapidly and live in harsh conditions due to their unicellular or simple multicellular structure [2].

The term algae has no formal understanding of taxonomy, as this term is used routinely to indicate a set of photosynthetic organisms artificially attached polyphyletic and not cohesive. This classification includes organisms that do not share a common origin, but continue multiple or independent lines of evolution. Moreover, there are relevant morphological similarities between them [3].

Microalgae are considered as primitive plants (thallophytes), i.e., lacking roots, stems and leaves, have no sterile covering of cells around the reproductive cells and have chlorophyll as their primary photosynthetic pigment. Microalgae structures are primarily focused on energy conversion without any development beyond cells, and their simple development allows them to adapt to prevailing environmental conditions and prosper in the long term [4].

Recently, microalgae have been exploited in different areas, such as, biofuels production, waste treatment, food supplements, carbon dioxide (CO_2) abatement, production of fine chemicals, immobilized systems for production of some extracellular compounds or biosorption of heavy metals [5]. Despite the large number of applications where microalgae are applied, there are some challenges that must be solved before they can represent a true solution in many industrial processes.

It is estimated that there are between one and ten million species of algae, and most of them are microalgae. According to the Smithsonian National Museum of Natural History [6], the algae can be classified as follows:

- **Bacillariophyta (diatoms):** They are unicellular organisms that are important components of phytopankton as primary sources of food for zooplankton in both marine and freshwater habitats. Most diatoms are planktonic, but some are bottom dwellers or grow on other algae or plants.

- **Charophyta(stoneworts):** These organisms are freshwater plants and generally grow anchored to the substratum by rhizoids with a shoot extending upward. These similarities have led some scientists to identify the charophytes as ancestors of the mosses. Their green color comes from chlorphylls a and b.

- **Chlorophyta (green algae):** Most chlorophytes are aquatic, but some green algae can live on the surface of snow, on tree trunks, in soils, or symbiotically with protozoans, hydras or lichen-forming fungi. The typical color of plants in the Chlorophyta, resulting from the dominant chlorophyll pigments, is some shade of apple or grass green, although certain species may appear yellow-green or blackish-green due to the presence of carotenoid pigments or high concentrations of chlorophyll. Chlorophytes appear more than a billion years ago in the fossil record.

- **Chrysophyta (golden algae):** These are photosynthetic, unicellular organisms that are abundant in freshwater and marine environments. Chrysophytes contain chlorophylls a and c, which are masked by the accessory pigment fucoxanthin, a carotenoid. In many ways, golden algae are, biochemically and structurally similar to brown algae. Both golden algae and brown algae store food outside of the chloroplast in the form of polysaccharide laminarin, or chrysolaminarin. In both groups, motile cells have unequal flagella of similar structure.

- **Cyanobacteria (blue-green algae):** Even though the Cyanobacteria are classified as bacteria (lacking a membrane-bounded nucleus) they are photosynthetic and are included among our algal collections. Cyanobacteria played a decisive role in elevating the level of free oxygen in the atmosphere of the early Earth. Cyanobacteria can change remarkably in appearance, depending on the environmental conditions. Cyanobacteria are different in many important ways from other photosynthetic prokaryotes. Instead of the bacteriochlorophylls found in purple and green bacteria, blue-greens contain chlorophyll a, as in eucaryotic phototrophs, and, produce free oxygen as a byproduct of photosynthesis.

- **Pyrrophyta/Dinoplhyta (dinoflagellates):** The division Pyrrophyta (from the Greek "pyrrhos" meaning flame-colored) comprises a large number of unusual algal species of many shapes and sizes. There are about 130 genera in this group of unicellular microorganims, with about 2000 living and 2000 fossil species described so far. The name "dinoflagellate" refers to the forward- spiraling swimming motion of these organisms. They are free-swimming protists (unicellular eukaryotic microorganisms) with two flagella, a nucleus with condensed chromosomes, chloroplasts, mitochondria, and Golgi bodies. Biochemically, photosynthetic species possess green pigments, chlorophylls a and c, and golden brown pigments, including peridinin.

- **Phaeophyta (brown algae):** The Phaeophyta are almost entirely marine, frequently dominating rocky shores in cold and temperate waters throughout the world. The giant kelp, Macrocystis pyrifera , forms expansive seaweed forests off the west coast of North America and provides habitat and shelter for many other organisms. Tropical waters have fewer species of brown algae, although genera such as Sargassum and Turbinaria can dominate in some areas to form small-scale forests.

- **Rhodophyta (red algae):** Of the approximately 6000 species, most red algae are marine; only a few occur in freshwater. Rhodophytes are usually multicellular and grow attached to rocks or other algae, but there are some unicellular or colonial forms. They do not have flagellated cells, are structurally complex, and have complex life cycles divided into three phases. Many red algae feature pit connections between the cells, and their cell walls include a rigid component composed of microfibrils and a mucilaginous matrix. Agar and carrageenin are two red algal mucilages that are widely used for gelling and thickening purposes in the food and pharmaceutical industries.

2. Microalgae Utility

As noticed, microalgae have a lot of potential utilities. However, in recent years one of the most studied application is the biofuels production. Fossil fuels are widely used in dayly life and their consumption is increasing every day, especially in transportation industry. However, they are a non-renewable energy source, so they are becoming scarcer causing the price rising by high demand and limited supply. Added to this, the use of these resources leads to increased emissions of air pollutants from different sources such as oil refining to fuel application.

To overcome the problems associated with the use of fossil fuels, some people have sought new forms of energy, in particular for biofuels, which are capable of reducing production costs, plus they can contribute to reducing CO_2 emissions from transportation industry and private cars utilization [7].

Based on the feedstock used for production and the technologies used to convert feedstock into fuel, biofuel technologies can be classified into two groups: biofuels of first- and second-generation. Technologies that normally utilize the sugar or starch fraction of plants as feedstock to produce ethanol and those utilizing oilseed crops to produce biodiesel are known as first-generation biofuels. On the other hand, biofuels produced using technologies that convert lignocellulosic biomass are called second-generation biofuels. These are biofuels produced from advanced feedstock [7].

Nowadays, increasing interest in second-generation biofuels is observed. For example, second generation ethanol is produced through the conversion of lignocellulosic biomass, which their sources can be waste seed husks, stalks and fast-growing grasses and trees. Making use of those sources, its production does not compete with food supply. Lignocellulosic biomass is comprised of polysaccharides, which are converted into carbohydrates through hydrolysis or other chemical process; the sugars are then fermented into ethanol using existing fermentation technology.

Besides, the production of high added-value products, such as carotenoids, phycobilis, fatty acids, medication delivery platforms, pharmaceuticals and bioplastics can be conducted in microalgae. These organisms are promising sources considering that they represent a safe way to develop these products, because microalgae are innocuous. However, there are some challenges as the way that they are produced and their harvest.

As one can notice, there are different algae applications that have remarkable diverse characteristics. Therefore, the requested conditions to reproduce or grow them are varied, including aspects such as the type and amount of nutrients, the lighting conditions, the medium pH, turbidity, salinity, temperature among others. So, one of the main challenges in the cultivation of microalgae is the choice of the most suitable system to grow microalgae. These systems are known as photobioreactors and there are mainly two types: closed systems and open systems.

To accomplish the growing needs of each microalgae, it is important to develop a particular system required to keep regulated the conditions to promote their growth. Photobioreactors are considered a specific class of bioreactors which are devices designed and constructed to support a biologically active environment. In particular, these photobioreactors consider the light as a variable to be controlled in order to optimize the microalgae growth.

Photobioreactors classified as open show poor mass transfer rates yielding to low biomass productivity [5]. Different forms of open photobioreactors can be distinguished: these may be natural waters such as lakes and ponds, waterfall type inclined, circular ponds, and rough raceways systems [8]. Open systems can be easily constructed and operated. These are the two major reasons because they are the most widely used commercially. However, they have also important problems such as poor light utilization by cells, evaporative losses of liquid medium, diffusion of CO_2 to the atmosphere, requirement of large areas of land, contamination by predators and other fast growing heterotrophs.

Closed systems reduce contamination risks and increase the possibility of implementing control systems on the culture conditions. They can be found as flat plate systems, bubble columns, air-lifts, stirred tanks, toroidals, among others. In these photobioreactors, several different instruments can be setup to promote the culture medium recirculation, the exchange of $\mathrm{CO_2}$, etc. Since the illumination is essential for growing microalgae, closed photobioreactors are designed with transparent containers. Despite all their aforementioned benefits, closed systems have the major disadvantage of the high cost of construction as well as operation and maintenance.

Independently of the class of bioreactor used for the microalgae culture, the regulation of their operation conditions is very important for increasing the biomass concentration with respect to the one that can be obtained in nature. This seems to be a relevant factor because when microalgae is used for biodiesel production, the most important challenges include the large-scale implementation of an integrated algae production system and the harvesting of algae in a way that downstream processing to produce bioproducts of high added-value is enforced [9].

This regulation must consider the type of photobioreactor to be used: the operation conditions of open systems are more difficult to be regulated because these reactors are exposed to contaminants or external factors such as dust or rain. On the opposite, in a closed system, to variables regulation is easier, however, the technical/economical ratio can be no so attractive.

One of the main factors that increases the reactor operation cost are the measurement systems. Photobioreactors usually work in very specific and time varying conditions. This fact underlies the necessity of having specific and reliable sensors to monitor different variables. Usually these sensors must have a wide operating range which usually increases the cost of each sensor or simply, these sensors do not exist at all. One possibility for decreasing costs of a closed or open system is through the use of virtual sensors which allow to estimate variables that can result in economic (costly or technically) measurements , from the measurements that are feasible to obtain.

In terms of control engineering, virtual sensors are called state estimators or observers. One of the most attractive aspects of using state observers is the possibility of virtually measuring internal variables of the photobioreactor, i.e., substrate concentrations, a subproduct or an intermediate compound of the reaction. These measurements can be obtained as result of computer algorithm that uses reduced information from the photobioreactor: an available information regarding the microorganism metabolism such as biomass and the input used to feed the reactor.

State estimation for bioreaction systems has been an active field for many years [10], [11], [12]. Many different methodologies were proposed to solve this problem in last decades [13], [14], [15]. In general, these processes are defined by the concentration of several compounds with chemical or biochemical nature that usually can not be measured directly or that can not be measured independently [16]. Moreover, the majority of state observers for bioreactors have been designed to converge asymptotically. An additional limiting issue of this class of observers is the approximate estimation of actual states due to the effect of uncertainties and disturbances that may affect the dynamics of the photobioreactor.

In principle, if the variables of the photobioreactor can be recovered from the output information in finite time (even under the effect of the same uncertainties and perturbations), the potential application (as source of information for controller or software sensor) of the associated observer would be much better done [17]. One option to enforce the finite time convergence of the estimation error is the application of discontinuous terms of the output error in the observer design. The most advanced application of such terms has been developed using the tools provided by the sliding mode theory [18]. The so-called sliding mode observers are recognized by their robustness against bounded perturbations and uncertainties in the model of the system to be estimated. However, their states still preserve the usual chattering that is considered as the most undesired consequence of using discontinuous terms on the observer design. In the last three decades, the emergence of the so-called high-order sliding modes (HOSM) has provided new algorithms (such as super-twisting, twisting, suboptimal, etc) that are less affected by the chattering.

Different observer models based on HOSM have been applied to design full-state feedback controllers, numeric differentiators, output based controllers and state observers [19], [20], [21]. The proposed observers forced the finite-time stability in the error estimation. Moreover, a new generation of recently introduced Lyapunov functions allows us to determine the time of convergence exactly [22], [23]. In general, these observers were designed for second order systems obeying the chain of integrators form. A different strategy based on a step-by-step structure was useful to extend the application of the super-twisting algorithm to systems with higher number of states that were still described with a chain of integrators form. However, the switch for each step was defined by a state dependent function that must be calculated on-line [24].

This chapter is organized as follows: the section 3 describes the mathematical model used to describe the microalgae growing dynamics including all parameters used for testing the observer proposed in this study. The section 4 contains the detailed description of the finite time state estimator used to reconstruct the variation of all the variables included in the model described in section 3 using just the biomass measurements. Section 5 presents the simulated results for both the dynamic model as well as the estimated states produced by the fixed time observer. Section 6 describes a feasible strategy to apply the proposed identifier in real photobioreactors. Finally, section 7 contains some concluding remarks.

3. Mathematical Model for Microalgae Culture

The Droop model [25] is a simple and widely used model that can represents the growth of microalgae. This model describes the dynamic evolution of biomass, the internal quota of nitrogen and the substrate used to feed microalgae. The model is formally presented as:

$$\begin{aligned}
\dot{x}(t) &= x(t)\left(\mu\left(Q_N(t)\right) - D\right) \\
\dot{Q}_N(t) &= \rho\left(s(t)\right) - \mu\left(Q_N(t)\right) Q_N(t) \\
\dot{s}(t) &= D\left(s^{in}(t) - s(t)\right) - \rho\left(s(t)\right) x(t)
\end{aligned} \tag{1}$$

where the time varying parameters $\rho(s(t))$ and $\mu(Q_N(t))$ are:

$$\rho(s(t)) = \rho_m \frac{s(t)}{s(t) + K_s}$$

$$\mu(Q_N(t)) = \bar{\mu}\left(1 - \frac{K_Q}{Q_N(t)}\right)$$

The biomass concentration is denoted as $x\left[\mu mol L^{-1}\right]$, the internal quota $Q_N\left[\mu mol L^{-1}\right]$ which is defined as the quantity of intracellular nitrogen per unit of biomass and the substrate concentration is denoted by $s\left[\mu mol L^{-1}\right]$. The substrate uptake rate is represented as $\rho\left[L\,\mu m^{-3} d^{-1}\right]$, while $\mu\left[L\,\mu m^{-3} d^{-1}\right]$ is the specific growth rate and $D\left[d^{-1}\right]$ is the constant dilution rate.

The selection of x and s_{in} as output and input respectively, provides the full relative degree equal to 3 according to the procedure described in [26]. Therefore, using the results given in the same reference, there exists a nonlinear transformation $z = T(X_b, S, O)$ ($z^\top = [z_1, z_2, z_3]$) given by

$$\begin{aligned}
z_1 &= y = x \\
z_2 &= \dot{x} = x\left[\bar{\mu}\left(1 - \frac{K_Q}{Q_N}\right) - D\right] \\
z_3 &= x\left[\frac{\bar{\mu} K_Q}{Q_N^2}\left(\frac{\rho_m s}{s + K_s} - \mu Q_N\right)\right] + \left\{x\left[\bar{\mu}\left(1 - \frac{K_Q}{Q_N}\right) - D\right]\right\}\bar{\mu}\left(1 - \frac{K_Q}{Q_N}\right)
\end{aligned} \tag{2}$$

such that, the new variables z obey the following dynamics

$$\begin{aligned}
\frac{d}{dt} z_1(t) &= z_2(t) \\
\frac{d}{dt} z_2(t) &= z_3(t) \\
\frac{d}{dt} z_3(t) &= f(z(t)) + g(z(t))\, u(t)
\end{aligned} \tag{3}$$

where the nonlinear function $f(z)$ is given by:

$$f(z) = \bar{\mu}K_Q \frac{z_1}{Q_N}\left[\frac{-\rho_m s\dot{Q}_N}{Q_N^2(s+K_s)} - \frac{\bar{\mu}K_Q\dot{Q}_N}{Q_N^2}\right]$$

$$+\bar{\mu}K_Q\left(\frac{\rho_m s}{Q_N(s+K_s)} + \bar{\mu}\left[\frac{K_Q}{Q_N} - 1\right]\right)\left(\frac{z_2}{Q_N} - \frac{z_1\dot{Q}_N}{Q_N^2}\right) \tag{4}$$

$$+\frac{z_2\bar{\mu}K_Q\dot{Q}_N}{Q_N^2} + z_3\left[\bar{\mu}\left(1 - \frac{K_Q}{Q_N}\right) - D\right] - \frac{\bar{\mu}K_QK_s\rho_m z_1}{Q_N^2(s+K_s)^2}\left(Ds + \frac{\rho_m sx}{s+K_s}\right)$$

Additionally, the input-associated function $g(z)$ obeys the following structure

$$g(z) = \frac{\bar{\mu}K_QK_s\rho_m D z_1}{Q_N^2(s+K_s)^2} \tag{5}$$

The non-linear functions $Q_N(z)$ and $s(z)$ are described as

$$Q_N(z) = \frac{-\bar{\mu}K_Q}{\frac{z_1}{z_2} - \bar{\mu} + D}$$

$$s(z) = \frac{\beta(z)Q_N(z)K_s}{\rho_m - \beta Q_N(z)} \tag{6}$$

$$\beta(z) = \frac{z_3 - z_2\left[\bar{\mu}\left(1 - \frac{K_Q}{Q_N(z)}\right) - D\right]}{z_1\left(\frac{\bar{\mu}K_Q}{Q_N(z)}\right)} + \bar{\mu}\left(1 - \frac{K_Q}{Q_N(t)}\right)$$

In general, measuring the internal quota and substrate is considered as an expensive and time-consuming procedure. However, the existence of the so-called full relative degree condition allows to use biomass as the sole required information to design an observer or software sensor. Therefore, this study was focused on developing an algorithm that serves as a software sensor to reconstruct the concentrations of Q_N and s [27]. As one can understand, microalgae biomass can be straightforwardly estimated by a simple optoelectronic sensor using a red light emitting led and the corresponding photodiode

4. Finite-time State Estimator for Microalgae Culture

The observer development requires a set of technical assumptions. These assumptions are:

Assumption 1. The nonlinear function f defined in (4) is Lipschitz with respect to its only one argument, that is

$$|f(z_a) - f(z_b)|^2 \leq L_1 \|z_a - z_b\|^2$$

with the constant L_1 a positive scalar and $z_a, z_b \in \mathbb{R}^3$.

Assumption 2. The function associated to the control action g provided in (5) is bounded with a bound L_2, that is

$$0 < G_1 \leq |g(z)|^2 \leq G_2, \quad G_1, G_2 \in \mathbb{R}^+$$

Finally, the control action must fulfill the following:

Assumption 3. The control action u belongs to the following admissible set:

$$U_{adm} = \left\{ u \mid \|u(t)\|^2 \leq u^+, \ t \geq 0, \ u^+ \in \mathbb{R}^+ \right\} \tag{7}$$

If all these assumptions are fulfilled, the proposed observer satisfies the following differential equations

$$\begin{aligned}
\frac{d}{dt}\hat{z}_1(t) &= \tilde{z}_2(t) + k_{11}\phi_{11}(e_1(t), \alpha_1) \\
\frac{d}{dt}\tilde{z}_2(t) &= k_{12}\phi_{21}(e_1(t), \alpha_1) \\
\frac{d}{dt}\hat{z}_2(t) &= E_2(\tilde{z}_3(t) + k_{21}\phi_{12}(e_2(t), \alpha_2)) \\
\frac{d}{dt}\tilde{z}_3(t) &= E_2(k_{22}\phi_{22}(e_2(t), \alpha_2)) \\
\frac{d}{dt}\hat{z}_3(t) &= E_3(f(\hat{z}(t)) + g(\hat{z}(t))u(t)) + E_3\left(\sum_{i=1}^{3} K_i e_i(t) + K_{SM} sign(e_3(t))\right)
\end{aligned} \tag{8}$$

with:

$$e_i = \hat{z}_i - \tilde{z}_i, \quad i \in \{1,2,3\} \tag{9}$$

The set of variables $\hat{z}_i$ represents the corresponding estimated trajectories of z_i. In particular, $\tilde{z}_1 = z_1$.

The observer gains represented by k_{11}, k_{12}, k_{21} and k_{22} are positive scalars that must be adjusted to force the convergence of the observer trajectories to the states of the uncertain system. The indicator function $E_i(t)$ is given by

$$E_i = \begin{cases} 0 & t < t_i^* \\ 1 & t \geq t_i^* \end{cases} \tag{10}$$

The switching time t_i^* is found as a result of the fixed-time converge obtained by the observer in the following section. The nonlinear functions $\phi_{1j}(e_j, \alpha_j)$ and $\phi_{2j}(e_j, \alpha_j)$ $(j = 1,2)$ were designed in agreement to the proposal given in [28]. Then, they are defined as

$$\begin{aligned}
\phi_{1j}(e_j, \alpha j) &= |e_j|^{1/2} sign(e_j) + \alpha_j |e_j|^{3/2} sign(e_j) \\
\phi_{2j}(e_j, \alpha_j) &= \frac{1}{2} sign(e_j) + 2\alpha_j e_j + \frac{3}{2}\alpha_j^2 |e_j|^2 sign(e_j)
\end{aligned} \tag{11}$$

where α_1 and α_2 are positive constants.

The convergence of the observer trajectories to the uncertain system states is proven in the following proposition.

Proposition: *Consider the nonlinear system (3) represented in the Brunovskii form and its corresponding proposed discontinuous observer given by (8). Suppose that functions f and g satisfied (4) and (5) respectively. If the control action u satisfies (7), then the observer is uniformly exact convergent if its gains belong to the set*

$$K = K_1 \cup K_2$$

where K_j is the subset of observer gains given by

$$K_j = \begin{array}{l} \left\{ (k_{j1}, k_{j2}) \in \mathbb{R}^2 \mid 0 < k_{j1} < 2\sqrt{L_j},\ \frac{k_{j1}}{4} + \frac{L_j^2}{k_{j1}^2} < k_{j2} \right\} \\ \cup \left\{ (k_{j1}, k_{j2}) \in \mathbb{R}^2 \mid 2\sqrt{L_j} < k_{j1}, 2L_j < k_{j2} \right\} \end{array} \tag{12}$$

and if

$$\begin{array}{l} K_{SM} = -2G_2 u^+ - \alpha,\ \ \alpha > 0 \\ K_i > 0 \end{array} \tag{13}$$

then there is a fixed-time T_M such that the estimation error $e = \begin{bmatrix} e_1 & e_2 & e_3 \end{bmatrix}^\top$ fulfils $e(t) = 0 \quad \forall t \geq T_M$ with the convergence time T_M satisfying the following identity $T_M = t_1^* + t_2^*$ where

$$t_i^* = \frac{4\lambda_{\max}^{1/2}\{P_i\}\eta_i^{1/2}}{\varepsilon_i} + 12\,(2C_{2i})^{7/6}\,(\varepsilon_i)^{-1/6}$$
$$\varepsilon_i \geq C_{1i}\left(2C_{3i} + 2\sqrt{C_{3i}^2 + C_{4i}}\right)$$

where $C_{1i}, C_{2i} C_{3i}, C_{4i}$ and η_i are positive constants. The method to obtain these gains is detailed in the appendix A and B of [28] and the matrices P_i are solutions of the following Linear Matrix Inequalities with $N_i \geq |z_{i+1}(t)|\ \forall t \geq 0$:

$$\begin{bmatrix} P_i A_i + A_i^\top P_i + \varepsilon_i I + 4N_i^2 C^\top C & P_i B \\ B^\top P_i & -1 \end{bmatrix} \leq 0$$

$$A_i = \begin{bmatrix} -k_{1i} & 1 \\ -k_{2i} & 0 \end{bmatrix},\ C^\top = \begin{bmatrix} 1 \\ 0 \end{bmatrix},\ B = \begin{bmatrix} 0 \\ 1 \end{bmatrix}$$

Proof. *The proof of this proposition can be straightforwardly obtained from the results presented in [27]. We omit all the details of the proof just to avoid unnecessary repetition of concepts regarding observer design.*

The following section describes the numerical implementation of the observer using the Drop model presented in (1) as data generator.

5. Numerical Results

The model of the photobioreactor is simulated using the following parameters:

Parameter	Value	Units
D	1.3	d^{-1}
ρ_m	9.3	$\mu mol\ \mu m^{-3}\ d^{-1}$
$\bar{\mu}$	2.0	d^{-1}
s^{in}	100.0	$\mu mol\ L^{-1}$
K_s	0.105	$\mu mol\ L^{-1}$
K_Q	1.8	$\mu mol\ \mu m^{-3}$

These parameters were estimated in [29]. Their values were taken as they were presented in the aforementioned study. Under the selected set of parameters, the simulated microorganism growing was not inhibited by the substrate nor the product. These conditions are usually considered in observer design for bioreactors. The transformation (2) of the system (1) presented in this study is simulated in Matlab/Simulink. The simulation is executed using a fixed step numerical integration algorithm (ODE1-Euler) with an integration step of 0.0001 hours.

Figure 1 shows the performance of the observer in the z-coordinates. The trajectories depicted in this Figure are obtained by using the set of discontinuous system described in (3). The individual Figures (one used for each variable of the transformed system) show also the step-by-step convergence of all the states. The first state z_1 converges before z_2 and it does it before z_3. The convergence of all the states are achieved in finite time and the estimation error converged before 0.65 h. This value almost coincides with the predicted time T (0.69 h) proposed in the theorem and it is considered relatively low with respect to the period of simulation (10.0 h).

For comparison purposes, a high-gain observer is also simulated. This observer structure is adapted from the results presented in [30].

$$\frac{d}{dt}\hat{z}_1(t) = \hat{z}_2(t) + l_1 e_1(t)$$

$$\frac{d}{dt}\hat{z}_2(t) = \hat{z}_3(t) + l_2 e_1(t)$$

$$\frac{d}{dt}\hat{z}_3(t) = f(\hat{z}(t)) + g(\hat{z}(t))u(t) + l_3 e_1(t)$$

The high-gain observer gains are selected following the method presented in [10]. The gains selected for developing the numerical simulation are $l_1 = 5,000$, $l_2 = 10,000$ and $l_3 = 100,000$.

The trajectories of this alternative observer converged to the true values of the system (1). However, the convergence is achieved asymptotically for the high gain observer. In this case, the transient process exhibited oscillations with big amplitude (Figure 1). This condition is improved by the fixed time discontinuous observer presented in this study. Moreover, the convergence of the uniform observer states is completed after 0.65 hours.

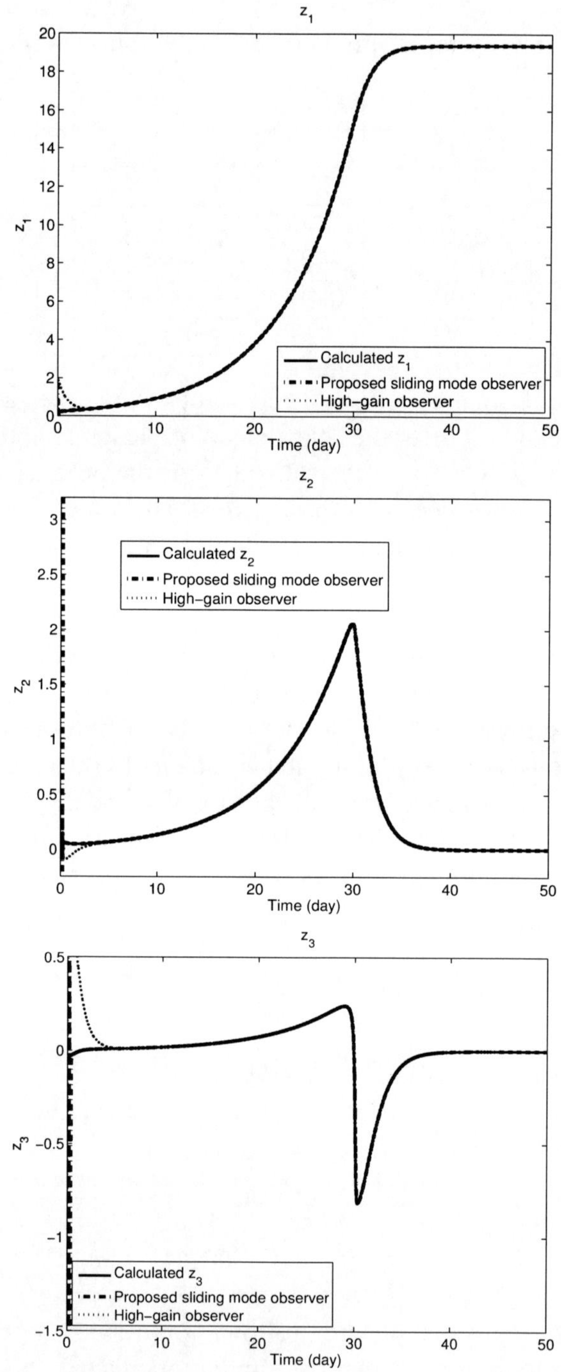

Figure 1: Comparison of the trajectories obtained for the transformed system, the observer proposed in this study as well as the high gain state estimator for the variables $a)z_1$, $b)z_2$ and $c)z_3$.

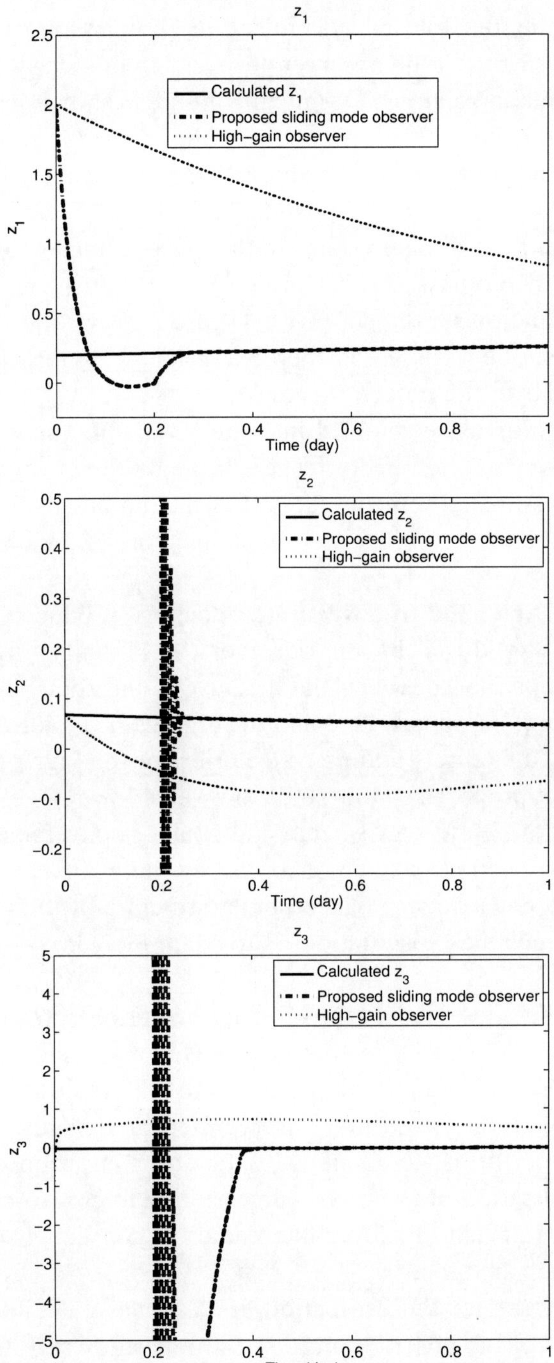

Figure 2: Detailed view of the comparison of the trajectories obtained for the transformed system, the observer proposed in this study as well as the high gain state estimator for the variable $a)z_1$, $b)z_2$ and $c)z_3$. These detailed views demonstrate the faster and better approximation of the variables used to describe the transformed system.

Figure 2 shows a detailed view of the trajectories of z_1, z_2 and z_3 obtained from the numerical simulation of the transformed system and its comparison with the estimated values obtained from the trajectories of the observer proposed in this study as well as the high gain and first order sliding mode versions. Despite this figure is showing high amplitude oscillations around the true trajectory, this is an effect of the scale. Moreover, one can notice that all the three variables are not reached by the estimated states generated by the high-gain observer.

The convergence of the estimated states in the $z-$coordinates forces the corresponding fixed-time estimation of biomass and substrate. These values are obtained as a solution of the inverse transformation presented in (6). Figure 3 shows the trajectories of both, the bioreaction system described in (1) and the observer obtained as the solution the inverse transformation applied over the trajectories of (3).

The reconstructed biomass is showed in Figure 3-a. One may notice that there is no evident difference between both trajectories, the biomass time evolution and the estimated state. The substrate is estimated some minutes after the biomass is fully reconstructed. This is an expected behavior considering the step-by-step nature of the observer proposed in this study.

Figure 3-b demonstrates the true substrate trajectory produced by the model (1) and the estimated substrate produced by the observer (8). Finally, Figure 3-c demonstrates the corresponding comparison between the actual Q_N and their corresponding estimates produced by the fixed time proposed here and the high gain version.

In the same figure, the corresponding trajectories for the high gain are also shown. In both cases, the observer proposed in this study showed a lower level of oscillations around the true trajectories of the bioreactor system and smaller convergence time. In particular, one may notice the big oscillations demonstrated by the estimated variable associated to the substrate. This is a consequence of the ladder structure of the transformed system. A feasible solution to reduce these oscillations is to adapt the gain of the fixed time observer introduced in this study.

Figure 4 depicts a detailed view of the biomass evolution obtained from the model (1), its estimate produced by the fixed-time observer as well as the state generated by the high-gain observer. One must notice that within the Figure only the true trajectory of the model and the one obtained with the fixed-time observer appeared. This is an additional confirmation of the superior performance obtained by the observer proposed in this study. Moreover, any possible application of the high-gain observer to construct an output close-loop controller may present relevant fails over the reactor as consequence of the big amplitude oscillations in the estimation process.

The mean square error for the estimation process was calculated for evidencing the fixed-time convergence of the observer proposed in this study and its comparison with the high-gain and the first order sliding mode observers. The time evolution of the error is depicted in Figure 5.

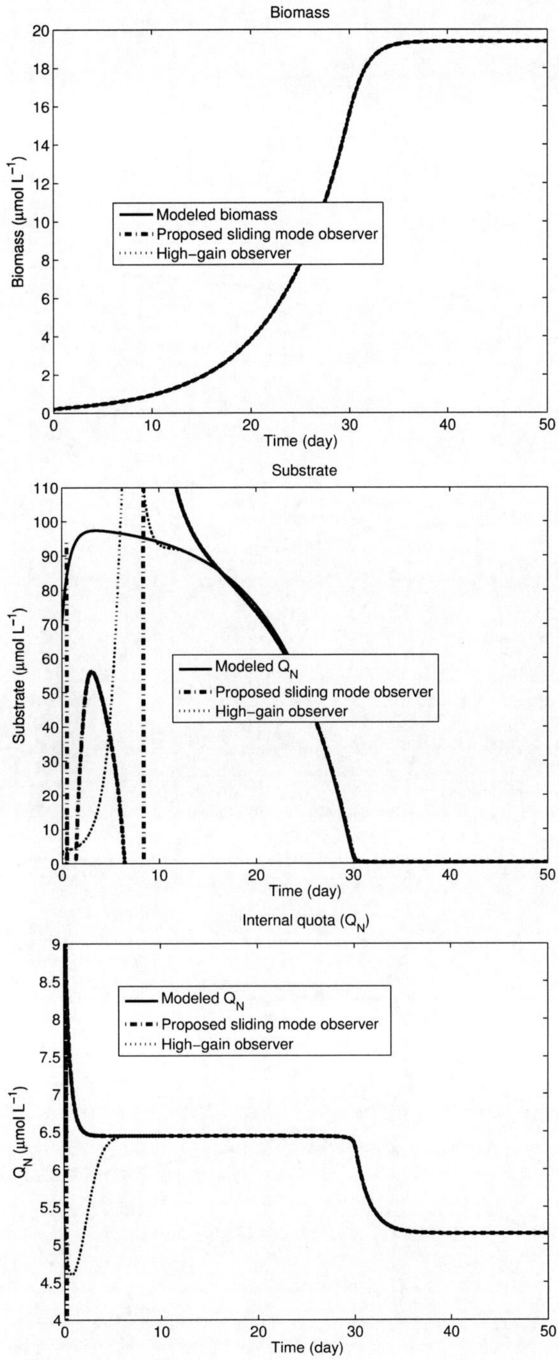

Figure 3: Comparison of the trajectories obtained for the bioreactor system, the observer proposed in this study as well as the high gain state estimator for the variables $a)X$, $b)S$ and $c)Q_N$.

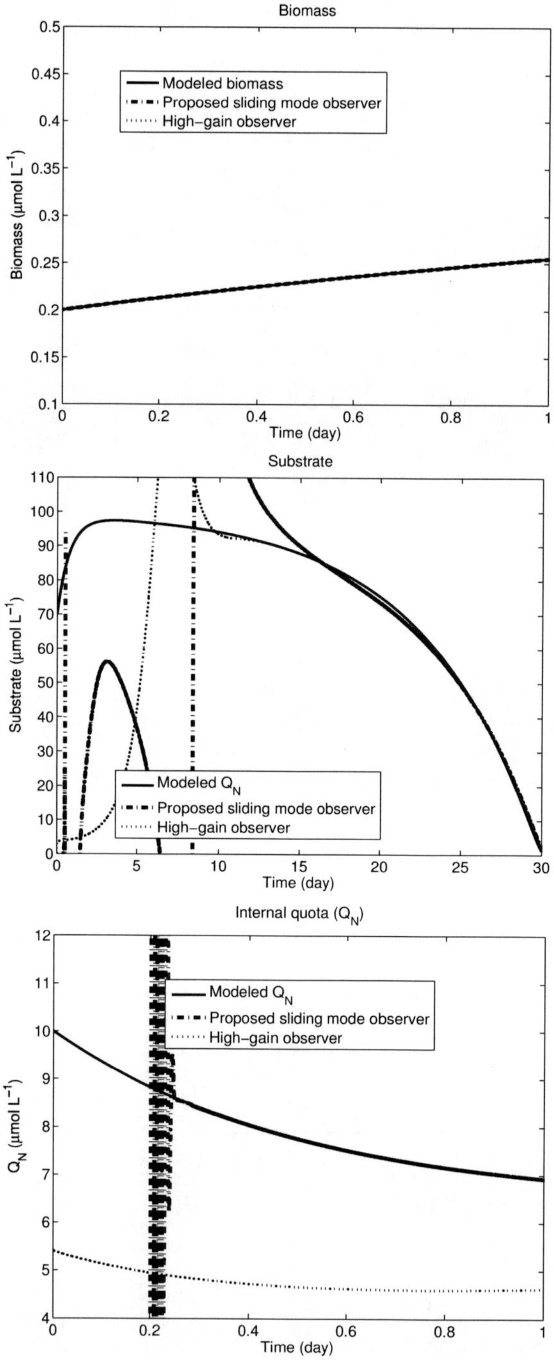

Figure 4: Detailed view of the comparison for the trajectories obtained for the bioreactor system, the observer proposed in this study as well as the high gain state estimator for the a) biomass, b) substrate. and c) the nitrogen quota.

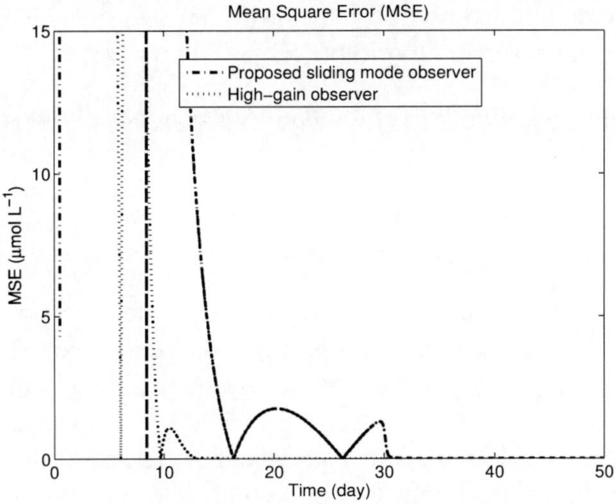

Figure 5: Comparison of the mean square error evaluated over the estimation error generated by the observer proposed in this study as well as the high gain state estimator

6. Real-time Implementations

The last stage on designing and applying the proposed state observer is how to implement it in actual photobioreactors. Considering the class of solution presented in this chapter, the problem can be solved as follows:

a) To develop an on-line sensor for the microalgae biomass. This can be done by a simple optoelectronic device that can measure the absorbance of a sample derived from the main flow of the reactor.

b) The absorbance value can be acquired by a digital device which can implement the corresponding function that transforms the absorbance value into the corresponding biomass concentration. This can be done using an standard calibration procedure.

c) The biomass concentration can be injected into a discretized version of the observer that can be implemented using a first order Euler approximation of the observer (8), that is

$$\hat{z}_1\left((k+1)T_s\right) = \hat{z}_1(kT_s) + T_s\tilde{z}_2(kT_s) + T_s k_{11}\phi_{11}\left(e_1(kT_s), \alpha_1\right)$$

$$\tilde{z}_2\left((k+1)T_s\right) = \tilde{z}_2\left(kT_s\right) + T_s k_{12}\phi_{21}\left(e_1\left(t\right), \alpha_1\right)$$

$$\hat{z}_2\left((k+1)T_s\right) = \hat{z}_2\left(kT_s\right) + T_s E_2\left(\tilde{z}_3\left(t\right) + k_{21}\phi_{12}\left(e_2\left(t\right), \alpha_2\right)\right)$$

$$\tilde{z}_3\left((k+1)T_s\right) = \tilde{z}_3\left(kT_s\right) + T_s E_2\left(k_{22}\phi_{22}\left(e_2\left(t\right), \alpha_2\right)\right)$$

$$\hat{z}_3\left((k+1)T_s\right) = \hat{z}_3\left(kT_s\right) + T_s E_3\left(f\left(\hat{z}\left(t\right)\right) + g\left(\hat{z}\left(t\right)\right)u\left(t\right)\right) + T_s E_3\left(\sum_{i=1}^{3} K_i e_i + K_{SM} sign\left(e_3\right)\right)$$

where T_s is the sampling time forced by the digital device and k is the discrete counter used to update the observer algorithm.

These three simple stages yields to the real time implementation of the observer proposed in this study.

Conclusion

This chapter introduces a class of nonlinear step-by-step state observer using a sliding mode based scheme that is able to converge on a predefined (fixed) time. This observer is working for a special type of systems that has been inspired by some sort of biotechnological system. The observer is proposed to have a class of step-by-step super-twisting like structure. The observer has been tested using a simulated model that fulfill the considered type of uncertain systems that corresponds to a photobioreactor system. The comparison with a well-known state observer (high gain) was developed to emphasize the contribution of the result developed in this study. The observer has shown remarkable characteristics to reconstruct the non-measurable states fulfilling the convergence time constrain imposed by the method proposed in this chapter. This kind of observers can be further implemented to solve the state estimation problem of systems with a big number of variables such as the microalgae genetic networks or their metabolic pathways. Moreover, the fixed time convergence can be used to design close loop output based controllers using the estimated states.

References

[1] Y. Chisti, "Biodiesel from microalgae," *Biotechnologies advances*, vol. 25, pp. 294–306, 2007.

[2] T. Mata, A. Martins, and N. Caetano, "Microalgae for biodiesel production and other applications: A review," *Renewable and sustainable energy reviews*, vol. 14, pp. 217–232, 2010.

[3] L. Barsanti and P. Gualtieri, *Algae. Anatomy, Biochemestry, and Biotechnology*, Taylor and F. Group, Eds. CRC Press, 2006.

[4] L. Brennan and P. Owende, "Biofuels from microalgae - a review of technologies for production, processing, and extractions of biofuels and co-products," *Renewable and sustainable energy reviews*, vol. 14, pp. 557–577, 2010.

[5] C. Ugwu, H. Aoyagi, and H. Uchiyama, "Photobioreactors for mass cultivation of algae," *Biore*, vol. 99, pp. 4021–4028, 2008.

[6] *Algae Research*, Smithsonian National Museun of Natural History. [Online]. Available: http://botany.si.edu/projects/algae/classification.htm

[7] G. Timilsina and A. Sherestha, "How much hope should we have for biofuels?" *Energy*, vol. 36, pp. 2055–2069, 2011.

[8] M. Borowitzka and N. Moheimani, Eds., *Algae for biofuels and energy.* Springer, 2013.

[9] L. Christenson and R. Sims, "Production and haversting of microalgae for wastewater treatment, biofuels, and bioproducts," *Biotechnologies advances*, vol. 25, pp. 686–702, 2011.

[10] J. P. Gauthier, H. Hammouri, and S. Othman, "A simple observer for nonlinear systems applications to bioreactors," *IEEE Transactions on Automatic Control*, vol. 37, no. 6, pp. 875–880, 1992.

[11] D. G. Luenberger, *Introduction to Dynamic Systems: Theory, Models, and Applications.* New York: John Wiley & Sons, 1979.

[12] S. V. Drakunov, "Sliding-mode observers based on equivalent control method," in *Proceedings of the 31st IEEE Conference on Decision and Control*, 1992.

[13] F. Lafont, E. Busvelle, and J. P. Gauthier, "An adaptive high-gain observer for wastewater treatment systems," *Journal of Process Control*, vol. 21, no. 6, pp. 893–900, 2011.

[14] M. Mihoub, A. S. Nouri, and R. B. Abdennour, "A second order discrete sliding mode observer for the variable structure control of a semi-batch reactor," *Control Engineering Practice*, vol. 19, pp. 1216–1222, 2011.

[15] T. Zhang and M. Guay, "Adaptive nonlinear observers of microbial growth processes," *Journal of Process Control*, vol. 12, no. 5, pp. 633–643, 2002.

[16] D. Theilliol, J. C. Ponsart, J. Harmand, C. Join, and P. Gras, "On-line estimation of unmeasured inputs for anaerobic wastewater treatment processes," *Control Engineering Practice*, vol. 11, no. 9, pp. 1007–1019, 2003.

[17] J. Moreno, J. Alvarez, E. Rocha-Cozatl, and J. Diaz-Salgado, "Super-twisting observer-based output feedback control of a class of continuous exothermic chemical reactors," in *Proceedings of the 9th International Symposium on Dynamics and Control of Process Systems (DYCOPS)*, Leuven, Belgium, 2010, pp. 727–732.

[18] V. Utkin, J. Guldner, and J. Shi, *Sliding Mode Control in Electro-Mechanical Systems*, second edition ed. Automation and Control Engineering, 2009.

[19] A. Levant, *Sliding Mode Control in Engineering.* Marcel Dekker, 2002, ch. High Order Sliding Modes.

[20] ——, "Robust exact differentiation via sliding mode technique," *Automatica*, vol. 34, no. 3, pp. 379–384, 1998.

[21] ——, "Sliding order and sliding accuracy in sliding mode control," *International Journal of Control*, vol. 58, no. 6, pp. 1247–1263, 1993.

[22] J. Moreno, "Lyapunov analysis of non homogeneous super-twisting algorithms," in *Proceedings of the 11th International Workshop on Variable Structure Systems (VSS)*, 2010, pp. 534–539.

[23] A. E. Polyakov and A. S. Poznyak, "Method of lyapunov functions for systems with higher-order sliding modes," *Automation and Remote Control*, vol. 72, no. 5, pp. 944–965, 2011.

[24] T. Floquet and J. P. Barbot, "Super twisting algorithm based step-by-step sliding mode observers for nonlinear systems with unknown inputs," *International Journal of System Sciece*, vol. 38, no. 10, pp. 803–815, 2007.

[25] M. Droop, "Vitamin b12 and marine ecology iv: The kinetics of uptake growth and inhibition in monochrisis lutheri," *Journal of the Marine Biological Association*, vol. 48, no. 3, pp. 689–733, 1968.

[26] A. Isidori, *Nonlinear control systems*, ser. Communications and control engineering series. Springer-Verlag, 1995, vol. 1.

[27] N. Martínez-Fonseca, I. Chairez, and A. Poznyak, "Uniform step by step pbserver for aerobic bioreactor based on super-twisting algorithm," *Bioprocess and Biosystems Engineering*, 2014.

[28] E. Cruz-Zavala, J. Moreno, and L. Fridman, "Uniform robust exact differentiator," *IEEE Transactions on Automatic Control*, vol. 56, p. 11, 2011.

[29] M. Khaksar Toroghi, G. Goffaux, and M. Perrier, "Output feedback passivity-based controller for microalgae cultivation," in *Preprints of the 12th IFAC Symposium on Computer Applications in Biotechnology, The International Federation of Automatic Control*, Mumbai, India, 2013.

[30] M. Z. Hadj-Sadok and J. L. Gouzé, "Estimation of uncertain models of activated sludge processes with interval observers," *Journal of Process Control*, vol. 11, no. 3, pp. 299–310, 2011.

In: Microalgae and Other Phototrophic Bacteria
Editor: Luis G. Torres Bustillos
ISBN: 978-1-63482-078-3

Chapter 8

PHOTOBIOREACTORS: IMPROVING THE BIOMASS PRODUCTIVITY

A. J. Martínez-Roldán* and R. O. Cañizares-Villanueva
Laboratorio de Biotecnología de Microalgas. Departamento de Biotecnología y Bioingeniería. Centro de Investigación y de Estudios Avanzados del Instituto Politécnico Nacional. Av. Instituto Politécnico Nacional, Col. San Pedro Zacatenco, Delegación Gustavo A. Madero, México

ABSTRACT

Microalgae are an extremely diverse group of microorganisms, the principal characteristic of which is the ability to conduct oxygenic photosynthesis. There are two arbitrary groups of microalgae: cyanobacteria (prokaryotes) and microalgae (eukaryotes) (Wijffels et al., 2013;Schmidt & Wilhelm, 2014). The commercial interest in producing microalgal biomass is due to the large number of high value-compounds that can be obtained from it, such as proteins, pigments, polysaccharides, fatty acids, and even substances with microbiocidal, antiviral and anticancer activities (Zittelli et al., 2013). In addition, microalgae were recently proposed as a feasible sustainable source of biofuels, based on the following properties:

a) Their capacity to fix the CO_2 in the flue gases of industrial processes.
b) Their capability to use the nitrogen and phosphorus present in wastewaters.
c) The great variety of biofuels that can be produced from microalgal biomasses, i. e., ethanol (through the fermentation of sugars) (John et al., 2011), biodiesel (through transesterification of oils) (Arias-Peñaranda et al., 2013), hydrogen (through photolysis and photofermentation) (Scoma et al., 2010) and methane (through anaerobic fermentation of the biomass) (Zittelli et al., 2013a).

---

* Corresponding author: martinezr@cinvestav.mx.

1. ALGAL CULTIVATION

The development of the first microalgal cultures, which occurred in the 1950s, was conducted in lagoons. The originally proposed utility of such cultures was the treatment of wastewaters (Oswald & Gotaas, 1957; Oswald & Golueke, 1960;Oswald, 1988). However, due to their selective culturing conditions, only a few genera can be cultivated in this type of photobioreactor (PBR) (Richmond, 2004). Nonetheless, despite all of the disadvantages, lagoons are the most employed technology for the production of microalgal biomasses.

Commercial-scale production of microalgal biomasses began in the 1960s and the first country to promote the production of *Chlorella* was Japan, in which circular ponds mixed using a rotating arm called Pivot. The boom in *Spirulina* production began in the 1970s in Mexico, with Sosa Texcoco S. A. using raceway ponds (Borowitzka, 1999). Currently, the global production of microalgal biomass is estimated to be approximately 20, 000 tons per year, most of which is employed in the production of nutritional and dietary supplements. The genera *Arthrospira*, *Chlorella* and *Dunaliella* are the most relevant in this regard (Richmond, 1999) because their selective culture conditions facilitate their growth in open systems, which considerably reduces the production costs.

2. PHOTOBIOREACTORS (PBRS)

Due to the complexities of microalgal cultivation and the need to satisfy the very particular requirements of a genus for production, many different configurations of PBRs have been developed (Carvalho et al., 2006). Today, the diverse microalgal cultivation systems are classified as follows:

1. Open systems. These are the oldest types of PBRs and have been employed for the last 50 years for biomass production. In these systems, not only is the control of the environmental conditions and consequently the process variables very difficult but the interaction between the culture and the environment is complete (Carvalho et al., 2006; Zittelli et al., 2013).
   The disadvantage of open systems is that they can be employed only to produce microalgae with very highgrowth rates or tolerance of selective growth media (e. g., high salinity or alkalinity, etc.) (Torzillo et al., 1986; Borowitzka & Moheimani, 2013).
2. Closed systems. These systems were developed due to the necessity to cultivate species with no selective growing conditions or with a low growth rate (μ). Employing this type of system ensures not only that the inoculated species will proliferate but also that all of the culture conditions, such as the temperature, pH, and concentration of CO_2, can be controlled (Tredici 2010). Within this group, there is a great diversity of configurations that meet the specific growth requirements of the microorganisms (Carvalho et al., 2006).

2.1. Open Culturing Systems

The first open culturing systems were developed in natural bodies of water that had specific characteristics. Under certain weather conditions, microalgal blooms occurred in lagoons. The biomass was collected by the natives and used as food; the iconic examples of this practice are the production of *Spirulina* in Texcoco Lake by the Aztecs in Mexico and the production of *Spirulina* in Chad, Africa (Hu, 2004). These types of systems are currently used in Myanmar, where *Spirulina* grows naturally in a group of volcanic craters and is collected and consumed by the natives (Borowitzka and Moheimani, 2013). However, these systems do not permit intensive biomass production because the culture is exposed to environmental variations (Zittelli et al., 2013).

There are also artificial open systems (Fig. 1), normally consisting of: (a) a circular pond mixed using a gyratory arm (b) water in open raceways that is mixed using a paddle wheel (Tredici, 2010; Zitelli et al., 2013) or (c) the thin-layer systems developed in the Czech Republic in the 1970s (Šetlik et al., 1970). In open systems, the culture is completely in contact with the environment and the risk of contamination with another species or with rotifers is very high. Therefore, these systems are suitable for cultivating species with a high μ, such as *Chlorella*, or species that grow under selective conditions, such as *Dunaliella* (high salinity) and *Spirulina* (high pH and high alkalinity) (Carvalho et al., 2006; Zittelli et al., 2013).

Mixing should not be rough because it can cause cellular lysis; the level of sensitivity to shear stress in the different groups of microalgae is as follows: green algae < cyanobacteria < diatoms < dinoflagellates (Bajpai et al., 2014). In lagoons and ponds, the culture cannot exceed 3-cm deep to ensure an efficient use of the light and should remain homogeneous through paddle-wheel mixing (Carvalho et al., 2014). In the raceway cultures, the flow velocity should be sufficient to maintain a homogeneous culture and avoid sedimentation of the cells (Del Campo et al., 2007).

As shown in Figure 1, in the thin-layer systems, the slope of the exposure zone (to light) causes an accelerated flow (0. 5 to 0. 65 m s–1), and at the end, the culture falls into a container (Doucha & Lívanský, 2006) ensuring no sedimentation of the biomass despite its high concentration (up to 50 g L-1) (Doucha & Lívanský, 2009). To ensure a correct supply of light to the culture, the depth of the liquid cannot be greater than 1 cm because the cell concentration can be very high and light can penetrate only a few millimeters (Masojídek et al., 2011).

In open systems, the loss of water due to evaporation is very significant; some authors estimated that the evaporation rate is up to 10 L m^{-2} d^{-1} (Cooney et al., 2011; Slade & Bauen, 2013), which could cause variation in the concentrations of nutrients in the culture medium. However, currently, this type of PBR is mostly employed due to its lower operation and maintenance costs (Tredici, 2004; Zitelli et al., 2013). In Table 1, some operations that employ this type of PBR are listed as well as the species cultured and the level of productivity obtained.

The inclined thin-layer systems deserve special mention because their configuration allows a very high biomass concentration to be reached (> 10 g L^{-1}) in comparison with that of the other types of open systems, and this property considerably reduces the biomass-harvesting costs (Tredici, 2004; Torzillo, 2010;Masojídek et al., 2011); however, only species with a cell wall sufficiently strong to tolerate the shear stress of the pumping process can be cultivated (Scarsella et al., 2010; Torzillo, 2010). In Table 1, some of the species that are cultivated using this configuration and the levels of biomass productivity are listed.

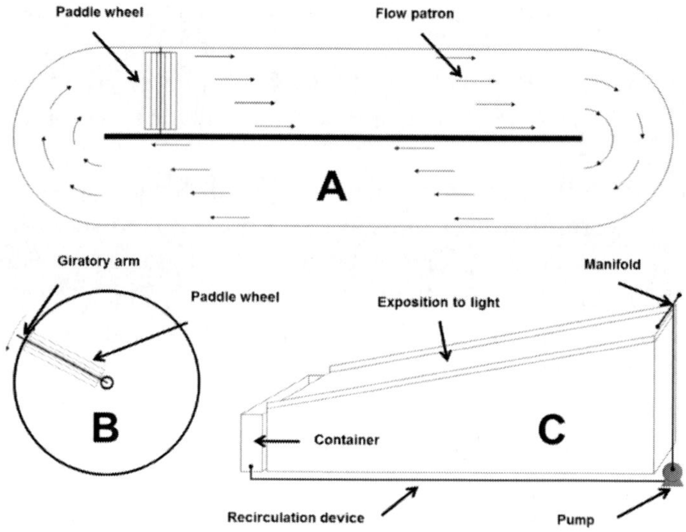

Figure 1. Open systems for microalgae culture. (A) Raceway. (B) Pivot. (C) Inclined thin layer system.

Table 1. Species cultivated in open systems and the values of concentration and productivity of biomass reached

Configuration	Specie	Maximal biomass concentration ($g\ L^{-1}$)	Maximal biomass Productivity ($g\ L^{-1}\ d^{-1}$)	Reference
Open pond	*Phaeodactylumtricornutum*	~1. 0	~0. 10	Silva-Benavides et al., 2013
	Phaeodactylumtricornutum	~0. 3	~0. 150	Torzillo et al., 2012
	Nannochloropsis sp.	--	~0. 16	Moazami et al., 2012
	Nannochloropsis oculata	~1. 0	--	Roselet et al., 2013
	C. vulgaris	--	0. 35	Shuwen et al., 2013
Inclined thin layer	*Chlorella* sp.	>40	~ 3. 0	Doucha & Lívanský, 2006
	Chlorella sp. P12	>40	3 – 4	Doucha & Lívanský, 2009
	Chlorella spp.	>40	1. 46	Masojídek et al., 2011
Waved-bottom thin layer cascade	*Nannochloropsis* sp.	>2. 0	0. 18	Torzillo et al., 2010

2.2. Closed Culturing Systems

Because few genera can grow under nonspecific conditions, different configurations of closed fermenters that are specialized for certain microalgal cultures have been developed, which are called photobioreactors. The main difference between these devices and the open systems is that the former have no contact with the environment because their designs allow efficient control of the operational conditions, ensuring that only the inoculated species proliferate (Zitelli et al., 2013).

To date, the closed system configurations that have been developed include pipes, panels, and tanks, etc. and the option chosen depends upon the specific requirements of the microorganism to be cultivated (e.g., light requirement and resistance to shearing stress, etc.), the culturing conditions (heterotrophy, mixotrophy or autotrophy) and the product to be obtained (Tredici, 2010).

2.2.1. Tubular PBRs

Tubular PBRs are the oldest type of closed system and the type most often employed to culture microalgae (Pirt et al., 1983; Lee & Pirt, 1984; Torzillo et al., 1986) because of the high ratio between the illuminated area and the culture volume (A/V) that they have, which can values of 40 to 170 m^{-1} (Watanabe & Hall 1996; Tredici & Chini-Zittelli, 1998;Suh & Lee, 2003; Scoma et al., 2012). A high A/V ensures good light transfer (Wang et al., 2012); however, because a high quantity of infrared radiation is absorbed by the culture, the temperature increases (Richmond, 1987) and a cooling system is required. Hence, the operational costs are significantly greater than that of the open systems.

To construct this type of PBR, transparent pipelines with a maximum diameter of 10 cm, composed of polyethylene, glass, polyvinyl chloride, polycarbonate or acrylic, etc. are employed (Posten et al., 2009; Wang et al., 2012). Although all of these materials meet the transparency needs for microalgal cultivation, the advantages and disadvantages must be evaluated prior to constructing a PBR.

Tubular PBRs can be operated as horizontal or vertical configurations, as shown in Figure 2. In horizontal tubular PBRs, the gaseous exchange is conducted in a degasser to which CO_2 is added and the O_2 produced by photosynthesis is removed. The movement of the liquid through the PBR can be conducted using pumps (Scoma et al., 2012) or by establishing an airlift system (Molina et al., 2001). The option selected will depend on the operational volume (e. g., the airlift system does not work well with high volumes) and the sensitivity of the cell to shear stress (pumping can cause cell death in some genera) (Scarsella et al., 2012).

The major inconvenience of the horizontal tubular PBRs is the accumulation of the oxygen produced by photosynthesis in the medium (Masojidek et al., 2010) because oxygen can be toxic to the cells when it reaches more than 400% of the saturation value (>32 mg L^{-1}). The toxicity of O_2reduces the growth rate and consequently, the level of biomass production (Molina et al., 2001; García-González et al., 2005; Ugwu et al., 2007). To avoid reaching inhibitory concentrations of oxygen, the most important design parameters of horizontal tubular PBRs are the diameter of the tube, the rate of growth rate, and the concentration of the biomass (Torzillo, 1997; Molina et al., 2001; Torzillo et al., 2003; Posten et al., 2009).

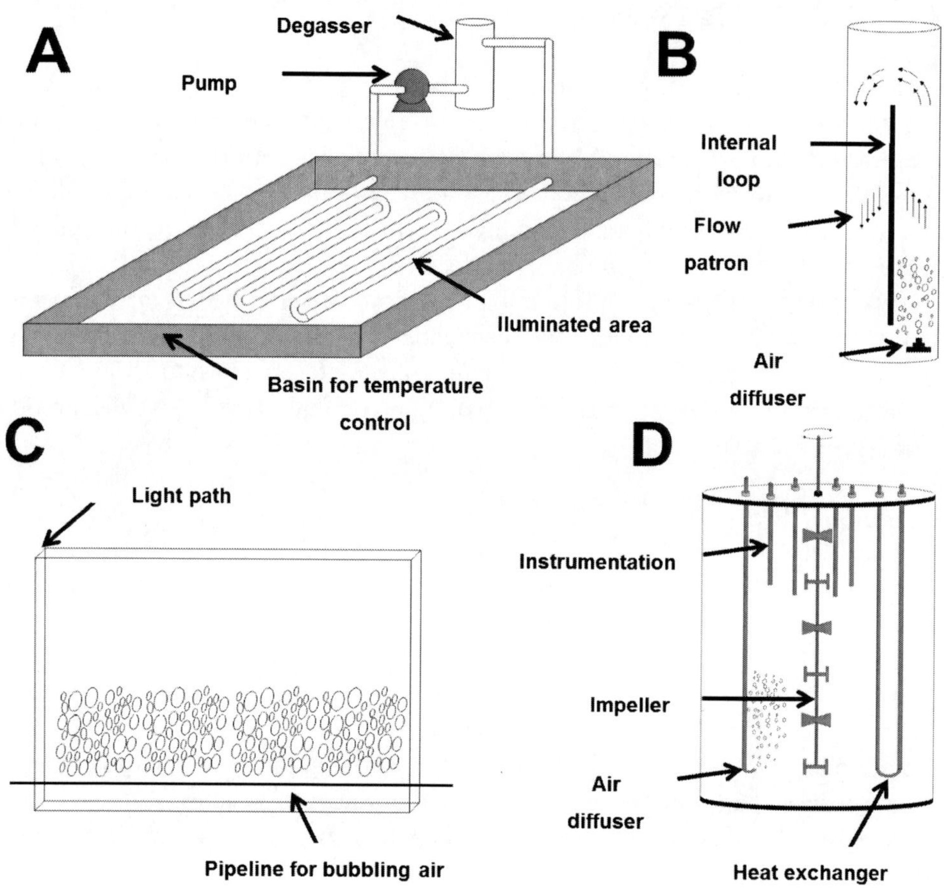

Figure 2. Closed systems. (A) Horizontal tubular. (B) Column. (C) Flat plate. (D) Stirred tank.

Despite the inconveniences of the horizontal tubular configuration, this type of PBR has been widely employed to grow different genera of microalgae (Table 2), particularly, green algae (Del Campo et al., 2001; García-González et al., 2005; Scoma et al., 2012), diatoms (Molina et al., 2001; Silva-Benavides et al., 2013) and cyanobacteria (Torzillo et al., 1986; Carlozzi & Torzillo, 1996).

In vertical tubular PBRs, the removal of the O_2 produced by photosynthesis is very efficient because of the large degree of turbulence in the liquid; however, the major limitation is in scaling up because if the tube diameter is wider than 10 cm, dark zones are generated in the center of the PBR, particularly in high-concentration cultures (Posten et al., 2009; Arias-Peñaranda et al., 2012). To avoid such an occurrence, annular PBRs were developed, which consist of two cylinders with different diameters concentrically placed to form an annular chamber (Zittelli et al., 2003; Chini-Zittelli et al., 2006). Because lower productivity is achieved using this type of PBR compared with that obtained in horizontal tubular PBRs (Dillschneider & Posten, 2013), they are employed to cultivate only a few genera of microalgae (Table 2).

Table 2. Species cultivated in tubular PBRs and the values of concentration and productivity of biomass reached

Configuration	Species	Maximal biomass concentration ($g\ L^{-1}$)	Maximal biomass Productivity ($g\ L^{-1}\ d^{-1}$)	Reference
Horizontal	*Phaeodactylumtricornutum*	2. 3	1. 2	Acién-Fernandez et al., 2001
	Porphyridium cruentum UTEX 161 w	3. 0	--	Camacho-Rubio et al., 1999
	Muriellopsis sp.	>2. 0	---	Del Campo et al., 2001
	D. salina UTEX 2538	~ 0. 60	0. 10	García-González et al., 2005
	P. tricornutum	1. 4	0. 3	Silva-Benavides et al., 2013
	S. platensis M2	--	0. 52	Torzillo et al., 1996
	P. tricornutum	~0. 6	~. 50	Torzillo et al., 2012
Horizontal with solar concentrators inside a greenhouse	*Spirulina platensis* M2	~2. 3	~0. 60	Masojídek et al., 2003
Strongly curved horizontal	*Spirulina*	---	1. 2	Carlozzi & Torzillo, 1996
Inclined	*Chlorella sorokiniana*	5. 0	>1. 0	Ugwu et al., 2002
Column	*Nannochloropsis* sp.	~1. 5	~0. 40	Martínez-Roldán et al., 2014
	Chlorella zofingiensis	~3. 0	~0. 27	Zhu et al., 2013
	Scenedesmus dimorphus	~5. 0	--	Jiang et al., 2013
Annular	*Tetraselmissuecica*	1. 71	0. 56	Chini-Zittelli et al., 2006
	Nannochloropsis sp.	--	>0. 20	Zittelli et al., 2003
Helical	*P. tricornutum*	> 3. 5	1. 40	Acién-Fernandez et al., 2003
Helical cone-shaped	*S. platensis*	1. 2	--	Watanabe & Hall, 1996

2.2.2. Flat panel PBRs

Flat-panel PBR development began in the 1980s when Ramos de Ortega & Roux (1986) used a rigid panel composed of polycarbonate for the production of *Chlorella* biomass; the operating mode was similar to that of a tubular PBR despite its shape. In the 1990s, the use of flat-panel PBRs boom escalated and they were employed by many research groups throughout the world (Qiang & Richmond, 1996). This phenomenon was due to the higher levels of biomass productivity that could be reached in PBRs with this configuration (Table 3) and because the shear stress values were very low, which prevented the destruction of very sensitive microalgal species, such as cyanobacteria (Gitelson et al., 1996; Hu et al., 1998; Zhang et al., 1999; Reyna-Velarde et al., 2010).

Characterization of the growth of different species of microalgae using PBRs of this configuration demonstrated an increased productivity (superficial and volumetric) and a better use of light compared with those of lagoons and tubular PBRs (Richmond, 1996). However, to make this possible, the light path must not exceed 10 cm because increasing this value considerably decreases the level of productivity (Zou & Richmond, 1999). This decrease is caused by the auto-shading effect that occurs when the cell concentration is so high that the light does not fall uniformly on all of the cells (Tamburic et al., 2011).

Table 3. Species cultivated in flat plate and the values of concentration and productivity of biomass reached

Configuration	Species	Maximal biomass concentration (g L^{-1})	Maximal biomass Productivity (g L^{-1} d^{-1})	Reference
15 – 22 mm light path solar tracked	*C. vulgaris*	~1. 0	--	Hindersin et al., 2013
	S. obliquus	~4. 0	--	Hindersin et al., 2013
1 cm light path	*Chlorococcum littorale*	~22	>9. 0	Hu et al., 1998
1. 2 cm light path	*Nannochloropsis*sp.	--	0. 85	Zittelli et al., 2000
1. 3 cm light path	*Nannochloropsis* sp.	8. 5	0. 85	Richmond & Cheng-Wu, 2001
2. 6 cm light path Bubble mixed	*S. platensis*	~8. 4	~2. 2	Hu et al., 1996a
2. 6 cm light path	*S. platensis*	~5. 0	>2. 4	Qiang & Richmond, 1996
10 cm light path	*Nannochloropsis* sp.	0. 24	0. 32	Cheng-Wu et al., 2001
	Nannochloropsis sp.	--	>1. 5	Zittelli et al., 2000
15 cm light path	*Spirulina* sp.	1. 81	0. 12	Reyna-Velarde et al., 2010

Table 4. Species cultivated in stirred tanks and the values of concentration and productivity of biomass reached

Configuration	Species	Maximal biomass concentration (g L^{-1})	Maximal biomass Productivity (g L^{-1} d^{-1})	Reference
Internally illuminated	*C. sorokiniana*	~1. 5	--	Ogbonna et al., 1996
3. 5 L externally illuminated	*Choricystis minor*	~2. 0	~0. 35	Mazzuca-Sobczuk & Chisti, 2010
2 L Externally illuminated	*Protoceratium reticulatum*	--	--	Gallardo-Rodríguez et al., 2010
2 L Externally illuminated Turbidostat mode	*Neochloris oleoabundans*	0. 4	0. 55-	Gallardo-Rodríguez et al., 2010
5 L volume. Heterotrophy	*Chlorella protothecoides*	~15. 0	~1. 9	Li et al., 2007
		~3. 5	~0. 60	Xu et al., 2006
2 L	*Chlorella saccharophila*	~1. 2	--	Isleten-Hosoglu et al., 2012
2. 5 L	*Chlorococcum* sp.	~18. 0	0. 18	Zhang & Lee, 2001

Flat-panel PBRs can be constructed with sheets of polyethylene, glass, polyvinyl chloride, polycarbonate or acrylic, etc. (Zou & Richmond, 1999; Zou et al., 2000; Reyna-Velarde et al., 2010; Zittelli et al., 2013). To ensure an effective supply of CO_2, pipelines must be placed in the bottom of the PBR and mixing must be performed via airlift to guaranty minimal shear stress (Reyna-Velarde et al., 2010).

This type of system offers the possibility of modular operation that facilitates the scaling-up processes compared with other PBR configurations, such as horizontal tubular systems (Zittelli et al., 2000). It is also possible to place flat-panel PBRs in a row as long as they are separated by 0. 25 to 1 m intervals. This arrangement increases the number of liters of culture per square meter of land (Zhang et al., 1999; Slegers et al., 2011;).

2.2.3. Stirred tank PBRs

Stirred tank PBRs comprise a glass or stainless steel tank equipped with a mechanical mixing system consisting of a series of paddles fixed in an axis (Fig. 2). Oxygen is supplied in sterile (filtered) air that is blown through a diffusor in the bottom of the reactor. The risk of contamination in this type of system is very low because is possible to sterilize the reactor using steam. However, because the majority of microalgal genera are sensitive to shear stress (the mixing must be moderate) and to the oxygen concentration (inhibitory levels must be avoided) (Carvalho et al., 2006), the operational conditions must be carefully controlled. This configuration has been widely used to culture species that produce compounds of high commercial value (Table 4) because their heterotrophy allows them to reach high concentrations and high biomass productivity.

Some of the first experimental cultures of microalgae were conducted using stirred tanks (Pruess et al., 1954); these PBRs are limited to autotrophic conditions because the large diameter of the vessel does not permit efficient light provision and hence the biomass productivity is very low. To overcome this problem, complex internal illumination systems using fluorescent lamps or optic fibers were developed (Ogbonna et al., 1996; Ugwu et al., 2008); however, the use of such illumination devices make scaling-up difficult.

Using stirred tank PBRs for autotrophic cultures presented showed some difficulties, such as the inefficient use of light, low cellular productivity and the unfeasibility of outdoor operation (Ugwu et al., 2008); however, this configuration has been very useful for the production of heterotrophic microalgae, by taking advantage of only the respiratory metabolic pathway (without light) (Barsanti & Gualtieri, 2014). This metabolic flexibility was demonstrated 50 years ago and provided new possibilities for microalgal biotechnology. Myer (1951) listed some of the studies in which the capability of certain species to grow in the dark while consuming an organic source of carbon was demonstrated. The principal advantages of stirred tank PBRs are that they can be sterilized and that filtered air can be blown into them (Ugwu et al., 2008).

3. Improving Biomass Productivity

3.1. Gaseous Exchange (CO_2/O_2)

The principal mass transfer processes that occur in a microalgal culture are the diffusion of CO_2 from the gas stream into the culture and the removal of the oxygen that is produced by the photosynthesis (Suh & Lee, 2003). The mass transfer inside PBRs is modified by the characteristics (e. g., density and viscosity, etc.) of the fluid and the operational conditions of the system (e. g., gas flow rate, liquid flow rate, configuration, and geometry of the PBR, etc.) (Guieysee et al., 2011).

The appropriate CO_2/O_2 balance is very important in microalgal cultures because ribulose-1, 5-bisphosphate carboxylase/oxygenase (Rubisco) has double functionality. At high concentrations of carbon dioxide, the Rubisco fixes CO_2 via the Calvin-Benson cycle (CBC), producing 2 molecules of 3-phosphoglycerate during the first step of the cycle, but under high O_2 tension, the Rubisco performs photorespiration. This process consists of inserting an oxygen molecule during the first step of the CBC to produce 2-phosphoglycolate, and it cannot be introduced to the CBC wasting a molecule of CO_2 (Sharkey & Weise, 2012).

Carbon is the most important nutrient for microalgae because it represents approximately 50% of the microalgal biomass (Carvalho et al., 2006). When CO_2 dissolves in the culture medium, three different inorganic species are generated and they eventually reach equilibrium (Eq. 1). An insufficient amount of CO_2 are often added to cultures, but the addition of CO_2 in excess causes a decrease in the pH and negatively affects the performance of the culture, diminishing the production of biomass. Therefore, it is very important to ensure that only the necessary amount of CO_2 is added to the culture.

$$CO_{2(aq)} + H_2O \leftrightarrow H_2CO_3 \xleftrightarrow{\text{pka} = 6.35} HCO_3^- + H^+ \xleftrightarrow{\text{pka} = 10.33} CO_3^{2-} + 2\,H^+ \qquad \text{(Eq. 1)}$$

In open systems, the accumulation of oxygen in the medium is not a problem because the large surface area permits efficient O_2 removal; however, the amount of CO_2 that diffuses into the medium is not sufficient to sustain concentrated cultures (Carvalho et al., 2014). For concentrated cell cultures (> 1 g L^{-1}), the simplest way to add CO_2 is direct bubbling; even the use of flue gasses was proposed because their CO_2 concentrations range from 10 to 15% $_{V/V}$ (Doucha & Lívanský, 2014).

The use of CO_2 in open systems is inefficient because the residence time of the gas and the cell concentration are very low. Some authors reported that the efficiencies of CO_2 use (the amount of CO_2 that is converted to biomass relative to the amount of CO_2 that is added to the culture) in this type of PBR ranged from 13 to 20% (Becker, 1994; Richmond & Becker 1986). However, in thin-layer systems, this efficiency increases considerably, reaching values of up to 70% because a high concentration of biomass can be achieved in this type of PBR (Doucha & Lívanský, 2006). In closed systems, the use of CO_2 is more efficient than in open systems because the residence time of the gas is higher, ensuring its fixation by the Rubisco (Kumar et al., 2010).

Many strategies have been proposed to favor the efficient utilization of CO_2, one of which is controlling the pH because the photosynthetic process (through the consumption of CO_2) increases the pH from 0. 1 to 0. 6 units, and it is necessary to start the flow of CO_2 only when the pH value is higher than 8 and suspend it when the pH reaches 7. 5 (Torzillo et al., 2012; Doucha & Lívanský, 2014).

The removal of the oxygen produced by photosynthesis is another factor that reduces the biomass productivity in microalgal cultures; it was estimated that a O_2 concentration of greater than 30 mg L^{-1} reduces the growth rate (Torzillo, 1997). The inhibitory effect of high O_2 concentrations on microalgal growth is caused by the increased generation of reactive oxygen species (ROS). When the amount of ROS produced exceeds the intracellular potential to neutralize them, the photoinhibitory process is triggered (Nishiyama et al., 2006). This process can reduce the biomass productivity rate by as much as 30% (Torzillo et al., 2003; Masojidek et al., 2010).

In open systems, the accumulation of oxygen is not as prevalent because the large surface area of the PBR and the low solubility of O_2 permit its efficient elimination to the atmosphere (Mendoza et al., 2013). However, in some configurations of closed systems, O_2 accumulates and this could considerably diminish the concentration of biomass that can be reached (Torzillo, 1997).

According to Torzillo (1997), in horizontal tubular PBRs, the production of oxygen is expressed by equation 2, in which μ is the growth rate (h^{-1}), X corresponds to the biomass concentration (g L^{-1}), T_c is the time it takes the liquid to pass through the PBR (h), and Y_0 is a dimensionless parameter that is the oxygen/biomass yield ($g_{oxygen}\ g_{biomass}^{-1}$), which is normally approximately 0. 50 (2 g of O_2 are generated by 1 g of biomass), although the value depends upon the growth conditions and must be obtained experimentally.

$$\frac{\text{Produced } O_2}{(\text{g L}^{-1})} = \frac{\mu X T_C}{Y_0} \qquad \text{(Eq. 2)}$$

The horizontal tubular PBRs that are equipped with an airlift system for mixing do not present oxygen-accumulation problems because the mixing device causes a high level of turbulence that ensures the efficient removal of O_2 from the culture medium (Camacho-Rubio et al., 1999).

The columnar type of PBR, whether annular or flat plate, does not have this problem (Zittelli et al., 2003; Chini Zittelli et al., 2006; Vega-Estrada et al., 2005; Reyna-Velarde et al., 2010), principally due to the good level of mixing that is possible, which can reach Reynolds-number values of greater than 4000, indicating a very turbulent regimen (Ratchford et al., 1992; Su et al., 2010).

3.2. Mixing

Mixing is necessary to prevent cell sedimentation inside the PBR and to avoid the generation of gradients of nutrient concentrations, gases, and light. Additionally, a turbulent regimen favors the exchange of the cells that are exposed to light, which increases the level of biomass productivity reached in the culture (Richmond, 2004; Grobbelaar, 2010; Marshall & Huang, 2010).

The extent of mixing and consequently the level of turbulence in a PBR depend on the amount of energy provided to the system; better mixing entails higher energy consumption and therefore an increase in the processing costs, which affects the economic viability of the process. In addition, if the objective of the process is the production of bioenergetics, the amount of energy added to the system should be minimized so that the energy contained in the biomass has a positive energy balance (Das & Obbard, 2011; Khoo et al., 2011).

Turbulence is very important in autotrophic cultures because it ensures that the incident light falls on all of the cells. This condition is particularly important in high-density cultures in which the mixing rate and the growth rate are directly proportionally (Markl, 1980) and consequently, so is the biomass productivity rate (Das & Obbard, 2011). Additionally, high turbulence eliminates the oxygen produced by photosynthesis; in cultures with low agitation rate, the O_2 concentration can reach values four times the saturation value, whereas in efficiently mixed cultures, the O_2 concentration can be less than three times the saturation value (Richmond, 2004), which diminishes the risk of photorespiration occurring (Sharkey & Weise, 2012).

In open systems (pivot and raceways), mixing is conducted using paddle wheels (Tredici, 2010; Zitelli et al., 2013) and a liquid flow rate of 0. 2 to 0. 5 m s^{-1} must be reached to ensure homogeneity (Del Campo et al., 2007). The lack of turbulence within the PBR causes death zones when cell sedimentation occurs; the sedimented cells die and can be used by protozoa and bacteria for their growth, endangering the culture (Richmond, 2004; Grobbelaar, 2010).

In columns and flat-plate PBRs, mixing is sufficient to prevent the accumulation of O_2 and the O_2 concentration rarely reaches values greater than 20 mg L^{-1} because the continuous bubbling of air allows O_2 removal (Qiang & Richmond, 1996; Chini-Zittelli et al., 2006).

In the case of the horizontal tubular PBRs in which mixing is performed using pumps, high O_2 concentrations can be reached, making it necessary to induce a high level turbulence within the degasser (Torzillo et al., 1998). A solution to this problem was proposed by Acién-Fernández et al., (2001), who added an airlift external-loop system to a horizontal tubular PBR to ensure the efficient removal of the oxygen produced by photosynthesis.

Another strategy to improve the degree of turbulence within horizontal tubular and flat-plate PBRs is incorporating static mixers (baffles) because they increase the fraction of retained gas and the rate of mass transfer and consequently, the level of biomass productivity reached (Ugwu et al., 2002; Ugwu et al., 2003; Ugwu et al., 2005; Huang et al., 2014).

Although biomass productivity is directly related to the turbulence of the culture, many species of microalgae are very sensitive to shear stress, so the type and vigor of mixing to use will depend on the species under cultivation (Grobbelaar, 2010).

3.3. Heat

The light supply under indoor conditions is generally obtained using cold white fluorescent lamps (CFWLs) and, more recently, using light emitting diodes (LEDs). The amount of heat produced by these types of lamps is very low, so it is not necessary to install a cooling device (Darko et al., 2014). However, under outdoor conditions, the situation is different for the following reasons:

1. The quantity of photosynthetically active radiation (PAR) received by a microalgal culture in a sunny day is 10 times greater than that needed for photosynthesis, which causes that the excess energy that is absorbed need to be dissipated as fluorescence (in almost negligible amounts) and heat (Melis et al., 1998).
2. The high amount of infrared radiation collected by the PBR increases the temperature of the culture (Goetz et al., 2011).

For the above-mentioned reasons, the temperature of the culture may increase to 10 to 30° C higher than the environmental temperature, which can considerably affect the productivity of the system (Wang et al., 2012). Moreover, in certain locations, the difference between the temperature during the day and at night is significant and regardless of the PBR configuration employed, it is necessary to install a temperature-control device (Cheng-Wu et al., 2001; Doucha et al., 2005; Masojídek et al., 2009; Hindersin et al., 2014). Controlling the temperature is important because it not only affects the biomass productivity but also the composition and quality of the biomass (De Oliveira et al., 1999).

The risk of the microalgal cultures overheating is not as great in open PBRs because most of the heat is dissipated (due the high surface area) or is eliminated via liquid evaporation. However, some microalgal species are extremely sensitive to temperature variations and thus a heatexchanger must be installed in the bottom of the PBR (Torzillo et al., 2012; Silva-Benavides et al., 2013). In closed PBRs, the risk of the temperature increasing is significant because of the high ratio of their illuminated area relative to the volume of the culture (Torzillo et al., 2003).

It is difficult to maintain horizontal tubular PBRs at a certain temperature because large gradients are generated in them, the magnitude of which depends on the length of the pipeline; moreover, the residence time in the degasser is too brief to allow the sufficient elimination of heat from the culture (Torzillo et al., 2003).

Various strategies to control the temperature in this type of PBR have been proposed. One of these strategies is installing a heat exchanger in a portion of the tube (Acién-Fernández et al., 2003); however, the area is too small to ensure efficient heat transference and the extent of heat removal is not very good. An alternative strategy is to increase the size of the heat exchanger; however, the technique causes a dark section in the PBR that could decrease the productivity of the process.

Another strategy that was tested was implanting a desalinized-water sprinkling system; however, this device caused considerable water waste because even though a collection device was added, a large fraction of water was lost through evaporation (Torzillo et al., 1986; Carlozzi & Torzillo, 1996). The most successful strategy demonstrated to date is to submerge the PBR in a basin containing desalinated-water at a controlled temperature, which ensures a large heat-transfer area (the surface of the tubes). This strategy has proven successful for growing a large number of microalgal species (Torzillo et al., 1991; Masojídek et al., 1999;Masojídek et al., 2000; Silva-Benavides et al., 2013)

Very different strategies to maintain a constant temperature of the cultures in flat-plate PBRs have been proposed. One of the first strategies to be employed was the addition of a system for sprinkling and recollecting desalinized-water. This system is very simple and minimal equipment is required but it is useful only in desert climates and requires the consumption of large amounts of water (Hu et al., 1996a and 1996b;Zou & Richmond, 1999; Cheng-Wu et al., 2001; Richmond & Cheng-Wu, 2001).

In climates in which it is not possible to add this type of refrigeration system, a heat exchanger is installed inside the PBR (Sierra et al., 2008), but this strategy is not widely used.

3.4. Light Transfer

The general photosynthetic reaction is shown in equation 3, in which in addition to water and CO_2, light is another nutrient required by all photosynthetic microorganisms; 9. 5 moles of photons is necessary to fix one mole of carbon dioxide (Melis, 2009).

$$CO_2 + H_2O \xrightarrow{\text{9.5 light photons}} \frac{1}{6} C_6H_{12}O_6 + O_2 \qquad \text{(Eq. 3)}$$

Light is provided to indoor cultures using different types of lamps. CFWLs emit light with wavelengths of 490 to 600 nm, which correspond to violet to red-orange. In contrast, LEDs are monochromatic, which is an advantage because the light emitted matches a specific wavelength. However, the quality of the light emitted by the two types of lamps is very different because the proportion of PAR in the light from LEDs is almost 80%, whereas that from CFWLs does not exceed 40% (Liu, 2008; Halonen et al., 2010). However, in both cases the amount of light emitted is sufficient to achieve biomass concentrations of greater than 1 g L^{-1} (Arias-Peñaranda et al., 2013; Martínez-Roldán et al., 2014).

Outside, the quantity of PAR that can be received by a microalgal culture on a sunny day can reach 2000 $\mu E\ m^{-2}\ s^{-1}$ (Melis, 2009; Hindersin et al., 2013); however, saturation of the photosynthetic apparatus occurs at approximately 300 $\mu E\ m^{-2}\ s^{-1}$ for green algae (Sousa et al., 2012) and at approximately 200 $\mu E\ m^{-2}\ s^{-1}$ for cyanobacteria (Torzillo & Vonshak, 1994; Vonshak et al., 2000). Thus, it is understandable that the productivity of outdoor cultures is decreased due to the cellular damage caused by the ultrahigh level of irradiation to which they are exposed (Melis et al., 1998; Torzillo et al., 2003; Vonshak & Torzillo, 2004). The extent of reduction in biomass productivity due to photoinhibition can range from 20 to 30% (Grobbelaar, 2007).

3.4.1. Photoinhibition (PI)

Photoinhibition is defined as the reduction in the photosynthetic rate when the culture is exposed to ultrahigh irradiation (Adir et al., 2003; Grobbelaar, 2007). This phenomenon occurs largely in cultures adapted to low light (Grobbelaar & Kurano, 2003). PI occurs because the cells cultivated indoors increase the size of their light-harvesting complexes (LHCs) to utilize all of the light that they receive, and when cells with large LHCs are exposed to high light levels, they absorb excessive photons (Falkowski & Chen, 2003). Despite the fact that these cells have the capacity to dissipate a fraction of the absorbed light via various processes, their intrinsic ability is often exceeded, causing oxidative stress (Adir et al., 2003).

The energy-dissipation systems utilized by the cells include the following: (a) the xanthophyll cycle, which consists of a reversible process in which a violaxanthin (a carotenoid) molecule loses its epoxide groups (two), first becoming anteraxanthin (one epoxide group) and finally becoming zeaxanthin (no epoxide groups). This mechanism is activated in a few minutes by a diminution of the luminal pH (Müller et al., 2001; Horton & Ruban, 2005). (b) structural modification of the LHCs, which reduces their assimilative capacity, ensuring the diminution of the amount of energy absorbed; however, this process takes several days (Melis, 2009).

3.4.2. Avoiding the PI in PBRs

There are various strategies to avoid exposing cultures to ultrahigh irradiation, thus preventing photoinhibition. The simplest strategy is to produce a high level of turbulence within the PBR, which is relatively easy in vertical tubular, annular and flat-panel PBRs because their airlift method of mixing can cause very high turbulence in the medium (Hu et al., 1996a; Zhang et al., 1999;Cheng-Wu et al., 2001; Zittelli et al., 2003;Chini-Zittelli et al., 2006; Reyna-Velarde et al., 2010). In contrast, in horizontal tubular PBRs, the layer that is in contact with the tube has a laminar regime (low turbulence); this layer is a few millimeters deep and carries a high risk of photoinhibition (Torzillo et al., 2003). To increase the turbulence inside this type PBR, large energy input is required. To avoid this energy consumption, static mixers have been employed (Ugwu et al., 2002; Ugwu et al., 2003; Ugwu et al., 2005); these mixers increase the level of turbulence but make it difficult to clean the system and increase the cost of constructing the PBR.

In thin-layer systems, the flow rate is greater and consequently so is the level of turbulence, which reduces the risk of photoinhibition (Doucha et al., 2005; Doucha & Livanský, 2006; Doucha & Livanský, 2009; Doucha & Livanský, 2014).

Another strategy to avoid photoinhibition is to employ high cell concentrations (Torzillo et al., 1996; Richmond, 2000) because this condition, coupled with a good mixing rate, favors the autoshading phenomenon. Both of these conditions permit changing the cells that are exposed to light (Doucha & Lívanký, 2006). Some authors suggested that to prevent the risk of photoinhibition, an optimal ratio of the amount of cells (or chlorophyll) and the quantity of light ($\mu E\ g^{-1}$ o $\mu E\ mg_{Chl}^{-1}$) must be reached. However, this value must be obtained experimentally because it is specific to each species (Richmond, 2000).

In thin-layer and flat-plate PBR systems, the light path can be modified to diminish the amount of photoinhibition and prevent autoshading but the necessary size of the light path is specific to the microalgae in culture (Richmond et al., 2003). Normally, the thickness of the light path cannot be greater than 10 cm in flat-panel PBRs and greater than 2 cm in thin-layer

PBRs (Zou & Richmond, 1999; Zhang et al., 1999;Richmond & Cheng-Wu, 2001; Richmond et al., 2003;Torzillo et al., 2010; Reyna-Velarde et al., 2010).

Finally, it is advisable to employ cultures that are adapted to high levels of irradiation, which will prevent photoinhibition and avoid the occurrence of a lag phase (Masojidek et al., 2003; Masojidek et al., 2009).

CONCLUSION

Throughout the history of modern phycology, different technologies for the production of large microalgal biomass have emerged due to the wide variety of compounds and products that can be obtained from them. To control the many factors that modify the behavior of microalgal cultures, various culture strategies (heterotrophic, mixotrophic or autotrophic) have been developed, different types of PBR configurations have been designed and modified, various types of construction materials have been developed, and numerous culture media have been designed and optimized.

Initially, the systems for microalgal cultivation were simple and unmechanized but resulted in low levels of biomass productivity. Over time, these systems became sophisticated such that now the values of the operational parameters, such as the pH, temperature, and concentrations of oxygen and carbon dioxide, and of the culture conditions (e. g., the feed rate, temperature, and nutrient supply) are controlled in real time. Nevertheless, the theoretically maximal level of biomass productivity has not yet been reached.

Despite the sustained effort applied over the last 60 years to develop PBR configurations ranging from simple devices (lagoons) to highly specialized closed system as well as the effort exerted to obtain a genetically modified "superalga", the reality is that at present, there are few cost-effective biomass production processes on a commercial scale. A combination of various aspects of PBRs is necessary to ensure the success of the process; therefore, instead of continuing to generate new PBR configurations, the right configuration should be chosen for each microalgal species, guided by the knowledge of its physiology and metabolism.

REFERENCES

Acién-Fernández, F. G., Hall, D. O., Cañizares-Guerrero, E., Krishna Rao, K. & Molina-Grima, E. (2003). Outdoor production of *Phaeodactylumtricornutum* biomass in a helical reactor. *Journal of Biotechnology*, *103*(2), 137-152.

Acién-Fernández, F. G., Fernández-Sevilla, J. M., Sánchez-Pérez, J. A., Molina-Grima, E. & Chisti, Y. (2001). Airlift-driven external-loop tubular photobioreactors for outdoor production of microalgae: assessment of design and performance. *Chemical Engineering Science*, *56*(8), 2721-2732.

Adir, N., Zer, H., Shochat, S. & Ohad, I. (2003). Photoinhibition–a historical perspective. *Photosynthesis Research*, *76*(1–3), 343-370.

Arias-Peñaranda, M. T., Cristiani-Urbina, E., Montes-Horcasitas, C., Esparza-García, F., Torzillo, G. & Cañizares-Villanueva, R. O. (2013). *Scenedesmusincrassatulus* CLHE-

Si01: A potential source of renewable lipid for high quality biodiesel production. *Bioresource Technology*, *140*, 158-164.

Arias-Peñaranda, M., Reyna-Velarde, R. & Cañizares-Villanueva, R. O. (2012). Hydrodynamic characterisation of an airlift reactor for domestic wastewater treatment. *International Journal of Environmental Engineering*. *4*(3), 171-182.

Bajpai, R., Zappi, M., Dufreche, S., Subramaniam, R. & Prokop, A. (2014). Status of algae as vehicles for commercial production of fuels and chemicals. In R. Bajpai, A. Prokop, & M. Zappi, (Eds.) *Algal Biorefineries:Cultivation of cells and products* (First Edition, 3 – 2 4). The Netherlands: Springer.

Barsanti, L. & Gualtieri, P. (2014). *Algae: anatomy, biochemistry, and biotechnology* (Second Edition). Boca Raton, FL: CRC.

Becker, E. W. (1994). *Microalgae-Biotechnology and microbiology*. Cambridge, Cambridge University Press.

Borowitzka, M. A. (1999). Commercial production of microalgae: ponds, tanks, tubes and fermenters. *Journal of Biotechnology*, *70*, 313-321.

Borowitzka, M. A. & Moheimani, N. R. (2013). Open pond culture systems. In M. A. Borowitzka & N. R. Moheimani (Eds.) *Algae for Biofuels and Energy*. 133-152. The Netherlands: Springer

Camacho-Rubio, F., Acién-Fernández, F. G., Sánchez-Pérez, J. A., García-Camacho, F. & Molina-Grima, E. (1999). Prediction of dissolved oxygen and carbon dioxide concentration profiles in tubular photobioreactors for microalgal culture. *Biotechnology and Bioengineering*, *62*, 71-86.

Carlozzi, P. & Torzillo, G. (1996). Productivity of *Spirulina* in a strongly curved outdoor tubular photobioreactor. *Applied Microbiology and Biotechnology*, *45*(1 – 2), 18-23.

Carvalho, A. P., Meireles, L. A. & Malcata, F. X. (2006). Microalgal reactors: a review of enclosed system designs and performances. *Biotechnology Progress*, *22*, 1490-1506.

Carvalho, J. C., Matsudo, M. C., Bezerra, R. P., Ferreira-Camargo, L. S. & Sato, S. (2014). Microalgae Bioreactors In R. Bajpai, A. Prokop, & M. Zappi, (Eds.) *Algal Biorefineries:Cultivation of cells and products* (First Edition, 83-126) The Netherlands: Springer.

Cheng-Wu, Z., Zmora, O., Kopel, R. & Richmond, A. (2001). An industrial-size flat plate glass reactor for mass production of *Nannochloropsis* sp. (Eustigmatophyceae). *Aquaculture*, *195*(1), 35-49.

Chini-Zittelli, G., Rodolfi, L., Biondi, N. & Tredici, M. R. (2006). Productivity and photosynthetic efficiency of outdoor cultures of *Tetraselmissuecica* in annular columns. *Aquaculture*. *261*(3), 932-943.

Cooney, M., J., Young, G. & Pate, R. (2011). Bio-oil from photosynthetic microalgae: case study. *Bioresource Technology*, *102*, 166-177.

Darko, E., Heydarizadeh, P., Schoefs, B. & Sabzalian, M. R. (2014). Photosynthesis under artificial light: the shift in primary and secondary metabolism. *Philosophical Transactions of the Royal Society B: Biological Sciences*, *369*(1640), 20130243.

Das, P. & Obbard, J. P. (2011). Incremental energy supply for microalgae culture in a photobioreactor. *Bioresource Technology*, *102*(3), 2973-2978.

De Oliveira, M. A. C. L., Monteiro, M. P. C., Robbs, P. G. & Leite, S. G. F. (1999). Growth and chemical composition of *Spirulinamaxima* and *Spirulinaplatensis* biomass at different temperatures. *Aquaculture International*, *7*(4), 261-275.

Del Campo, J. A., García-González, M. & Guerrero, M. G. (2007). Outdoor cultivation of microalgae for carotenoid production: current state and perspectives. *Applied Microbiology and Biotechnology*, *74*(6), 1163-1174.

Del Campo, J. A., Rodríguez, H., Moreno, J., Vargas, M., Rivas, J. & Guerrero, M. G. (2001). Lutein production by *Muriellopsis* sp. in an outdoor tubular photobioreactor. *Journal of Biotechnology*, *85*(3), 289-295.

Dillschneider, R. & Posten, C. (2013). Closed bioreactors as tools for microalgae production. In J. W. Lee (Ed.) *Advanced Biofuels and Bioproducts*. pp. 629-649. New York: Springer.

Doucha J., Lívanský, K. (2014). High density outdoor microalgal culture. In R. Bajpai, A. Prokop, & M. Zappi, (Eds.) *Algal Biorefineries:Cultivation of cells and productsAlgal Biorefineries*(First Edition, 147-173). The Netherlands: Springer.

Doucha, J. & Lívanský, K. (2006). Productivity, CO_2/O_2 exchange and hydraulics in outdoor open high density microalgal (*Chlorella* sp.) photobioreactors operated in a Middle and Southern European climate. *Journal of Applied Phycology*, *18*(6), 811-826.

Doucha, J. & Lívanský, K. (2009). Outdoor open thin-layer microalgal photobioreactor: potential productivity. *Journal of Applied Phycology*. *21*, 111-117.

Doucha, J., Straka, F. & Lívanský, K. (2005). Utilization of flue gas for cultivation of microalgae *Chlorella* sp. in an outdoor open thin-layer photobioreactor. *Journal of Applied Phycology*, *17*(5), 403-412.

Falkowski, P. G. & Chen, Y. B. (2003). Photoacclimation of light harvesting systems in eukaryotic algae. *Advances in Photosynthesis and Respiration*, *13*, 423-447.

Farrelly, D. J., Everard, C. D., Fagan, C. C. & McDonnell, K. P. (2013). Carbon sequestration and the role of biological carbon mitigation: a review. *Renewable and Sustainable Energy Reviews*, *21*, 712-727.

Gallardo Rodríguez, J. J., Sánchez Mirón, A., García Camacho, F., Ceron Garcia, M. C., Belarbi, E. H. & Molina Grima, E. (2010). Culture of dinoflagellates in a fed-batch and continuous stirred-tank photobioreactors: growth, oxidative stress and toxin production. *Process Biochemistry*, *45*(5), 660-666.

García-González, M., Moreno, J., Manzano, J. C., Florencio, F. J. & Guerrero, M. G. (2005). Production of *Dunaliellasalina* biomass rich in 9 *cis*-β-carotene and lutein in a closed tubular photobioreactor. *Journal of Biotechnology*, *115*(1), 81-90.

Gitelson, A., Qiuang, H. & Richmond, A. (1996). Photic volume in photobioreactors supporting ultrahigh population densities of the photoautotroph *Spirulinaplatensis*. *Applied and Environmental Microbiology*. *62*(5), 1570-1573.

Goetz, V., Le Borgne, F., Pruvost, J., Plantard, G. & Legrand, J. (2011). A generic temperature model for solar photobioreactors. *Chemical Engineering Journal*, *175*, 443-449.

Grobbelaar, J. U. (2007). Photosynthetic characteristics of *Spirulinaplatensis* grown in commercial-scale open outdoor raceway ponds: what do the organisms tell us?. *Journal of Applied Phycology*, *19*(5), 591-598.

Grobbelaar, J. U. (2010). Microalgal biomass production: challenges and realities. *Photosynthesis Research*, *106*(1-2), 135-144.

Grobbelaar, J. U. & Kurano, N. (2003). Use of photoacclimation in the design of a novel photobioreactor to achieve high yields in algal mass cultivation. *Journal of Applied Phycology*, *15*(2–3), 121-126.

Guieysse, B., Quijano, G., & Muñoz, R. (2011). Bioreactors-Design: Airlift bioreactors. In M. Moo-Young (Ed.), *Comprehensive Biotechnology*, (Second Edition, Volume 2, pp. 199-212). The Netherlands: Elsevier.

Halonen, L., Tetri, E. & Bhusal, P. (2010). *IEA Annex 45 – Guidebook on energy* efficient *electric lighting for buildings*. Finland: Aalto University.

Hindersin, S., Leupold, M., Kerner, M. & Hanelt, D. (2013). Irradiance optimization of outdoor microalgal cultures using solar tracked photobioreactors. *Bioprocess and Biosystems Engineering*, *36*(3), 345-355.

Hindersin, S., Leupold, M., Kerner, M. & Hanelt, D. (2014). Key parameters for outdoor biomass production of *Scenedesmusobliquus* in solar tracked photobioreactors. *Journal of Applied Phycology*. DOI:10. 1007/s10811-014-0261-2

Horton, P. & Ruban, A. (2005). Molecular design of the photosystem II light-harvesting antenna: photosynthesis and photoprotection. *Journal of Experimental Botany*, *56*(411), 365-373.

Hu, Q. (2004). Industrial production of microalgal cell-mass and secondary products-major industrial species. In A. Richmond (Ed.). *Handbook of Microalgal Culture: Applied Phycology and Biotechnology* (Second Edition, 264 – 272). Oxford, UK: Blackwell Science Ltd.

Hu, Q., Guterman, H. & Richmond, A. (1996a) A flat inclined modular photobioreactor for outdoor mass cultivation of photoautotrophs. *Biotechnology and Bioengineeering*, *51*, 51-60.

Hu, Q., Guterman, H. & Richmond, A. (1996b). Physiological characteristics of *Spirulina platensis* (Cyanobacteria) cultured at ultrahigh cell densities. *Journal of Phycology*, *32*, 1066-1070.

Hu, Q., Kurano, N., Kawachi, M., Iwasaki, I. & Miyachi, S. (1998). Ultrahigh-cell-density culture of a marine green alga *Chlorococcumlittorale* in a flat-plate photobioreactor. *Applied Microbiology and Biotechnology*. *49*(6), 655-662.

Huang, J., Li, Y., Wan, M., Yan, Y., Feng, F., Qu, X., Wang J., Shen G., Li W., Fan J. & Wang, W. (2014). Novel flat-plate photobioreactors for microalgae cultivation with special mixers to promote mixing along the light gradient. *Bioresource Technology*, *159*, 8-16.

Isleten-Hosoglu, M., Gultepe, I. & Elibol, M. (2012). Optimization of carbon and nitrogen sources for biomass and lipid production by *Chlorellasaccharophila* under heterotrophic conditions and development of Nile red fluorescence based method for quantification of its neutral lipid content. *Biochemical Engineering Journal*, *61*, 11-19.

Jiang, Y., Zhang, W., Wang, J., Chen, Y., Shen, S. & Liu, T. (2013). Utilization of simulated flue gas for cultivation of *Scenedesmusdimorphus*. *Bioresource Technology*, *128*, 359-364.

John, R. P., Anisha, G. S., Nampoothiri, K. M. & Pandey, A. (2011). Micro and macroalgal biomass: a renewable source for bioethanol. *Bioresource Technology*, *102*, 186-193.

Khoo, H. H., Sharratt, P. N., Das, P., Balasubramanian, R. K., Naraharisetti, P. K. & Shaik, S. (2011). Life cycle energy and CO_2 analysis of microalgae-to-biodiesel: Preliminary results and comparisons. *Bioresource Technology*, *102*(10), 5800-5807.

Kumar, A., Ergas, S., Yuan, X., Sahu, A., Zhang, Q., Dewulf, J., Malcata, F. J. & Van Langenhove, H. (2010). Enhanced CO_2 fixation and biofuel production via microalgae: recent developments and future directions. *Trends in Biotechnology*, *28*(7), 371-380.

Lee Y. K., & Pirt S. J. (1984). CO_2 absorption rate in an algal culture: effect of pH. *Journal of Chemical Technology and Biotechnology. Biotechnology*, *34*(1), 28-32.

Li, S., Luo, S. & Guo, R. (2013). Efficiency of CO_2 fixation by microalgae in a closed raceway pond. *Bioresourse Technology*, *136*, 267-272.

Li, X., Xu, H. & Wu, Q. (2007). Large-scale biodiesel production from microalga *Chlorellaprotothecoides* through heterotrophic cultivation in bioreactors. *Biotechnology and Bioengineering*, *98*(4), 764-771.

Liu, Y. S. (2008). An overview of the development of major light sources: from light bulbs to solid state lighting, III-Nitride devices and nanoengineering. In Z. C. Feng (Ed.). *III-Nitride Devices and Nanoengineering* (pp. 1-20) Taiwan: Imperial College Press.

Markl, H. (1980). Modeling of algal production systems. In G. Shelef, & C. J. Soeder. (Eds.), *Algae Biomass*. (First Edition, pp. 361–383). The Netherlands: Elsevier Press.

Marshall, J. S. & Huang, Y. (2010). Simulation of light-limited algae growth in homogeneous turbulence. *Chemical Engineering Science*, *65*(12), 3865-3875.

Martínez-Roldán A. J., Perales-Vela H. V., Cañizares-Villanueva R. O., Torzillo G. (2013). Physiological response of *Nannochloropsis* sp. to saline stress in laboratory batch cultures. *Journal of Applied Phycolog*, *26*(1), 115-121.

Masojídek, J., Kopecký, J., Giannelli, L. & Torzillo, G. (2011). Productivity correlated to photobiochemical performance of *Chlorella* mass cultures grown outdoors in thin-layer cascades. *Journal of Industrial Microbiology & Biotechnology*, *38*, 307-317.

Masojídek, J., Papáček, Š., Sergejevova, M., Jirka, V., Červený, J., Kunc, J., Korečko, J., Verbovikova, O., Kopecký, J., Štys, D. & Torzillo, G. (2003). A closed solar photobioreactor for cultivation of microalgae under supra-high irradiance: basic design and performance. *Journal of Applied Phycology*, 15(2–3), 239-248.

Masojídek, J., Sergejevová, M., Rottnerová, K., Jirka, V., Korečko, J., Kopecký, J., Zaťková I., Torzillo, G. & Štys, D. (2009). A two-stage solar photobioreactor for cultivation of microalgae based on solar concentrators. *Journal of Applied Phycology*, *21*(1), 55-63.

Masojídek, J., Torzillo, G., Koblížek, M., Kopecký, J., Bernardini, P., Sacchi, A. & Komenda, J. (1999). Photoadaptation of two members of the Chlorophyta (*Scenedesmus* and *Chlorella*) in laboratory and outdoor cultures: changes in chlorophyll fluorescence quenching and the xanthophyll cycle. *Planta*, *209*(1), 126-135.

Masojídek, J., Torzillo, G., Kopecký, J., Koblížek, M., Nidiaci, L., Komenda, J., Lukavská, A. & Sacchi, A. (2000). Changes in chlorophyll fluorescence quenching and pigment composition in the green alga *Chlorococcum* sp. grown under nitrogen deficiency and salinity stress. *Journal of Applied Phycology*, *12*(3-5), 417-426.

Masojídek, J., Vonshak, A. & Torzillo, G. (2010). Chlorophyll fluorescence applications in microalgal mass cultures. In D. J. Suggett, M. A. Borowitzka, O. Prášil (Eds.) *Chlorophyll a Fluorescence in Aquatic Sciences: Methods and Applications*(First Edition, pp. 277-292). The Netherlands: Springer.

Mazzuca Sobczuk, T. & Chisti, Y. (2010). Potential fuel oils from the microalga *Choricystisminor*. *Journal of Chemical Technology and Biotechnology*, *85*(1), 100-108.

Melis, A. (2009). Solar energy conversion efficiencies in photosynthesis: minimizing the chlorophyll antennae to maximize efficiency. *Plant Science*, *177*(4), 272-280.

Melis, A., Neidhardt, J. & Benemann, J. R. (1998). *Dunaliellasalina* (Chlorophyta) with small chlorophyll antenna sizes exhibit higher photosynthetic productivities and photon use efficiencies than normally pigmented cells. *Journal of Applied Phycology*, *10*(6), 515-525.

Mendoza, J. L., Granados, M. R., de Godos, I., Acién, F. G., Molina, E., Heaven, S. & Banks, C. J. (2013). Oxygen transfer and evolution in microalgal culture in open raceways. *Bioresource Technology*, *137*, 188-195.

Moazami, N., Ashori, A., Ranjbar, R., Tangestani, M., Eghtesadi, R. & Nejad, A. S. (2012). Large-scale biodiesel production using microalgae biomass of *Nannochloropsis*. *Biomass and Bioenergy*, *39*, 449-453.

Molina, E., Fernández, J., Acién, F. G. & Chisti, Y. (2001). Tubular photobioreactor design for algal cultures. *Journal of Biotechnology*, *92*(2), 113-131.

Müller, P., Li, X. P. & Niyogi, K. K. (2001). Non-photochemical quenching. A response to excess light energy. *Plant Physiology*, *125*(4), 1558-1566.

Myers, J. (1951). Physiology of the algae. *Annual Reviews in Microbiology*. *5*(1), 157-180.

Nishiyama, Y., Allakhverdiev, S. I. & Murata, N. (2006). A new paradigm for the action of reactive oxygen species in the photoinhibition of photosystem II. *Biochimica et Biophysica Acta-Bioenergetics*, *1757*(7), 742-749.

Ogbonna, J. C., Yada, H., Masui, H. & Tanaka, H. (1996). A novel internally illuminated stirred tank photobioreactor for large-scale cultivation of photosynthetic cells. *Journal of Fermentation and Bioengineering*, *82*(1), 61-67.

Oswald, W. J. (1988). Micro-algae and waste-water treatment. In M. A. Borowitzka & L. J. Borowitzka (Eds.). *Micro-algal Biotechnology*. (First Edition, pp 305–328). Cambridge, UK: Cambridge University Press.

Oswald, W. J. & Golueke C. G. (1960). Biological transformation of solar energy. *Advances in Applied Microbiology*, *2*, 223-262.

Oswald, W. J., Gotaas, H. B., Golueke, C. G. & Kellen, W. R. (1957). Algae in waste treatment. *Sewage and Industrial Wastes*, *29*, 437-457.

Pirt, S. J., Lee, Y. K., Walach, M. R., Pirt, M. W., Balyuzi, H. H., Bazin, M. J. (1983). A tubular bioreactor for photosynthetic production of biomass from carbon dioxide: design and performance. *Journal of Chemical Technology and Biotechnology. Biotechnology*, *33*(1): 35-58.

Posten, C. (2009). Design principles of photo-bioreactors for cultivation of microalgae. *Engineering in Life Sciences*, *9*(3), 165-177.

Pruess, L., Arnow, P., Wolcott, L., Bohonos, N., Oleson, J. J. & Williams, J. H. (1954). Studies on the mass culture of various algae in carboys and deep-tank fermentations. *Applied Microbiology*, *2*(3), 125–130.

Qiang, H. & Richmond, A. (1996). Productivity and photosynthetic efficiency of *Spirulinaplatensis* as affected by light intensity, algal density and rate of mixing in a flat plate photobioreactor. *Journal of Applied Phycology*, *8*(2), 139-145.

Ramos de Ortega, A. & Roux, J. C. (1986). Production of *Chlorella* biomass in different types of flat bioreactors in temperate zones. *Biomass*, *10*(2), 141-156.

Ratchford, I. A. & Fallowfield, H. J. (1992). Performance of a flat plate, air-lift reactor for the growth of high biomass algal cultures. *Journal of Applied Phycology*, *4*(1), 1-9.

Reyna-Velarde, R., Cristiani-Urbina, E., Hernández-Melchor, D. J., Thalasso, F. & Cañizares-Villanueva, R. O. (2010). Hydrodynamic and mass transfer characterization of a flat-panel airlift photobioreactor with high light path. *Chemical Engineering and Processing: Process Intensification*, *49*(1), 97-103.

Richmond, A., Becker, E. W. (1986). Technological aspects of mass cultivation—A general outline. In A. Richmond (Ed.) *Handbook of microalgal mass culture* (pp 245-264). Boca Raton, FL: CRC Press.

Richmond, A. (1987). The challenge confronting industrial microagriculture: high photosynthetic efficiency in large-scale reactors. In M. A. Ragan, & C. J. Bird (Eds.), *Twelfth International Seaweed Symposium*(First Edition, pp. 117-121). The Netherlands: Springer.

Richmond, A. (1996). Efficient utilization of high irradiance for production of photoautotropic cell mass: a survey. *Journal of Applied Phycology*, *8*(4-5), 381-387.

Richmond, A. (1999). Physiological principles and modes of cultivation in mass production of photoautotrophic microalgae. In Z. Cohen (Ed.) *Chemicals from Microalgae*. (First Edition, pp. 353–386). London, UK: Taylor & Francis.

Richmond, A. (2000). Microalgal biotechnology at the turn of the millennium: a personal view. *Journal of Applied Phycology*, *12*(3), 441-451.

Richmond, A. (2004). Biological principles of mass cultivation. In A. Richmond (Ed.). *Handbook of microalgal culture: applied phycology and biotechnology* (Second Edition, pp. 125–177). Oxford, UK: Blackwell Science Ltd.

Richmond, A. & Cheng-Wu, Z. (2001). Optimization of a flat plate glass reactor for mass production of *Nannochloropsis* sp. outdoors. *Journal of Biotechnology*, *85*(3), 259-269.

Richmond, A., Cheng-Wu, Z. & Zarmi, Y. (2003). Efficient use of strong light for high photosynthetic productivity: interrelationships between the optical path, the optimal population density and cell-growth inhibition. *Biomolecular Engineering*, *20*(4), 229-236.

Roselet, F., Maicá, P., Martins, T. & Abreu, P. C. (2013). Comparison of open-air and semi-enclosed cultivation system for massive microalgae production in sub-tropical and temperate latitudes. *Biomass and Bioenergy*, *59*, 418-424.

Scarsella, M., Torzillo, G., Cicci, A., Belotti, G., De Filippis, P. & Bravi, M. (2012). Mechanical stress tolerance of two microalgae. *Process Biochemistry*, *47*, 1603-1611.

Schmidt, M. & Wilhelm, C. (2014). Green Algae. In M. F. Hohmann-Marriott (Ed.). *The structural basis of biological energy generation.* (First Edition, pp. 309-333). The Netherlands: Springer.

Scoma, A., Giannelli, L., Faraloni, C. & Torzillo, G. (2010). Solar light driven H_2 production with the microalga *Chlamydomonasreinhardtii* in an outdoor photobioreactor. *Journal of Biotechnology*, *150*, 168.

Scoma, A., Giannelli, L., Faraloni, C. & Torzillo, G. (2012). Outdoor H_2 production in a 50-L tubular photobioreactor by means of a sulfur-deprived culture of the microalga *Chlamydomonasreinhardtii*. *Journal of Biotechnology*, *157*(4), 620-627.

Šetlík, I., Šust V. & Málek, I. (1970). Dual purpose open circulation units for large scale culture of algae in temperate zones. I. Basic design considerations and scheme of a pilot plant. *Algological Studies*, *1*, 111-164.

Sharkey, T. D. & Weise, S. E. (2012). Autotrophic carbon dioxide fixation. *Advances in Photosynthesis and Respiration*, *34*, 651-674

Sierra, E., Acién, F. G., Fernández, J. M., Garcia, J. L., González, C. & Molina, E. (2008). Characterization of a flat plate photobioreactor for the production of microalgae. *Chemical Engineering Journal*, *138*(1), 136-147.

Silva-Benavides, A. M., Torzillo, G., Kopecký, J. & Masojídek, J. (2013). Productivity and biochemical composition of *Phaeodactylumtricornutum* (Bacillariophyceae) cultures grown outdoors in tubular photobioreactors and open ponds. *Biomass and Bioenergy*, *54*, 115-122.

Slade, R. & Bauen, A. (2013). Micro-algae cultivation for biofuels: Cost, energy balance, environmental impacts and future prospects. *Biomass and Bioenergy*, *53*, 29-38.

Slegers, P. M., Wijffels, R. H., Van Straten, G. & Van Boxtel, A. J. B. (2011). Design scenarios for flat panel photobioreactors. *Applied Energy*, *88*(10), 3342-3353.

Sousa, C., De Winter, L., Janssen, M., Vermuë, M. H. & Wijffels, R. H. (2012). Growth of the microalgae *Neochlorisoleoabundans* at high partial oxygen pressures and sub-saturating light intensity. *Bioresource Technology*, *104*, 565-570.

Su, Z., Kang, R., Shi, S., Cong, W. & Cai, Z. (2010). Study on the destabilization mixing in the flat plate photobioreactor by means of CFD. *Biomass and Bioenergy*, *34*(12), 1879-1884.

Suh, I. S. & Lee, C. G. (2003). Photobioreactor engineering: design and performance. *Biotechnology and Bioprocess Engineering*, *8*(6), 313-321.

Tamburic, B., Zemichael, F. W., Crudge, P., Maitland, G. C. & Hellgardt, K. (2011). Design of a novel flat-plate photobioreactor system for green algal hydrogen production. *International Journal of Hydrogen Energy*, *36*(11), 6578-6591.

Torzillo G., Pushparaj B., Bocci F., Balloni W., Materassi R. & Florenzano G. (1986). Production of *Spirulina* biomass in closed photobioreactors. *Biomass*, *11*(1), 61-74.

Torzillo, G. (1997). Tubular bioreactors. In A. Vonshak (Ed.)*Spirulina platensis (Arthrospira): Physiology, Cell-Biology and Biotechnology* (First Edition, pp. 101-115). London:Taylor & Francis.

Torzillo, G. & Vonshak, A. (1994). Effect of light and temperature on the photosynthetic activity of the cyanobacterium *Spirulinaplatensis*. *Biomass and Bioenergy*, *6*(5), 399-403.

Torzillo, G., Accolla, P., Pinzani, E. & Masojidek, J. (1996). In situ monitoring of chlorophyll fluorescence to assess the synergistic effect of low temperature and high irradiance stresses in *Spirulina* cultures grown outdoors in photobioreactors. *Journal ofApplied Phycology*, *8*(4–5), 283-291.

Torzillo, G., Faraloni, C., Silva, A. M., Kopecký, J., Pilný, J. & Masojídek, J. (2012). Photoacclimation of *Phaeodactylumtricornutum* (Bacillariophyceae) cultures grown outdoors in photobioreactors and open ponds. *European Journal of Phycology*, *47*(2), 169-181.

Torzillo, G., Giannelli, L., Martínez-Roldán, A. J., Verdone, N., De Filippis, P. & Scarsella, M. (2010). Microalgae culturing in thin-layer photobioreactors. *Chemical Engineering Transactions*, *20*, 265-270.

Torzillo, G., Pushparaj, B., Masojidek, J. & Vonshak, A. (2003). Biological constraints in algal biotechnology. *Biotechnology and Bioprocess Engineering*, *8*(6), 338-348.

Torzillo, G., Sacchi, A., Materassi, R. & Richmond, A. (1991). Effect of temperature on yield and night biomass loss in *Spirulinaplatensis* grown outdoors in tubular photobioreactors. *Journal of Applied Phycology*, *3*(2), 103-109.

Tredici, M. R. (2004). Mass production of microalgae: photobioreactors. In A. Richmond (Ed.). *Handbook of Microalgal Culture: Applied Phycology and Biotechnology* (Second Edition, pp 178-214). Oxford, UK: Blackwell Science Ltd.

Tredici, M. R. & Zittelli, G. C. (1998). Efficiency of sunlight utilization: tubular versus flat photobioreactors. *Biotechnology and Bioengineering*, *57*(2), 187-197.

Tredici, M. R. (2010). Photobiology of microalgae mass cultures: understanding the tools for the next green revolution. *Biofuels*, *1*, 143-162.

Ugwu, C. U., Aoyagi, H. & Uchiyama, H. (2007). Influence of irradiance, dissolved oxygen concentration, and temperature on the growth of *Chlorellasorokiniana*. *Photosynthetica*, *45*(2), 309-311.

Ugwu, C. U., Aoyagi, H. & Uchiyama, H. (2008). Photobioreactors for mass cultivation of algae. *Bioresource Technology*, *99*, 4021-4028.

Ugwu, C. U., Ogbonna, J. C. & Tanaka, H. (2003). Design of static mixers for inclined tubular photobioreactors. *Journal of Applied Phycology*, *15*(2-3), 217-223.

Ugwu, C., Ogbonna, J. & Tanaka, H. (2002). Improvement of mass transfer characteristics and productivities of inclined tubular photobioreactors by installation of internal static mixers. *Applied Microbiology and Biotechnology*, *58*(5), 600-607.

Ugwu, C. U., Ogbonna, J. C. & Tanaka, H. (2005). Light/dark cyclic movement of algal culture (*Synechocystisaquatilis*) in outdoor inclined tubular photobioreactor equipped with static mixers for efficient production of biomass. *Biotechnology Letters*, *27*, 75-78

Vega-Estrada, J., Montes-Horcasitas, M. C., Domínguez-Bocanegra, A. R. & Cañizares-Villanueva, R. O. (2005). *Haematococcuspluvialis* cultivation in split-cylinder internal-loop airlift photobioreactor under aeration conditions avoiding cell damage. *Applied Microbiology and Biotechnology*, *68*(1), 31-35.

Vonshak, A., Cheung, S. M. & Chen, F. (2000). Mixotrophic growth modifies the response of *Spirulina* (*Arthrospira*) *platensis* (Cyanobacteria) cells to light. *Journal of Phycology*, *36*(4), 675-679.

Vonshak, A., Torzillo, G., 2004. Environmental stress physiology. In A. Richmond, A. (Ed.) *Handbook of Microalgal Culture*. (First Edition, pp. 57-82). Oxford: Blackwell Publishers.

Wang, B., Lan, C. Q. & Horsman, M. (2012). Closed photobioreactors for production of microalgal biomasses. *Biotechnology Advances*, *30*(4), 904-912.

Watanabe, Y. & Hall, D. O. (1996). Photosynthetic production of the filamentous cyanobacterium *Spirulinaplatensis* in a cone-shaped helical tubular photobioreactor. *Applied Microbiology and Biotechnology*, *44*(6), 693-698.

Wijffels, R. H. & Barbosa, M. J. (2010). An outlook on microalgal biofuels. *Science (Washington)*, *329*(5993), 796-799.

Wijffels, R. H., Kruse, O. & Hellingwerf, K. J. (2013). Potential of industrial biotechnology with cyanobacteria and eukaryotic microalgae. *Current Opinion in Biotechnology*, *24*, 405-413.

Xu, H., Miao, X. & Wu, Q. (2006). High quality biodiesel production from a microalga *Chlorellaprotothecoides* by heterotrophic growth in fermenters. *Journal of Biotechnology*, 126(4), 499-507.

Xu, L., Weathers, P. J., Xiong, X. R. & Liu, C. Z. (2009). Microalgal bioreactors: challenges and opportunities. *Engineering in Life Sciences*, *9*, 178-189.

Zhang, D. H. & Lee, Y. K. (2001). Two-step process for ketocarotenoid production by a green alga, *Chlorococcum* sp. strain MA-1. *Applied Microbiology and Biotechnology*, *55*(5), 537-540.

Zhang, K., Kurano, N. & Miyachi, S. (1999). Outdoor culture of a cyanobacterium with a vertical flat-plate photobioreactor: effects on productivity of the reactor orientation, distance setting between the plates, and culture temperature. *Applied Microbiology and Biotechnology*, *52*(6), 781-786.

Zhu, L., Wang, Z., Shu, Q., Takala, J., Hiltunen, E., Feng, P. & Yuan, Z. (2013). Nutrient removal and biodiesel production by integration of freshwater algae cultivation with piggery wastewater treatment. *Water Research*, *47*(13), 4294-4302.

Zittelli, G. C., Biondi, N., Rodolfi, L. & Tredici, M. R. (2013). Photobioreactors for mass production of microalgae. In A. Richmond (Ed.). *Handbook of Microalgal Culture: Applied Phycology and Biotechnology* (Second Edition, pp. 225-266). Oxford, UK: Blackwell Science Ltd.

Zittelli, G. C., Pastorelli, R. & Tredici, M. R. (2000). A modular flat panel photobioreactor (MFPP) for indoor mass cultivation of *Nannochloropsis* sp. under artificial illumination. *Journal of Applied Phycology*, *12*(3-5), 521-526.

Zittelli, G. C., Rodolfi, L. & Tredici, M. R. (2003). Mass cultivation of *Nannochloropsis* sp. in annular reactors. *Journal of Applied Phycology*, *15* (2-3), 107-114.

Zittelli, G. C., Rodolfi, L., Bassi, N., Biondi, N. & Tredici, M. R. (2013). Photobioreactors for microalgal biofuel production. In M. A. Borowitzka & N. R. Moheimani (Eds.) *Algae for Biofuels and Energy*. (First Edition, pp. 115-131). The Netherlands: Springer.

Zou, N. & Richmond, A. (1999). Effect of light-path length in outdoor flat plate reactors on output rate of cell mass and of EPA in *Nannochloropsis* sp. *Journal of Biotechnology*, *70*(1), 351-356.

Zou, N., Zhang, C., Cohen, Z. & Richmond, A. (2000). Production of cell mass and eicosapentaenoic acid (EPA) in ultrahigh cell density cultures of *Nannochloropsis* sp. (Eustigmatophyceae). *European Journal of Phycology*, *35*(2), 127-133.

In: Microalgae and Other Phototrophic Bacteria
Editor: Luis G. Torres Bustillos
ISBN: 978-1-63482-078-3

Chapter 9

MEASURING SUSTAINABLE DEVELOPMENT IN BIOFUEL PRODUCTION FROM MICROALGAE

Mariana Ortega Ramírez, Mario Giraldi Díaz and Gemma Cervantes Torre-Marín*
Research Group on Industrial Ecology (GIEI),
UPIBI, Instituto Politécnico Nacional, Mexico

ABSTRACT

It is a fact that a focus on sustainability must be taken into account in all processes, including those addressed in this book. This chapter gives a broad overview of the current environmental crisis humanity is facing, putting special emphasis on the impact of biofuel production according to a classification based on generations. Considering this, the potential benefits of microalgae biofuels are discussed. A definition for sustainable development is given. Industrial ecology and ecological economics are described as two approaches for sustainability implementation in the context of any production system. Several tools for sustainability measurement are also described in the context of biofuel production; these are: life cycle assessment, ecological footprint, carbon footprint, indicators sets and cost benefit analysis. As an example of the application of such tools, a case study on biodiesel production from microalgae is presented.

1. INTRODUCTION

Overexploitation of natural resources confront humanity with a variety of issues never experienced before by our species on a global scale. Soil loss and degradation, fresh water quantity and quality problems, decrease in forest areas, inefficient use of materials such as metals, depletion of non renewable resources such as petroleum, and an increased emission of pollutants are only some of the environmental impacts related to global issues such as climate change and ozone layer depletion, which are being experienced and left for generations to

* Corresponding author. gemma.cervantes@gmail.com.

come. In fact, most energy analyst expect world energy to continue rising in the ecological scenario by 50% from 2000 to 2030 (Meadows et al. 2004). This will imply a greater environmental impact being global warming one of the most dangerous ones.

Although climate change has been quoted as the greatest threat humanity is facing, the true solution lies not only in measuring and reducing greenhouse gas emissions, but in adopting a sustainable approach in all production processes that are being carried out. For that, this chapter gives an overview on sustainability subjects and tools for measurement that should be taken into account while environmentally evaluating biofuel production.

The subjects of study considered in this book can be grouped into three general areas: (i) cultivation and production of microalgae, (ii) energy exploitation, and (iii) derivative products with added value. In the case of energy supply, large quantities of raw materials and fuels are consumed due to the high global demand. Hence the importance of analyzing the environmental impact of such processes and the indicators associated to its sustainability.

In the actual context of alternative energy consumption to mitigate climate change and to reduce dependence on conventional energy, a massive production of microalgae as a source of supply for a future diversified energy market will be important. However, microalgae biofuels are not immune to the potential environmental impacts from their production chain. Therefore, the study of sustainability is an important complement to the research and technological development for microalgae as raw material of future biofuels and other biotechnological products.

In this text sustainable development is defined and the first Sustainable Development (SD) definition revised. The different measurement tools for SD are explained, with special emphasis on Life-Cycle Assessment (LCA). Also, some approaches for sustainable production are described, such as industrial ecology and ecological economics. A general description of biofuels based on generation and its impacts is shown in point 3. and the impacts are assessed. The potential of biofuels from microalgae is remarked, showing what is being done in research and production and also the perspectives for this type of biofuel. Finally, as part of this analysis, a case study on LCA about the biodiesel production from the microalgae *Chlorococcum humicola* at Instituto Politécnico Nacional in Mexico, is included in section 4, emphasizing the quantification of greenhouse gasses (GHG) emissions compared to other results of similar studies on biofuels.

2. Measuring Sustainable Development

Firstly, a definition for sustainable development is given, then different tools for its measurement are described, and finally two approaches for sustainable production are suggested.

a. Definitions

The Brundtland definition for Sustainable development (SD) is very well-known: "Sustainable development is the development that meets the needs of the present without compromising the ability of future generations to meet their own needs" (WCED 1987).

There are four main ideas inside the Brundtland definition: a) development, b) intragenerational equity, c) intergenerational equity and d) needs.

However the Brundtland definition is no longer sustainable. This concept has been criticized by a lot of contemporary authors (Martinez Alier, 2010; Stahel, 2011; Quijano, 2012; Schneider, 2011 and Zaccai, 2012). One of the main reasons for the unsustainability of Brundtland's definition is that it's based in the fulfillment of needs. Needs cover a wide range, they are numerous and are continuously growing. The way of fulfilling these needs (Max Neef, 1998) isn't universal, but rather depends on the individual, his historical context, and the region he inhabits. Another reason for the unsustainability of Brundtland's definition is the accelerated use of resources and the impact on Earth that meeting of our needs is causing (GFN 2011). Climate Change shows us that this style of living and rate of consumption will lead to the important depletion of natural resources in the Earth (CO2now 2014). Another warning, now from Humanity, comes from inequalities. Inequalities are growing in most countries (CITINOise, 2008). Therefore it is not possible to live according to the rate of our needs. Humanity needs to find another way of living (Cervantes & Ortega 2013).

It is necessary to redefine the concept of SD so that it is genuinely sustainable. This new definition cannot be based on an infinite and subjective concept –needs–, but on a countable and objective one –the available capitals–. These capitals are: natural capital (natural resources, biodiversity or the ecosystem and its services), human and/or social capital (knowledge, skills, multiculturalism, cohesion, participation, etc.), and economic capital (financial, infrastructure, etc.). These capitals may be non-substitutable, as Daly proposed (Daly 1996).

A new SD definition was stated: "Sustainable development is the kind of development that uses the available capitals for the present generations without compromising the ability of future generations to use the same capitals" (Xercavins 2005). This definition still proposes the intergenerational and intragenerational equity and includes the idea of limits, as capitals are finite. In a way it is close to degrowth because this would be an important strategy in order to keep capitals for future generations. The meaning of SD proposed in this definition is closer to sustainable degrowth than to the classic meaning of SD.

b. Measurement Tools

There are several tools to measure sustainability or concepts related to it. Here is a description of some of them:

Life Cycle Assessment

The so-called industrial metabolism concept, used in industrial ecology, refers to the whole set of processes that transform raw materials and energy plus workforce into finished products and waste output (van Berkel, 2006). This is also related to the importance of efficiency in production cycles when interacting with biogeochemical cycles, since disturbances could be caused to the ecosystems, both natural and human.

Every product (or activity) has a "life" that begins with its design and development, followed by the extraction of resources, production of supplies and manufacturing that gives it added value. Afterwards come its use and consumption, followed by the processes of waste management and its final disposal (Rebitzer et al., 2004); this concept is shown in Fig.1. Each of these stages cause an environmental impact in the form of resource depletion, emissions and effluents, and diverse phenomena that modify the environment (noise, radiation, heat, etc.). With this premise, the Life Cycle Assessment is defined as the "compilation and evaluation of the inputs, outputs and the potential environmental impacts of a product system throughout its life cycle" (ISO 14040/44; ISO, 2006). As linkage between processes and environmental impacts, different mathematical correlations (environmental impact models) have been developed to determine how an output generates an impact on the environment, and thereby quantify it.

The main goal of the LCA methodology is to lessen the environmental impact of products and services by guiding the decision-making process. Throughout the life cycle, material use intensity and energy consumption are identified in a production chain, both up and down the supply line. Therefore, for companies, designers, and governments, this tool represents a decision-making aid for implementing sustainable development. The information obtained from a life cycle assessment can also influence environmental policies and regulations. The LCA complements other tools to approach this conceptualization, such as cleaner production, eco-efficiency & eco-design, and development mechanisms.

Indicators

There are some indicators that will help understand the environmental impact of biofuels from microalgae. Some of them are called indexes and they are constituted by various variables. This is the case of Ecological Footprint, Carbon Footprint or Water Footprint. Others may be formed by fewer elements and may be organized as a group in an indicator set.

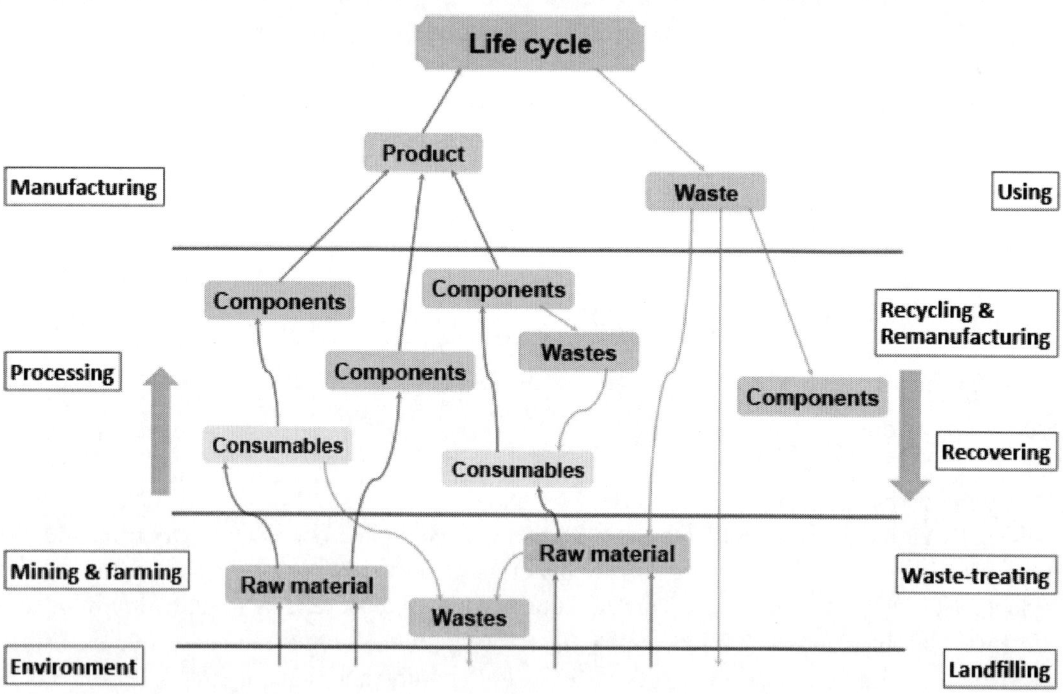

Figure 1. Approach to the concept of the life cycle of a product.

Biocapacity	– Ecological Footprint	=	ECOLOGICAL DEFICIT
Available area	– Consumed Area	=	NATURAL CAPITAL OVEREXPLOTATION

Figure 2. Defining ecological deficit. Source: the author adapted from Hernández and López 2004.

Ecological footprint

The Ecological Footprint was defined in 1996 by Wackernagel and Ress as "the area of ecologically productive land (crops, pastures, forests and aquatic ecosystems) required to produce the resources used, and to assimilate the wastes produced by a given population with a specific way of life, indefinitely" (Wackernagel & Rees 1996).

It is worth looking closely at the main elements in this definition: a) The footprint is measured using the area, i.e., indicates the consumed territory; b) It does not refer to any territory but to ecologically productive land; c) the land is required to produce resources; d) the territory is also required to assimilate waste; e) it refers to a group of people – to a specific population- that has a specific way of life, level of consumption and habits; f) the territory has to perform these functions indefinitely, or continuously over time.

The ecological footprint (EF) is related to biocapacity, which is the productive capacity of the territory expressed in global hectares per capita. The EF is the territory that is consumed, and the biocapacity is the territory or land that is available. When the EF was defined in 1996, globally we had already consumed more than one planet Earth (GFN 2014), Therefore the difference between biocapacity and EF was already a negative number. For this reason, the difference between these two variables is defined as ecological deficit (Figure 2).

In order to calculate the ecological footprint (Chambers et al., 2000) it is necessary to evaluate what type of food, housing, transportation, consumer goods and services the population on which the EF is calculated is using. The Global Footprint Network (GFN) regularly publishes guidelines for the calculation of the ecological footprint (Kitzes et al., 2010).

The ecological footprint is a comprehensive indicator of the environmental impact of a population, so it may be very valuable for evaluating the impact of the production or use of a biofuel. It could be a widely used indicator for comparing the impact of different biofuels.

Carbon Footprint

There is much confusion and disparity in the scientific and not scientific communities in the definition and use of this indicator. Most of the scientific community and specialized agencies consider the carbon footprint of a product, process or territory, as the quantification of emissions of greenhouse gases (GHGs) in the life cycle of the process, product or territory (Weidema et al, 2008) (Hammond, 2007). However, other entities and authors, although considering the GHG, do not consider the life cycle: "The carbon footprint is a measure of the impact of all GHG produced by our activities (individual, collective, and of the products) in the environment" (Schneider & Samaniego, 2009); others only consider CO_2 emissions or a few gases (Wright et al., 2011) or both. Other organizations consider that there are differences in carbon footprints that result in different types: organizational, from the value chain, and from the product (Carbon Trust, 2012).

This involves a wide disparity of ways to calculate this footprint. For those who consider the life cycle, the way to calculate it is through a Life Cycle Assessment (LCA) or an Input-Output Analysis (IO) (Hertwich & Peters 2009). For those who do not consider the life cycle, it can be calculated using the IPCC methodology (Eggleston et al. 2009), and the World Resource Institute protocols (WBCSD &WRI 2004).

ISO14064 builds on the life cycle methodology contained in ISO 14040 and 14044 and considers Carbon Footprint for all the lifecycle (ISO14064-1:2006, 2006).

Therefore, it is an interesting indicator to compare processes, products or also the behavior of various industrial ecosystems. Many companies and organizations already calculated their GHG emissions or carbon footprint. In Mexico there is the Mexico GHG Program (SEMARNAT & CESPEDES 2014) sponsored by CESPEDES that provides businesses with a calculation methodology and training to calculate their GHG emissions, based on the methodology of WRI.

A lot of research groups have calculated the carbon footprint of the production of a biofuel. The comparison among the carbon footprint of biofuels produced from different raw materials is used to analyze which crop has the best environmental performance.

Indicator set

In order to find out what degree of progress on sustainable development (SD) a system has, quantification and measurement of this progress are needed. SD indicator sets are an important tool to measure the degree of progress. There are many SD indicator sets, created by international and national bodies, for cities, regions and countries (UN-CSD 2014) (UPC 2014), but very few for assessing processes or smaller systems. If we want to assess the SD performance of a biofuel it is important to create and calculate the value of the SD indicator set for this biofuel.

In order to design indicators (Cervantes 2007) it is first necessary to determine the objectives of SD in the social, economic and environmental areas, based on the criteria of SD (Figure 3). Determining objectives is not trivial, but very important, since they guide and point out a direction and tendency towards SD. Subsequently, in line with each determined objective, one or more issues or themes associated to them must be established. Themes must be neutral, short, and help organize the different aspects involved in the SD objective. Finally metrics must be created. These may consist of a variable or a relationship between two or more variables, thus called indexes or macroindicators.

There is some work regarding SD indicators for enterprises (Tyteca 1998) or products (Cunningham et al., 2004), others regarding environmental indicators for industrial ecology (Dewulf and Van Langenhove 2005) (Zvolinschi et al., 2007), or economic indicators (Cleveland & Ruth 1998), and very few regarding creation and implementation of SD indicators for eco-industrial systems (Cervantes 2010 and Tiejun 2010) and for biofuel production. The European MESVAL project (MESVAL 2010) and the Research Group on Industrial Ecology (GIEI) from IPN (GIEI 2014) created a system of SD indicators to assess the level of global sustainable development of an industrial ecosystem, and also to compare the degree of SD of two different valuations of the same waste (ECOSIND 2006).

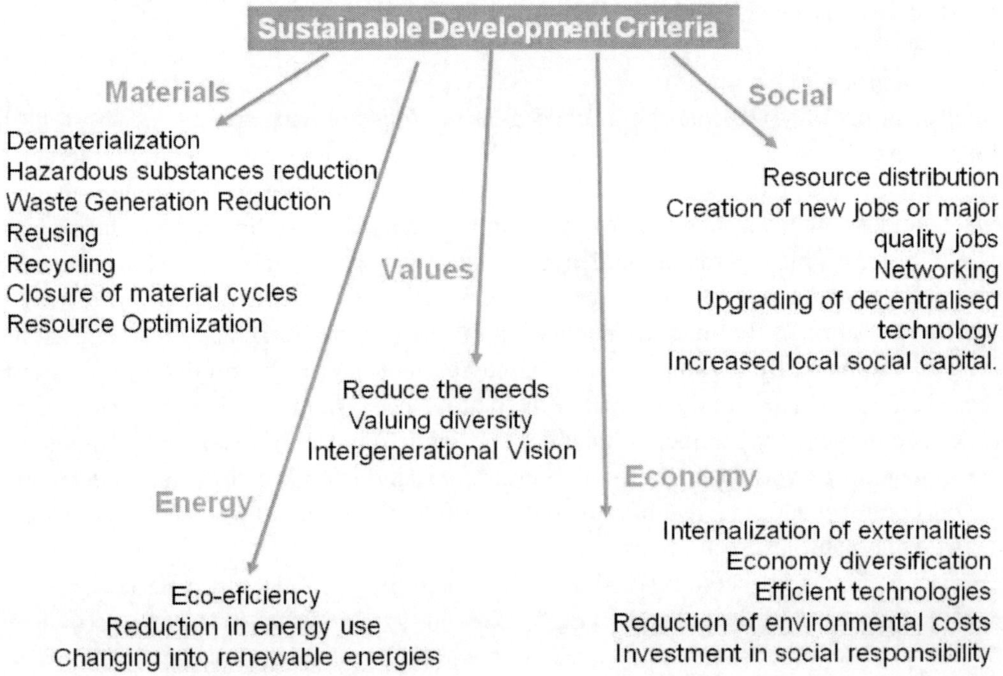

Figure 3. Sustainable development criteria.

Cost-benefit analysis

The Cost-Benefit Analysis (CBA) is an economical technique used to evaluate a set of projects or a set of options within a project. It is based in monetizing all the costs and all the benefits associated to the project, so that if the ratio between the two implies greater benefit, the project or project alternative is better. The information taken into account for the calculation has to include not only installation, operation and maintenance costs, or benefits from sales, but also all the related environmental and social externalities that come with the project.

An externality is defined as a third party of detrimental or beneficial effect for which no price is exacted and has to be estimated for the analysis. Pollution is a clear example of an externality. In a CBA approach, the profits of the polluter would be weighed against the damage done to the environment and to the social sphere, each measured in money terms. Only if profits exceed the damage would the polluter´s activities be efficient. It is important to emphasize that while it is often ignored in practice, a properly executed CBA should show the costs and benefits accruing to different social groups of beneficiaries and losers (Pearce et al., 2006).

Regarding the above, it has been argued that the best way in which policy-makers can contribute to sustainability is by selecting the best projects, where "best" is defined relative to a standard cost-benefit test. Conducting a proper state-of-the-art CBA on projects, thus considering that intangible environmental and social impacts are taken into account, could have a favorable impact on prospects for sustainable development (OECDE, 2006). Although CBA is already widely used to evaluate public policy on biofuels (Charles et al, 2013), it would be very important as well to assess biofuel production systems under this approach.

c. Approaches for Sustainable Production

Industrial ecology

Industrial ecology (IE) can be defined as a multidisciplinary approach, whose ultimate goal is to have industrial systems operate like natural ecosystems by the mutual interaction in cycling matter and increasing process efficiency of industries, society and nature. IE innovates and promotes a new way of thinking by expanding the limits from the firm to the industrial system. This is a major challenge for today's world and also one of the few ways in which industries can strive towards a more sustainable development.

One of the aims of the mutual cooperation of industries is to achieve zero emission/zero waste. This can be partly accomplished by having an industry use by-products and waste from another, similar to the cycling of matter in natural ecosystems. This approach potentially brings economic and environmental benefits that include not only resource savings, but also the minimization of waste generation and emissions. Industrial ecology can also improve a firm's corporate image and result in a more harmonic cooperation between industries and their social and natural surroundings.

However, industrial ecology is more than an approach to closing material cycles (also known as "industrial metabolism" and "industrial symbiosis"). In fact, it provides the foundation for a number of economic, environmental, and social approaches and tools which aim at the reduction of environmental impact, leading to higher efficiency and more sustainable development (Cervantes, 2013). IE can be applied to firm's entities or processes, including production process of biofuels, if each process operation is seen as an entity and reuse of some residual flows is considered (Carrillo 2013).

Ecological economics

Ecological economics arose during the 1980s among a group of scholars who realized that improvement in environmental policy & management and protecting the well-being of future generations depended on bringing these aspects together (Constanza et al., 2007). In this sense, ecological economics aim to improve and expand economic theory to integrate the earth's natural systems, human values, human health and well-being. It is based on strong sustainability, thus considering that capitals are not exchangeable.

In contrast with conventional economics, whose primary goal is to increase goods and services produced by human industries (built capital), assuming that ever-increasing GDP is desirable, possible, and that it benefits everyone, ecological economics take on a broader perspective. They recognize that human well-being also implies health and education (human capital), friends and family (social capital) and the contribution of the earth and its biological and physical systems (natural capital). Their goal is to develop a deeper scientific understanding of the complex associations between human and natural systems, and to use this understanding to develop effective policies that will lead to sustainability. In this regard, ecological economics seek true economic efficiency and good economic decision making through the consideration of all cost and benefits within the price of products and services, with the inclusion of externalities (Sustainable Path Foundation, 2008).

Ecological economics are still developing with the contribution of multidisciplinary professionals concerned with strong sustainability. While ecological economists are certainly diverse, the largest "cluster" works from the initial premise that the earth has a limited capacity for sustainably supporting people and their artifacts, determined by combinations of resource limits and ecological thresholds. To keep the economy operating sustainably within these limits, specific environmental policies need to be established (Constanza et al., 2007).

3. Biofuels and Microalgae Promise

It is accepted that using fossil fuels has caused global warming, whose effects include a potential increase in sea level and subsequent submerging of lowlands, deltas and islands, changing weather and extreme climatic events, as well as disturbances to the ecosystem patterns (Stocker et al. 2013). Therefore, global energy supply should be diversified, and where possible replaced with renewable and clean energy production processes to reduce GHG emissions. Bioenergy is the largest source of renewable energy that can provide heat and electricity, as well as transport fuels. The biomass-based energy is the oldest source of consumed energy known to mankind, and still accounts for roughly 13% (2011) of total primary energy supplied. Biofuels and waste energy recovery account for about 76%. Only 3.4% of this percentage is equivalent to liquid fuels such as biodiesel and ethanol, but it has been increasing 10% annually from 1990 to 2011 (IEA, 2013), which indicates that there is still potential for development in the biofuel production sector.

a. Classification

The term biofuel includes all fuels generated from potentially renewable resources, which can be derived from forest products, agriculture, aquaculture and waste. They are classified according to their physical state as: solid, such as wood pellets or charcoal; liquids such as biodiesel and ethanol; or as biogas. Another classification divides them into primary or secondary and unprocessed or processed. In turn, processed fuels are classified into those of first, second, and third generation. Figure 4 shows a classification according to raw materials and production process.

First Generation Biofuels

First generation biofuels are produced using conventional fermentation technology and transesterification. Common raw materials for these are seeds, whole grains, and crop plants such as corn, sugarcane, rapeseed, wheat, and sunflower, among others.

The main disadvantage of using first generation crops in biofuel production is that they compete with the demand for food (Timilsina 2010). On the other hand, it has been reported that first-generation biofuels result in low reduction levels of GHG emission compared with fossil fuels. They also have high production costs and in the long run could replace fossil fuels only to a modest level due to the high requirement of land for their production (Cherubini et al., 2009).

Currently, vegetable oils using conventional conversion technology that have attained profitable levels of production are 1st generation biofuels. In fact, edible vegetable oil has been the main source for biodiesel production. More than 95% of the world's biodiesel is produced on large scale from the agricultural industry. This has raised questions and controversies both for its possible impact on global food security and its competition in the use of productive soils (Balat et al, 2011). It has been estimated that about 1% of the world's available arable land is used for the production of biofuels, accounting for 1% of global transport fuels. This implies that to cover 100% of the supply of transport fuels with first generation biodiesel is unrealistic owing to the large areas of production land required. Thus, it is considered that raw materials of second and third generation represent better options for biofuel production.

Second Generation Biofuels

Second generation biofuels are produced from a variety of sources and inedible lignocellulosic biomass waste such as stalks of wheat, corn stover, wood, jatropha, castor beans, among others. Lignocellulose (cellulose, hemicellulose and lignin) is the main and most abundant component in the biomass produced by photosynthesis. Lignocellulosic materials have a more complex structure, and unlike the first generation, they require a special treatment before becoming biofuels. The technologies involved in the production of 2nd generation biofuels are the same as those used in the 1st generation, that is, biochemical conversion routes as hydrolysis and anaerobic digestion, but also include thermochemical routes such as gasification and pyrolysis (Singh Nigam 2010). Some advantages of the second generation materials with respect to the first are:

- They do not compete with land use for food production, since waste from lignocelullosic food crops is used.
- Increased efficiency, since they use the whole crop products and not just the seeds.
- A flexible technology can be used, which allows processing more sources of raw materials, many of which are not dependent on climate conditions as is the case of first-generation sugar crops.

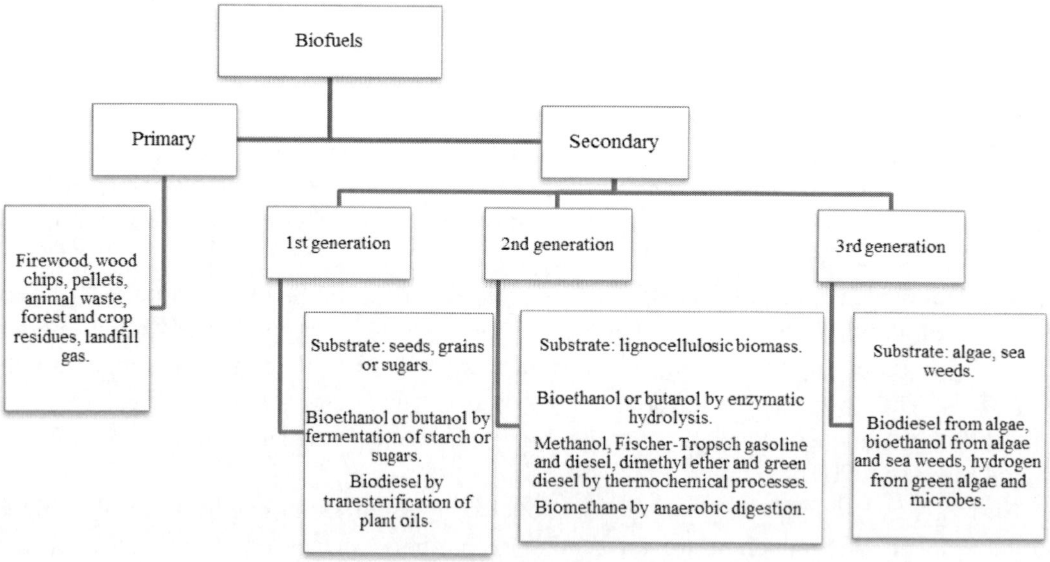

Figure 4. Classification of biofuels.Source: Singh Nigam, (2010).

While, second generation biofuels have clear advantages over first generation ones, the conversion technology is limited to small sectors, and it is not currently applied in large-scale commercial exploitations (Brennan et al., 2010).

Third Generation Biofuels

Finally, third generation energy sources are products specifically designed or adapted by various techniques to improve the conversion of biomass. Microalgae are an example of third generation raw materials. From these microorganisms, biodiesel, ethanol, hydrogen, and biogas can be readily produced (Vieira 2010).

It has been reported that a given area of cultivation of microalgae can produce about 10 to 100 times more lipids than any other oilseed crop. Whereas a terrestrial crop cycle takes from 3 months to 3 years to be exploited, algae begin to produce lipids between days 3 and 5 of culture, and can therefore be harvested daily. It has been reported that microalgae are the only raw material for generation of oils that could, according to its productivity, replace fossil-based diesel (Chisti, 2008).

The algal biomass is cultivated for processing with biological and thermochemical methods. Thermochemical methods include direct combustion to generate electricity, heat, and mechanical power. Biological methods include fermentation to produce energy carriers such as hydrogen, ethanol, and biogas, or the extraction of oils to produce biodiesel (Vieira 2010). The final stage of any process is, of course, combustion, which generates usable energy and CO_2, the latterto be again absorbed in the biomass growth cycle.

Several authors (Chisti 2007, Lardon et al. 2009, Stephenson et al. 2010, Campbell et al. 2011, Lee, 2013) have proposed the cultivation of microalgae as a system for CO_2 uptake as well as for the production of biodiesel and co-products based on several assertions about these microorganisms: a) photosynthetic efficiency greater than that of terrestrial plants, b) high growth rate, biomass doubling between 8 and 24 h, c) high lipid content, 20-70% in certain strains, and a higher proportion of triacylglycerol, essential for efficient biodiesel production, d) high performance in lipid production per hectare, e) direct biofixation of CO_2, thus being able to generate synergies of anthropic CO_2 consumption, f) potential large-scale production throughout the year, g) no competition with terrestrial plants for food production, h) generation of valuable co-products, and i) possibility of using waste water to provide nutrients for the crop, reducing process consumables and helping to treat domestic or industrial wastewater.

It is worth emphasizing that microalgae can be grown both in salt, fresh, waste and brackish water and on non-productive land, so there is great interest in promoting the use of algae for biofuel generation and provide an end to the debate on the cultivation of food and energy crops. Besides, microalgae culture does not require herbicides or pesticides under the conditions studied thus far and they produce beneficial valuable co-products such as proteins and polyunsaturated fatty acids. From biodiesel production, "waste to energy" technology can be used for residual biomass from extracted oil cake, both by direct thermal exploitation, and by producing added value biofuels as ethanol or methane from fermentation processes (Demirbas & Fatih., 2010; Lee, 2013).

Microalgae are emerging as one of the more promising sources for biodiesel production given several of their properties such as photosynthetic efficiency, their lack of competition with food crops, and their capacity to use multiple sources of carbon such as the CO_2 from industrial emissions (Collet et al. 2010).

b. Impact from the Production Process

This section contains an overview on environmental, social and economic impact, both global and local, concerning biofuel production.

It is worth mentioning that according to Rennings (2000), the most common problem areas in sustainability concepts are greenhouse gas effect, ozone layer depletion, acidification, eutrophication, toxic impact on ecosystems, toxic impact on humans, loss of biodiversity, use of soil or land and resource use. Within these areas, the sectors mainly involved are: energy, mobility and waste. Most of these concepts are related to environmental studies, although sustainability concerns take a broader approach considering also social and economic issues. However, among 35 social, economic and environmental criteria related to bioenergy, indicators on greenhouse gases balance and energy balance rank as the most important for expert on the subject (Markevicius et al., 2010).

Environmental

First and second generation raw materials for biofuel are mostly associated with impacts for the agriculture sector, derived from the production of raw materials. Such impacts relate to biodiversity loss, land use change, water management and use of chemical fertilizers and pesticides, all of which in some way have a relation to climate change and resource depletion.

When establishing a large area of cropland dedicated to grow bioenergy crops in large commercial plantations priority is given to monocultures, which can have negative effects on the local ecosystem, diminishing the production of local crop varieties and thus impacting agrobiodiversity.

Land use change can also cause loss of biodiversity when, for example, grassland is converted to cropland, having an effect on the quantity of carbon stored in the soil and in the above and below ground biomass that accompanies it. The former contributes to the release of carbon and nitrogen to the atmosphere and thus to climate change. It has been reported that production of biofuels at large scale could lead to the increase of GHG emissions. For example, conversion of peat forest to cropland for production of palm oil could emit 3,452 ton CO_2/ha, thus requiring 423 years to "pay the carbon debt" (Fargione et al., 2008).

Water management can also be an important issue, since bioenergy crops, as any crop, could imply pressure on aquifer levels leading to salinization or groundwater contamination from inadequate fertilization practices. It is expected that water requirements for bioenergy will increase. According to Gerbens-Leenes et al. (2012), in many countries biofuel water footprints, thus the direct and indirect water use over the entire supply chain, significantly contribute to surface and ground water scarcity. The same authors refer that water footprint of energy from biomass is nearly 70 to 700 times larger than that of fossil fuel.

Concerning the conversion stage, energy use becomes a very important factor to determine efficiency indicators. The energy balance of a biofuel production system can be quantified based on the comparison between the amount of energy needed in every production stage and the energy embedded in the biofuel (Khoo, et al., 2011). A general rule should be that the production of a unit of biofuel in terms of usable energy has to imply the consumption of less than that unit from fossil derived energy. Impacts related to the production of fossil derived energy are, at least, atmospheric emissions from the burning of fuels and the depletion of non-renewable resources.

Concerning third generation biofuel raw material, including microalgae, a potential impact is whether genetically engineered microalgae would be able to survive in the wild and multiply. Biologists worry that any transgenes that genetically modified strains contain to enhance their growth and strength could be transferred to other species in a way that could upset a fragile ecosystem (Caldwell, 2012). Another issue surrounding biofuel production from microalgae is the large use of water for its culture.

Until now, impacts have been described from a negative view; however all of them can be diminished given local circumstances. Specifically for first and second generation biofuels in the context of agricultural production, means of management based on ecosystem approaches can even enhance soil health and water efficient use. As an example, it has been reported that bioenergy crops cultivated under zero-tillage practice can lead to carbon storage (Smith, et al., 2007). Recently, the concept of Climate Smart Agriculture (CSA) has been proposed as a means of reducing agriculture´s impact on ecosystems parallel to improving yields (FAO, 2013). Concerning water use for third generation biofuels, quantifying the direct component of water demands can help determine the impacts of biofuel production on regional water resources, and therefore is an important criterion for evaluating optimal locations for algal cultivation (Zaimes & Khanna, 2013). Another way of diminishing the impact would be to use water resources with few competing uses, such as seawater and brackish water from aquifers, and re-circulating water. The later has the potential to also reduce nutrient loss, but comes with a greater risk of infection and inhibition from bacteria, fungi and viruses which are found in greater concentrations in recycled waters, along with non-living inhibitors such as organic and inorganic chemicals and remaining metabolites from destroyed algae cells (Slade & Bauen, 2013).

Social and economical

According to the Food and Agriculture Organization of the United Nations, although production of first generation biofuels could promote energy access in rural communities, providing support to economic growth for producers and long-term food security, there is also a risk that it could increase the cost of food and severely affect the poorest of the world (FAO, 2012). Concerning the impact from third generation biofuels, no studies on the subject have been published at this moment. This is a subject of potential and necessary analysis.

c. Potential of Biofuels from Microalgae

The use of microalgae as raw material for large scale biodiesel production is still under development (Sander et al. 2010, Campbell et al. 2011, IPCC 2011). The biodiesel production process from microalgae implies their cultivation and the extraction of their oils, and both processes are currently under research at a worldwide level to improve technical efficiency (Torres, L. Bandala E. (eds), 2014).

At the research level there are different technologies for biofuel production from microalgae under development:

- Co-processes in microalgae production: Bio-mitigation of CO_2 emissions with microalgae, waste water treatment with microalgae, and waste water use in microalgae culture (Brenan, 2010).

- Types of Microalgae culture systems: Open and closed-culture systems; Batch and continuous operation systems (Mata, 2010)
- Recovery of microalgal biomass (Brenan, 2010):
  - Harvesting methods: flocculation and ultrasonic aggregation; harvesting by flotation; gravity and centrifugal sedimentation and biomass filtration.
  - Extraction and purification of microalgal biomass: dehydration processes.
- Algal biofuel conversion technologies (Brenan, 2010):
  - Thermochemical conversion: gasification; thermochemical liquefaction; pyrolysis and direct combustion.
  - Biochemical conversion: anaerobic digestion; alcoholic fermentation; photobiological hydrogen production, etc.

At plant production level, currently there is only one closed tubular photobioreactor in the world operating commercially and producing high value (>$100,000/t) biomass. It is in Israel and is working with *H. pluvialis* (Benemann 2013).

The difficulties associated with efficient biodiesel production from microalgae are related to finding a strain with high lipid content, high speed of growth, easy harvesting, and positive cost-effectiveness (Demirbas & Demirbas., 2010). The problem is that nowadays there isn't a strain that gathers all the requirements (Klein-Marcuschamer 2013). There are other problems for the large-scale production such as insufficient low-cost concentrated CO_2 availability and inability of sunlight to penetrate most of the volume of a culture pond. Besides these, other challenges are finding a CO_2 emitting facility in a climatic region suitable for growing algae and with the necessary land and water resources required for algal culture, implementing an efficient nutrient recycling process (specially for P and N), and recovering energy from the spent biomass (Chisti 2013).

Although there are still a broad issues to overcome, the versatility of biofuel production from microalgae may provide answers to both the economic hurdles and the technological resolution of large-scale production. As an example, the combined biorefinery concept can be applied synergetically, linking the different stages of production and exploitation from microalgae, along with their added value co-products, and energy recovery from waste throughout the production chain.

4. Case Study: Life Cycle Assessment of Biodiesel from Microalgae

This case study is based on the production of biodiesel from microalgae that is currently being developed in Mexico at Instituto Politécnico Nacional. This section includes the description of the system studied, the methodology used to do the LCA and results. Although these were focused on GHG emissions, other impact categories were added to improve the environmental profile of the product, as shown in section (c).

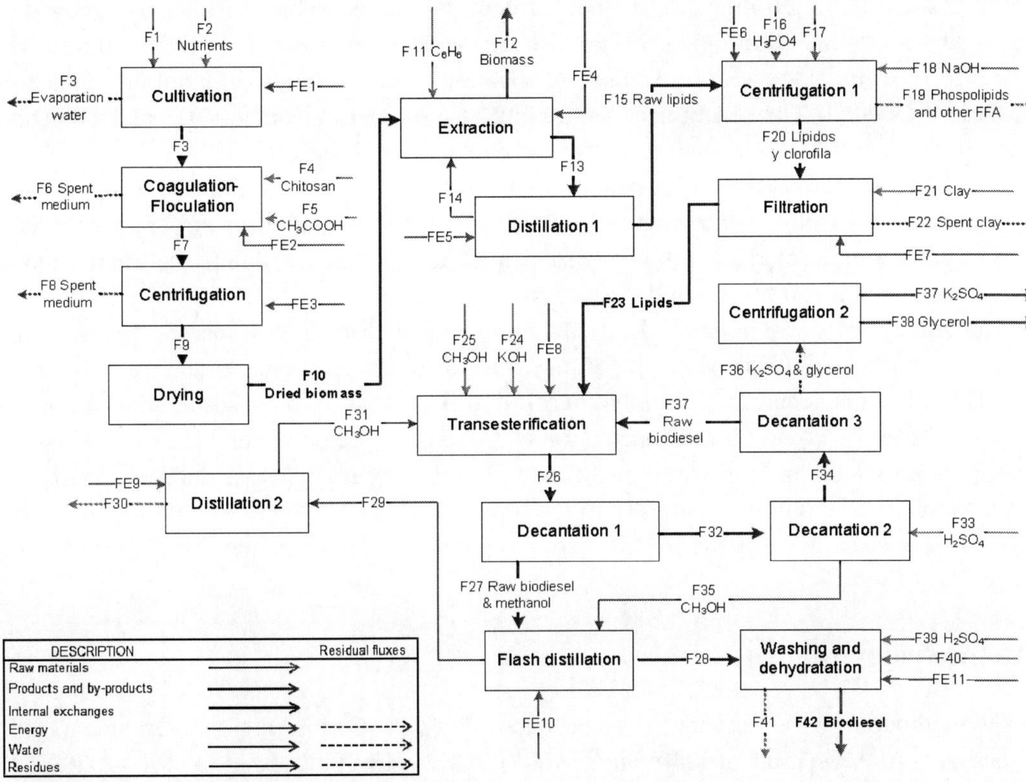

Figure 5. Process diagram of biodiesel production from microalgae.

a. Process Description

The system under study is presented in Figure 5. Unit processes, and both inflows and outflows of raw materials, waste and energy can be observed. Biodiesel is the main product, while residual biomass after extraction, potassium sulfate and glycerol are considered by-products.

During the culture stage, biomass is grown in a raceway reactor inside a greenhouse. The inoculum is previously grown in smaller reactors at laboratory level. Water loss from evaporation is constantly replaced during this stage. When biomass reaches the desired concentration, along with the spent medium, it is submitted to a coagulation-flocculation stage adding chitosan and acetic acid. When the flocculated biomass settles, the spent medium is discarded and biomass is centrifuged to reduce water content. After that it is dried in a conventional electric furnace.

Dried biomass is subjected to lipid extraction using hexane. The solution is then distilled to recycle remaining hexane. On the other hand, extracted lipids are mixed with water and phosphoric acid to remove hydrophilic phospholipids. The aqueous phase is then separated from the mixture by centrifugation and discarded. Subsequently, lipids are treated with sodium hydroxide to eliminate remaining free fatty acids, which are also removed by centrifugation. The resulting lipids still are filtrated with clay in order to remove chlorophyll.

The refined lipids are sent to a transesterification reactor where methanol and potassium hydroxide are added. The reaction is carried out at a temperature of 60 ° C and 60-80 bar. Produced raw biodiesel is decanted to obtain two phases: biodiesel with methanol on one side, and a mixture of catalyst, glycerol, methanol, biodiesel and free fatty acid on the other. The crude biodiesel phase with methanol is sent to a flash distillation process in order to obtain a purer biodiesel that is then washed and dehydrated to obtain the final product. Methanol is distilled and reused in the process.

Concerning the second phase from the first decantation after transesterification, it is treated with sulfuric acid to neutralize the potassium hydroxide and remaining unesterified free fatty acids, thus generating more biodiesel. In this process, two phases are created: a light one that is mainly composed of methanol (F35), and a heavy one composed of water, glycerol, biodiesel and solid potassium sulfate (F34). The heavy phase is decanted in order to recover more crude biodiesel to be sent to the transesterification reactor, while the mixture of glycerol and potassium sulfate is centrifuged to separate both by-products.

b. Methodology

The methodology used in this LCA was based on two ISO (International Organization for Standardization) standards (14040 and 14044) (ISO 14040 2006; ISO 14044 2006) in addition to the references of Guinee (2004), and guidance from the USEPA (2006). The steps involved are: (i) defining the scope and purpose, (ii) inventory analysis, (iii) environmental impact evaluation, and (iv) interpretation. The initial phase establishes what is to be achieved with the LCA, which is stated in the objectives of the study. Subsequently, three concepts are defined: (i) the functions of the system, which identifies the characteristics of the product and its scope, (ii) the functional unit is defined as the quantification of the identified functions (performance characteristics) of the product, this provide a reference to which the inputs and outputs are related (ISO 14040), and (iii) the limits of the system under study, and the criteria that define them.

The implementation of the LCA should account for all inputs, including upstream processing, and analysis completed by computing each of the possible outputs of the system, which may include the use and end of product life. The completion of this stage is concreted with the life cycle inventory (LCI). This is an iterative process, which can be adjusted as necessary to complete it successfully. The data, once validated, are linked to the functional unit to proceed after the computation of common variables. With the completed database, it continues with the evaluation of the life cycle impact assessment (LCIA). The different models proposed for the impact categories result in conversion factors that are applied to the values of the LCI, called characterization, and the obtained value is defined as an impact category indicator (see Figure 6).

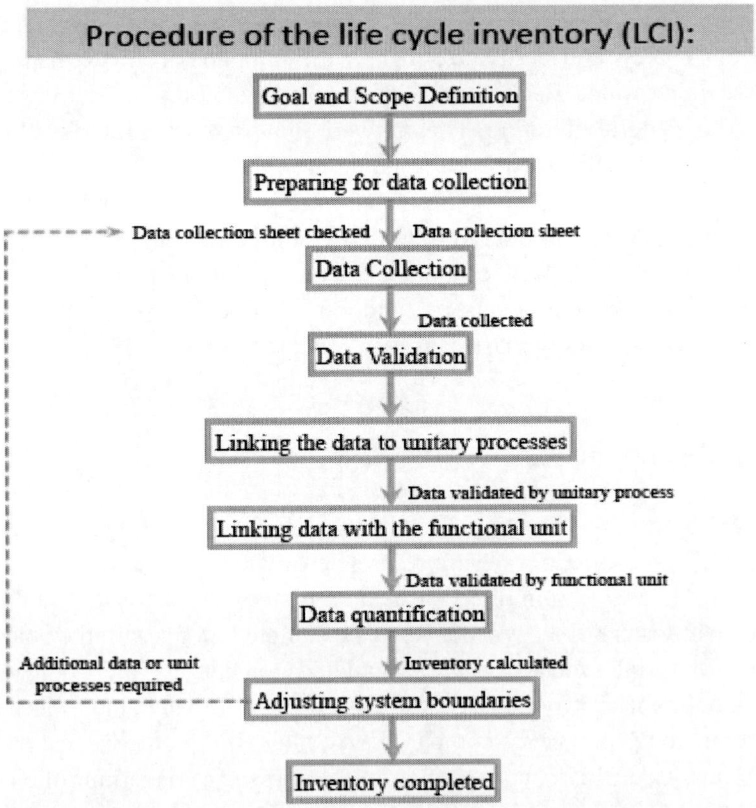

Figure 6. Descriptive procedure of the actual LCA (adapted from ISO 14044).

Below are described the characteristics of the LCA study according to the life cycle methodology:

Scope and goal. The scope of this study was to quantify the environmental impact associated to the biodiesel production from algae oil by a transesterification process, and to compare the results with other LCA studies on biodiesel production technologies.

Functional unit. The system function was set as the production of biodiesel at the gate of plant; therefore, the functional unit (FU) was defined as 100 MJ of energy produced by biodiesel considering a calorific value of 38 MJ kg^{-1}.

System boundaries and allocation. The boundaries for the production system were established on the basis of a cradle-to-gate principle according to the following criteria:

- The LCI was compiled and quantified from the stages of: (i) microalgae cultivation and oil extraction, (ii) mining, milling, and pre-manufacturing of other raw materials, (iii) upstream energy flow supplied to the system, and (iv) biodiesel production plant, based on conventional technology of transesterification process.
- The use and end of product life were not included. Both the supplies transport and capital goods were left out of the scope of the study.

- This study was focused on primary data provided by specialized databases such as Ecoinvent (v2.2), and cultivation data were obtained from experimental studies of our research group. Background information is compiled based on information provided in general literature. The software tool used as an interface between the LCI and LCIA was the Simapro® (v.7.3) software. The method used was EDIP 2003.
- Our study applied "load allocations", defined as the distribution of flows, out of a process, between the product system under study, and one or more different product systems. The ISO (2006), establishes separate inputs and outputs of the system between its different products or functions that reflect physical causality between them. For this case study mass causality was used.

c. Results and Discussion (2)

As mentioned before, the functional unit of the study was defined as 100 MJ of energy produced from the biodiesel. Considering reference flows, the required amount of cultivated biomass was estimated as 24 kg in dried base and which corresponds to 46 m^3 of culture (wet biomass and nutrient water). Energy balance was calculated along with the characterization of impacts. Carbon footprint scenarios were calculated considering the improvement of lipid concentration and biomass productivity for *Chlorococcum humicola* microalgae. Finally, results on carbon footprint were compared to other literature values. Although other categories were analyzed, this chapter cover only the results concerning the climate change category.

Energy balance

From the inventory construction stage it was determined that 2,112 MJ are needed to obtain the functional unit of study. Considering this, the balance between used and generated energy is clearly negative and thus inefficient. For the balance to be positive, the ratio between consumed and produced energy must be equal to 1. Presently, this value is 0.047. In Figure 7, the distribution of energy use is presented. It is observed that over 95% of it comes from the culture, dowstreaming and drying stages.

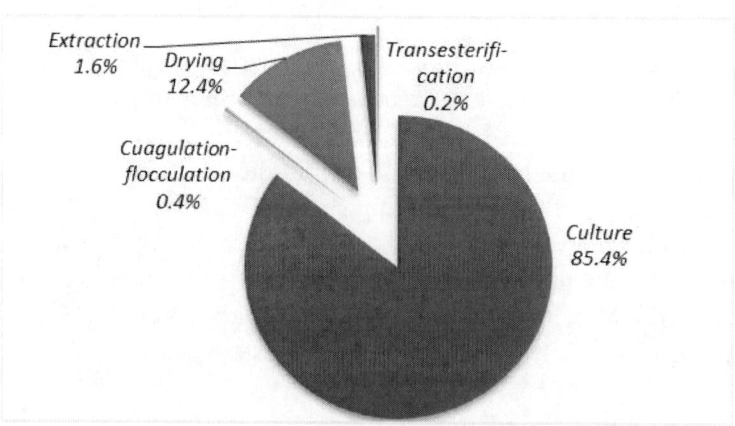

Figure 7. Percentage distribution of energy consumption in the production process.

Table 1. Impact caraterization from the production of 100 MJ of biodiesel

Impact Category	Associated impact		Impact Category	Associated impact	
Global warming 100a	0.55	kg CO_2 eq	Ecotoxicity water (chronic)	5.96	m^3
Ozone depletion	1.31E-07	kg CFC11 eq	Ecotoxicity water (acute)	-9.88E-03	m^3
Acidification	-1.23	m^2	Ecotoxicity soil (chronic)	64.82	m^3
Terrestrial eutrophication	-8.83E-05	m^2	Hazardous waste	17.10	Kg
Aquatic eutrophication EP(N)	-1.44E-03	kg N	Slags/ashes	-6.89E-01	Kg
Aquatic eutrophication EP(P)	-1.01E-02	kg P	Bulk waste	7.58E-02	Kg
Human toxicity air	-3.11E-05	m^3	Radioactive waste	7.27E-05	Kg
Human toxicity water	1.46E-04	m^3	Resources (all)	-3.78E-03	Kg
Human toxicity soil	-104.52	m^3			

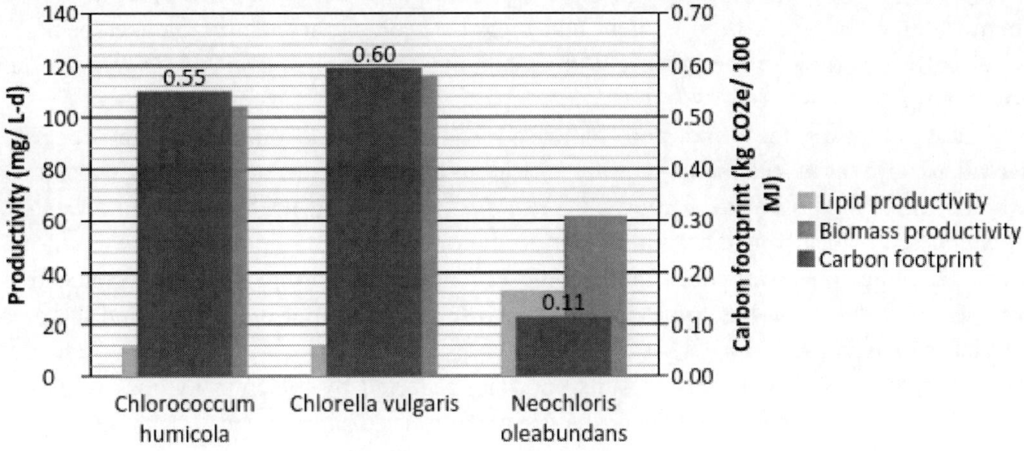

Figure 8. Microalgae lipid and biomass productivity relation to carbon footprint.

Energy consumption in the culture stage is related to motor efficiency of the raceway ponds used to move the paddlewheels and provide aeration. Currently, as the process is being developed at pilot scale, equipment is over-dimensioned and thus energy efficiency is low because the primary function is to evaluate strain growth. The amount of biomass that has to be recovered per unit of biofuel produced also affects the amount of energy employed, since it depends on lipid productivity. A higher lipid concentration results in less needed biomass, and thus less energy consumption per unit of product. Most LCA studies consider high lipid productivity between the ranges of 25-70%. The value considered in this study is 11% and it comes from real experimental data. As a way to improve energy balance, it was determined that it is necessary to increase the efficiency of the motors used and the lipid productivity.

Impact characterization

Results for the impact characterization of the production process are shown in Table 1. Negative values correspond to avoided impacts on the categories and come from the generation of by-products (glycerine and potassium sulphate from the transesterification stage) that substitute the production of this same products from virgin raw materials.

It is important to mention that most of the impact comes from the use of electricity. In the culture stage, which supposes the highest energy input, 67% of the GHG emissions come from this source. For the calculation, the energetic power mix from Mexico was used. This implies that the fossil fuel sources are high; therefore GHG emissions increase in all production stages. Use of mineral salts for the culture medium is the second source of impact in this stage. In this sense, in order to minimize the impact, IE criteria must be applied to valorize the spent culture medium from the coagulation-flocculation stage and diminish the use of salts. To achieve this, several technical barriers must be overcome since recirculation of spent medium supposes the increase of culture contamination.

Carbon footprint scenarios

Considering the importance of lipid productivity for the improvement of the process, three scenarios were considered in carbon footprint estimation. Different but real experimental values from lipid content and biomass productivity of microalgae species that are currently grown with the same culture medium as *Chlorococcum humicola* were taken into account.

Figure 8 shows the lipid and biomass productivity for three different species of microalgae grown in the same culture medium and with different lipid concentration: *Clorococcum humicola* (11% w/w), *Chlorella vulgaris* (10% w/w) and *Neochloris oleabundans* (53%). It was determined that lipid productivity and lipid content were the most influential indicators over the final biodiesel carbon footprint. As lipid concentration decreases, more biomass has to be cultivated and thus more input is required and the carbon footprint increases.

A comparison between these results and data reported by different authors is shown in Figure 9. Large differences between studies are observed since results from LCA greatly depend on system limits and scope. However, as in the former study, most LCA studies confirm the potential of microalgae as an energy source but highlight the crucial inevitability of decreasing the energy and fertilizer consumption (Singh & Gu, 2010).

In this study, environmental impact is reduced mostly because of the consideration of generation of byproducts (glycerine and potassium sulphate) which substitute the usage of virgin materials. However, it is necessary to make adjustments in the models using more experimental data and considering an IE approach, since it has been reported that the successful commercial implementation of algal biofuel shall depend on the development of high-value co-products such as renewable polymers or pigments (Singh & Gu, 2010).

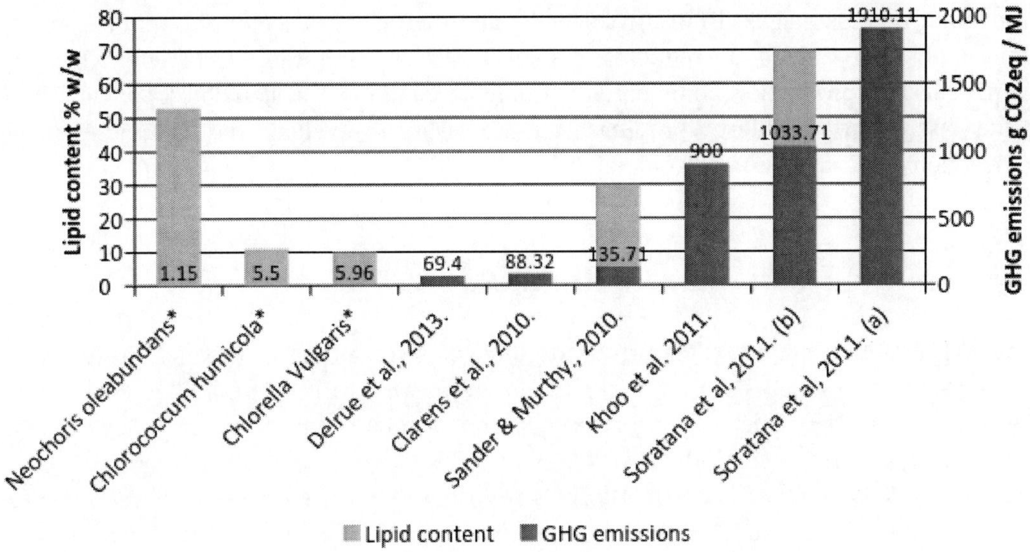

* Data from this work

Figure 9. GHG emissions from biodiesel production from microalgae.

CONCLUSION

Sustainability measurement tools have become important issue to assess finished products, commodities, services and including waste management. This has allowed promoting "green products", as well as to businesses and consumers can make more informed decisions in the marketplace. Likewise, the sustainability measurement can contribute in decision making to the energy analysts and the ecological economists to analyze the energy sources and the pathway to low-carbon economy. On the subject of study, the important factors in making a decision to establish a prospective large-scale bioenergy supply should focus on feedstocks that do not compete with edible crops but that contribute to diversifying energy sources and reduce stress on the depletion of conventional energy resources. The exploitation of microalgae as an energy source is not novel, nevertheless, with the increasing price of petroleum, the public aversion to nuclear power and the implementation of mitigation actions against global warming, there has been an upsurge in its research as a new source of renewable energy in the context of global supply.

Throughout the chapter it was stated that microalgae have a potential to produce biodiesel because of their many advantages as a sustainable feedstock. However, more research on the production stage to make the culture of microalgae efficient has to be developed, as was shown in the LCA study. The energy consumption represents a focal point where the process could still be unfeasible for industrial implementation in the current conditions. From the environmental perspective, there is an opportunity for researchers in this field to explore and to quantify other potential advantages of microalgae biomass and to further diversify more added value products to compensate, in a way, environmental loads generated in the production process stages. Both governments and the business sector should be involved in policies that are aimed at reducing capital cost and operating costs and which make microalgae fuel production commercially viable considering a sustainable approach. Finally, in the case studies of energy products, it is necessary to evaluate the use phase for better comparison with other products on equal functions.

REFERENCES

Balat, M. (2011). Potential alternatives to edible oils for biodiesel production–A review of current work. *Energy Conversion and Management, 52*(2), 1479-1492.

Benemann, J. (2013). Microalgae for Biofuels and Animal Feeds (Viewpoint). *Biotechnology and Bioengineering, 110*(9), 2319-2320.

Brennan, L. & Owende, P. (2010). Biofuels from microalgae—A review of technologies for production,processing, and extractions of biofuels and co-products. *Renewable and Sustainable Energy Reviews, 14*(2), 557-577.

Caldwell, E. / Ohio State University. Genetically Engineered Algae For Biofuel Pose Potential Risks That Should Be Studied [online]. 2012 [2014 August 28]. Available from: http://researchnews.osu.edu/archive/GEalgae.htm

Campbell, P., Beer, T. & Batter, D. (2011). Life cycle assessment of biodiesel production from microalgae in ponds. *Bioresource Technology, 102*, 50-56.

Carbon Trust. Carbon footprinting. The next step to reducing your emissions. 2012 [2014 August 28]. Available from: http://www.carbontrust.com/resources/guides/carbon-footprinting-and-reporting/carbon-footprinting.

Carrillo, G. (coord.) (2013). *La Ecología Industrial en México*. México DF: EdiUAM.

Cervantes, G. (2007). *Ecologia Industrial*. Barcelona: Fundació Pi I Sunyer.

Cervantes, G. (2013). Industrial Ecology for Climate Change Mitigation in *Energy and Environment: Management, Technology and Conflicts in a Warming World* (1st. Edition). EdiUOC: Barcelona.

Cervantes, G. & Ortega, M. (2013). Comparación de la huella de carbono de la producción de biodiesel a partir de diferentes microalgas. *Proceedings of the Primer Congreso Interamericano de Cambio Climático*. Viña del Mar, Chile, 28-30 octubre 2013.

Cervantes, G., Arce, J. & Mazza, L. (2010). Sustaintability Indicators for Industrial Ecology: Industrial ecosystems assessment. *Proceedings of the ISIE Asia-Pacific Meeting & ISIE MFA-ConAccount Meeting*. Tokyo (Japón), 7-9 noviembre 2010.

Chambers, N., Simmons, C. & Wackernagel, M. (2000). *Sharing Nature's Interest. Ecological Footprints as a indicator of sustainability*. New York: Earthscan.

Charles C., Gerasimchuk I., Bridle R., Moerenhout T., Asmelash E., Laan T./Global Subside Initiative. Biofuels-At what cost? A review of cost and benefits of EU biofuels policies [online]. 2013 [2014 August 28]. Available from: http://www.iisd.org/gsi/sites/default /files/biofuels_subsidies_eu_review.pdf.

Cherubini, F., Bird, N., Cowie, A., Jungmeier, G., Schlamadinger, B. & Woess-Gallash, S. (2009). Energy –and greenhouse gas- base LCA of biofuel and bioenergy systems: Key issues, ranges and recommendations. *Resources, conservation and recycling, 53*, 434-447.

Christy, Y. (2007). Biodiesel from microalgae. *Biotechnology Advances, 25*, 294-306.

Christy Y. (2008). Biodiesel from microalgae beats bioethanol. *Trends in biotechnology, 2*, 126-131.

Chisti Y. (2013) The Problems With Algal Fuels (Viewpoint). *Biotechnology and Bioengineering, 110 (9)*, 2318-2319.

Citynoise. (2008). Gini coefficients of selected countries, from publicly available data from the World Bank, Nationmaster, and the US Census Bureau [online]. 2008 [2014 August 28]. Available from: http://commons.wikimedia.org/wiki/File:Gini_since_WWII.gif.

Cleveland, C. & Ruth, M. (1998). Indicators of Dematerialization and the Materials Intensity of Use. *Journal of Industrial Ecology, 2*(3), 15–50.

Collet, P., Arnaud, H., Lardon, L., Ras, M., Goy, R. & Steyer, J. (2010). Life cycle assessment of microalgae culture coupled to biogas production. *Bioresource technology, 102*, 207-214.

Constanza, R., Norgaard, R., Daly, H., Goodland, R. & Cumberland, J. (2007). An Introduction to Ecological Economics : Chapter 2 [online]. [2014 August 28]. Available from: http://www.eoearth.org/view/article/150041/.

Cunningham, B., Battersby, N., Wehrmeyer, W. & Fothergill, C. (2004). A Sustainability Assessment of a Biolubricant. *Journal of Industrial Ecology, 7*(3-4), 179-192.

CO_2Now.org. 2014 [2014 August 28]. Available from: http://co2now.org/.

Daly, H. (1996). *Beyond Growth: The economics of sustainable development*. Boston: Beacon.

Demirbas, A., Demirbas, M. (2010). *Algae energy. Algae as new source of biodiesel*. London: Springer.

Dewulf, J. & Van Langenhove, H. (2005). Integrating industrial ecology principles into a set of environmental sustainability indicators for technology assessment. *Resources, Conservation and Recycling, 43*, 419–432.

ECOSIND project. *ECOSIND guide of recommendations for the planning and management of industrial zones through industrial ecology (2006*). Barcelona: Departament de Medi Ambient (Generalitat Catalunya).

Eggleston, H. S., Buendia, L., Miwa, K., Ngara, T. & Tanabe, K. (eds). 2006 IPCC Guidelines for National Greenhouse Gas Inventories [online]. 2009 [2014 August 28]. Available from: http://www.ipcc-nggip.iges.or.jp/public/2006gl/.

FAO. (2012). *A compilation of tools and methodologies to assess the sustainability of modern bioenergy*. Roma: Food and Agriculture Organization of the United Nations.

FAO. (2013). *Climate-smart agriculture sourcebook.* [without city]: Food and Agriculture Organization of the United Nations.

Fargione, J., Hill, J., Tilman, D., Polasky, S. & Hawthorne, P. (2008). Land clearing and the biofuel carbon debt. *Science, 319*, 1235-1238.

Gerbens-Leenes, P. W., van Lienden, A. R., Hoekstra, A. Y. & van der Meer Th. H. (2012). Biofuel scenarios in a wáter perspective: the global blue and green water footprint of road transport in 2030. *Global environmental change 22*, 764-775.

GFN Global Footprint Network. 2014 [2014 June 30]. Available from: http://www.footprintnetwork.org/en/index.php/GFN/..

GFN, Global Footprint Network. The National Footprint Accounts [online]. 2011 [2014 June 5]. Available from: http://www.footprintnetwork.org/en/index.php/GFN.

GIEI, Grupo de Investigación en Ecología Industrial. 2014 [2014 June 5]. Available from: http://gieiupibi.wordpress.com.

Guinée, J. B., Bruijn, H. de, Van Duin, R., Huijbregts, M. A. J., Lindeijer, E., Roorda, A. A. H., Van Der Ven, B. L. & Weidema, B. P. (2004). *Handbook on life cycle assessment: operational guide to the ISO standards*. Dordrecht: Kluwer Academic Publishers.

Hammond, G. (2007). Time to give due weight to the carbon footprint issue. *Nature, 445 (7125)*, 256– 256.

Hernández., E. & López, F. (2004). Cambio temporal en la huella ecológica de la región de Murcia y su uso como indicador de desertificación. *Papeles de geografía, 40*, 95-110, 2004.

Hertwich, E. & Peters, G. (2009). Carbon Footprint of Nations: A Global, Trade-Linked Analysis. *Environ. Sci. Technol., 43*, 6414–6420.

IEA [International Energy Agency] (2013). *Renewables Information (2013 Edition).* Paris: Organisation for Economic Co-operation and Development.

IPCC. Fuentes de energías renovables y mitigación del cambio climático. Resumen para responsables de políticas y resumen técnico [online]. 2011 [2014 August 28]. Available from: https://www.ipcc.ch/pdf/special-reports/srren/srren_report_es.pdf.

ISO 14040 (2006). Environmental management–Life cycle assessment–Principles and framework, International Organization for Standardization.

ISO 14044 (2006). Environmental management– Life cycle assessment–Requirements and guidelines, International Organization for Standardization.

ISO14064-1 (2006). Greenhouse gases -- Part 1: Specification with guidance at the organization level for quantification and reporting of greenhouse gas emissions and removals. International Organization for Standardization.

Khoo, H., Sharratt, P., Das, P., Balasubramanian, R., Naraharisetti, P. & Shaik, S. (2011). Life cycle energy and CO2 analysis of microalgae-to-biodiesel: preliminary results and comparisons. *Bioresource technology, 102*, 5800-5807.

Kitzes, J., Galli, A., Reed, A., Rizk, S. M., Ewing, B. & Wackernagel, M. (2010). *Guidebook to the National Footprint Accounts: 2010 Edition*. Oakland: Global Footprint Network.

Klein-Marcuschamer D. (2013). The Fine Print of Microalgal Biofuels (Viewpoint). *Biotechnology and Bioengineering, 110*(9), 2317.

Lardon, L., Hélias, A., Sialve, B., Steyer, J. & Bernard, O. (2009). Life cycle assessment of biodiesel production from microalgae. *Environmental science and Technology, 43*, 6475-6481.

Lee, J. W. (2013). *Advanced biofuels and bioproducts*. New York: Springer.

Martinez-Alier, A., Pascual U., Vivien F. D. & Zaccai, E. (2010). Sustainable de-growth:mapping the context, criticisms and future prospects of an emergent paradigm. *Ecological Economics, 69*(9),1741-1747.

Mata, T., Martins, A. & Caetano, N. (2010). Microalgae for biodiesel production and other applications: A review. *Renewable and Sustainable Energy Reviews, 14*, 217–232.

Markevicius, A., Katinas, V., Perednis, E. & Tamasauskiene, M. (2010). Trends and sustainability criteria of the production and use of liquid biofuels. *Renewable and sustainable energy reviews, 14*, 3226-3231.

Max-Neef, M. (1998). *El desarrollo a escala humana* (2nd ed.). Barcelona: Nordan e Icaria.

Meadows, D., Randers, J. & Meadows, D. (2004). Limits to growth. The 30 year update. Chealsea Green. 337.

MESVAL Proyecto, http://www.cimne.com/mesval/, consultado el 7 de diciembre de 2010.

Pearce, D., Atkinson G., Mourato, S. (2006). *Cost-Benefit Analysis and the environment: recent developments.* Paris: OECD.

Quijano, A. (2012). "Bien vivir": entre el “desarrollo” y la descolonialidad del poder [online]. 2012 [2014 August 28]. Available from: http://vientosur.info/IMG/pdf/ VS122_A _Quijano_Bienvivir---.pdf

Rebitzer, G., Ekvall, T., Frischknecht, R., Hunkeler, D., Norris, G., Rydberg, T., Schmidt, W. P., Suh, S., Weidema, B. P. & Pennington, D. W. (2004). Life cycle assessment Part 1: Framework, goal and scope definition, inventory analysis, and applications, *Environ.Int., 30*(5), 701-720.

Rennings, K. (2000). Redefining innovation - eco-innovation research and the contribution from ecological economics. *Ecological Economics, 32*(2), 319-332.

Sander, K. & Murthy, G. (2010). Life cycle analysis of algae biodiesel. *International Journal of Life Cycle Assessment, 15*, 704-714.

Schneider, F., Martinez-Alier, J. & Kallis, G. (2011). Sustainable degrowth. *Journal of Industrial Ecology, 15*, 654-656.

Schneider, H. & Samaniego, J. (2009). *La huella del carbono en la producción, distribución y consumo de bienes y servicios*. Santiago de Chile: CEPAL.

SEMARNAT, CESPEDES. Programa GEI México [online]. 2014 [2014 June 20]. Available from: http://www.geimexico.org/.

Singh, A., Singh Nigam, P. & Murphy, J. (2010). Renewable fuels from algae: An answer to debatable land base fuels. *Bioresource Technology, 102*, 10-16.

Singh, J. & Gu, S. (2010). Commercialization potential of microalgae for biofuels production. *Renewable and sustainable energy reviews, 14*, 2596-2610.

Smith, P., Martino, D., Cai, Z., Gwary, D., Janzen, H., Kumar, P. & Sirotenko, O. (2007). Agriculture. In B. Metz, O. R. Davidson, P. R. Bosch, R. Dave, & L. A. Meyer, *Climate Change 2007: Mitigation. Contribution of Working Group III to the fourth Assessment Report of the Intergovernmental Panel on Climate Chnage* (498-540). Cambridge: Cambridge University Press.

Slade R. & Bauen A. (2013). Micro-algae cultivation for biofuels: cost, energy balance, environmental impacts and future prospects. *Biomass and Bioenergy, 53*, 29-38.

Stahel, A. & Cendra, J. (2011). Desarrollo Sostenible: ¿sabemos de que estamos hablando?. *Revista Internacional de Sostenibilidad, Tecnología y Humanismo, 6*, 37-57.

Stephenson A., Kazamia E., Dennis J., Howe C., Scott S. & Smith A. (2010). Life Cycle Assessment of Potential Algal Production in d United Kingdom: A Comparison of Raceways and Air-Lift Tubular Bioreactors. *Energy Fuels, 24*, 4062-4077.

Stocker, T. F., Qin, D., Plattner, G. K., Tignor, M., Allen, S. K., Boschung, J., Nauels, A., Xia, Y., Bex, V., Midgley, P. M. (eds.). (2013). *Climate Change 2013: The Physical Science Basis. Contribution of Working Group I to the Fifth Assessment Report of the Intergovernmental Panel on Climate Change*. New York: Cambridge University Press.

Sustainable Path Foundation. (2008). Sustainable Path Foundation. White Papers [online]. 2008 [2014 August 28]. Available from: http://sustainablepath.org/wp-content/uploads /2008/11/what-is-ecological-economics-one-pag-pdf-080123.pdf

Tiejun, D. (2010). Two quantitative indices for the planning and evaluation of eco-industrial parks. *Resources, Conservation and Recycling 54*, 442–448, 2010.

Timilsina Govinda R. & Shrestha A. (2010). How much hope should we have for biofuels? *Energy, 36*, 2055-2069.

Torres, L. & Bandala, E. (coed.) (2014) Energy and Environment Nowadays. New York: Nova Science Publishers Inc. e-book.

Tyteca, D. (1998). Sustainability Indicators at the Firm Level. *Journal of Industrial Ecology 2 (4),* 61-77.

UN-CSD Commission for Sustainable Development. 2014 [2014 June 28]. Available from: http://sustainabledevelopment.un.org/csd.html.

UPC, Portal de Sostenibilidad. 2014 [2014 June 28]. Available from:, http://portal sostenibi lidad.upc.edu/.

USEPA [U.S. Environmental Protection Agency] (2006): Life cycle assessment: principles and practice, Report EPA/600/R-06/060.

Van Berkel, R. Industrial ecology & Cleaner production and eco-efficiency. In Marinova, D.; Annandale, D. Phillimore, J. *The international handbook on environmental technology management.* Cheltenham: Edward Elgar Publishing, 2006.

Vieira Costa, J. A. & Greque de Morais, M. (2010). The role of biochemical engeneering in the production of biofuels from microalgae. *Bioresource Technology, 102*, 2-9.

Wackernagel, M. & Rees, W. (1996). *Our Ecological Footprint: Reducing Human Impact on the Earth.* Gabriola Island: New Society.

WCED, World Commission on Environment and Development. Our Common Future. Published as Annex to General Assembly document A/42/427 [online]. 1987 [2014 August 28]. Available from: http://www.un-documents.net/wced-ocf.htm.

Xercavins, J., Cayuela, D., Cervantes, G. & Sabater, A. (2005). *Desarrollo Sostenible.* Barcelona: Edicions UPC.

Weidema, B., Thrane, M., Christensen,P., Schmidt, J. & Løkke, S. (2008). Carbon Footprint. A Catalyst for Life Cycle Assessment?. *Journal of Industrial Ecology, 12*(1), 3-6.

World Business Council for Sustainable Development (WBCSD) & World Resources Institute (WRI). The Greenhouse Gas Protocol. A Corporate Accounting and Reporting Standard. Revised Edition [online]. 2004 [2014 August 28]. Available from: http://www.ghgprotocol.org/files/ghgp/public/ghg-protocol-revised.pdf.

Wright, L., Kemp, S. & Williams, I. (2011). Carbon footprinting': towards a universally accepted definition. *Carbon Management, 2*(1), 61–72.

Zaccai, E. (2012). Over two decades in pursuit of sustainable development: influence, transformations, limits. *Environmental Development 1*, 79-90.

Zaimes, G. G. & Khanna V. (2013). Microalgal biomass production pathways: evaluation of life cycle environmental impacts. *Biotechnology for biofuels, 6*(88), 1-11.

Zvolinschi, A., Kjelstrup, S., Bolland, O. & van der Kooi, H. (2007). Exergy Sustainability Indicators as a Tool in Industrial Ecology. *Journal of Industrial Ecology, 11*(4), 85–98.

In: Microalgae and Other Phototrophic Bacteria
Editor: Luis G. Torres Bustillos
ISBN: 978-1-63482-078-3

Chapter 10

TECHNOLOGIES FOR THE PRODUCTION OF BIODIESEL FROM MICROALGAE: BIOMASS PRODUCTION THE MAIN CHALLENGE

Citlally Ramirez Lopez and Luis Fernandez Linares [*]
Unidad Profesional Interdisciplinaria de Biotecnología
del Instituto Politécnico Nacional, Mexico

ABSTRACT

Biodiesel is a renewable substitute for diesel fuel, it is considered CO_2-neutral because the amount of CO_2 produced during its combustion is the same that vegetable species absorb during their growth. Biodiesel is a mixture of fatty acid alkyl esters of long chain fatty acids obtained by alcoholysis or transesterification of triacylglycerols. The economically significant production of biodiesel from microalgae has been hailed as the ultimate alternative to depleting resources of petro-diesel due to its high cellular concentration of lipids. Microalgae present high solar energy yields, which leads to a superior lipid productivity, they present rapid growth rates and short generation times. In addition, the lipids produced are generally neutral.

There are technological, economical, supply, storage and safety related, as well as policy barriers linked to the development of biofuel production. Several conditions affect simultaneously the microalgae cultivation, such as light avaiability and intensity, land topography, climatic conditions, water supply and access to the carbon source and other nutrients. Furthermore, biomass yield also depends on the mode of cultivation (photoautotrophic, mixotrophic and heterotrophic production), types of culture (open or closed systems), culture strategies (batch or continuous culture), inhibitors concentration, mixing, dilution rate, depth and harvest frequency.

In this text, we present an overview of the factors that affect the production of the microalgal biomass yield and lipid production, both important factors to potentiate the feasibility of using microalgae as a source of oils to produce biodiesel.

[*] Corresponding author: lfernand36@gmail.com.

1. Introduction

The question of using potential food crops to produce biofuels has led to discussions about 'food *vs.* fuel' and the limitations of available agricultural land and fresh water. However, new generation of raw mateials for biofuels includes lignocellulosic residues and microalgae, which have received increasing attention net energy balance, more water related efficiency and and less or no requirement for a due to the advantages that these biofuel production systems present, such as a higher rable land (Chisti, 2007; Chen et al., 2012; Handler et al., 2012; Aguirre et al., 2013; Fon Sing et al., 2013; Ríos et al., 2013).

The proposal to use microalgae for biofuels production is not new and was first suggested in the 1940s-1950s. In 1970s, the large-scale cultivation of microalgae for producing sustainable liquid fuels was previously investigated and in the 1990s extensive research was carried out, especially at the Solar Energy Research Institute in Golden, Colorado, USA (Chisti, 2007; Schenk et al., 2008; Aguirre et al., 2013; Fon Sing et al., 2013). Specifically, microalgae production offers the potential for the production of high value compounds, wastewater treatment, nitrogen fixing, CO_2 mitigation and bioenergy production (Moheimani and Borowitzka, 2006; Chisti, 2007; Schenk et al., 2008; Salis et al., 2010; Das et al., 2011; Hadj-Romdhane et al., 2012; Mendoza et al,. 2013 a and b). In this manner, these microscopic organisms have been taken into consideration as a feedstock for the production of renewable biofuels such as bioelectricity, methane produced by anaerobic digestion of the algal biomass, biohydrogen produced under anaerobic conditions, bioethanol (sugar fermentation) and bioediesel derived from microalgal oil (Schenk et al., 2008; Feng et al., 2011; Mutanda et al., 2011; Hadj-Romdhane et al., 2012; Aguirre et al., 2013; Fon Sing et al., 2013).

The feasibility of using microalgae as row material for biodiesel depends on the culture systems, growth conditions, strains, productivities and oil content in microalgae. One of the main challenges in research and development of biodiesel from microalgae is the obtention of biomass at low cost and with a high oil content.

2. Microalgae and Biodiesel

2.1. Algae Classification

Algae are a polyphyletic group of organisms (including organisms that do not have the same origin, but are multiple, independent evolutionary lines) that comprises both, unicellular and multicellular forms and both, prokaryotes and eukaryotes, which are able to capture light energy through pigments (such as chlorophylls, carotenoids, anthocyanins, and phycoerythrinsphycobilins) that are necessary to carry out photosynthesis (Barsanti and Gualtieri, 2006; Salis et al., 2010). The classification of algae has five principal branches: chromista (brown algae, golden brown algae, and diatoms), red algae, dinoflagellates, euglenids, and green algae (Salis et al., 2010). Microalgae can be eukaryotic (Chlorophyta, Rhodophyta, Bacillariophyta) or prokaryotic (Cyanophyta) (Aguirre et al., 2013). Microalgae are a large group of microscopic algae considered primary producers on a global scale, and involved in all marine and freshwater ecosystems (Hernandez et al., 2009).

2.2. Composition of Microalgal Lipids

A fatty acid (FA) molecule consists of a hydrophilic carboxylate group attached to one end of a hydrophobic hydrocarbon chain (Figure 1). Fatty acids are constituents of lipid molecules neutral and polar, and are designated based on their two most important features 'the total number of carbon atoms in the hydrocarbon chain: the number of double bonds along the hydrocarbon chain'. Saturated fatty acids have no double bond, while unsaturated fatty acids consist of at least one double bond. When the carboxylate end of the fatty acid molecule is bonded to an uncharged head group (glycerol), a neutral lipid molecule calledtriacylglycerol is formed. On the other hand, the association of a fatty acid molecule to a charged head group (glycerol and phosphate complex) forms a phospholipid, a polar lipid molecule (Halim et al., 2012).

Lipids are any biological molecule that is soluble in an organic solvent. As mentioned above, most lipids contain fatty acids and can be classified into two categories based on the polarity of the molecular head group: neutral lipids, which comprise acylglycerols and free fatty acids (FFA) and polar lipids, which can be further sub-categorized into phospholipids (PL) and glycolipids (GL). Neutral lipids are primarily used in the microalgal cells as energy storage, while polar lipids pack in parallel to form bilayer cell membranes. Acylglycerol consists of fatty acids ester-bonded to a glycerol backbone and is categorized according to its number of fatty acids: triacylglycerols (TG), diacylglycerols (DG), monoacylglycerols (MG). FFA is a fatty acid bonded to a hydrogen atom. However, there are also some types of neutral lipids that do not contain fatty acids, such as hydrocarbons (HC), sterols (ST), ketones (K), pigments (carotenes and chlorophylls). Although these lipid fractions are soluble in organic solvents, they are not convertible to biodiesel (Halim et al., 2012).

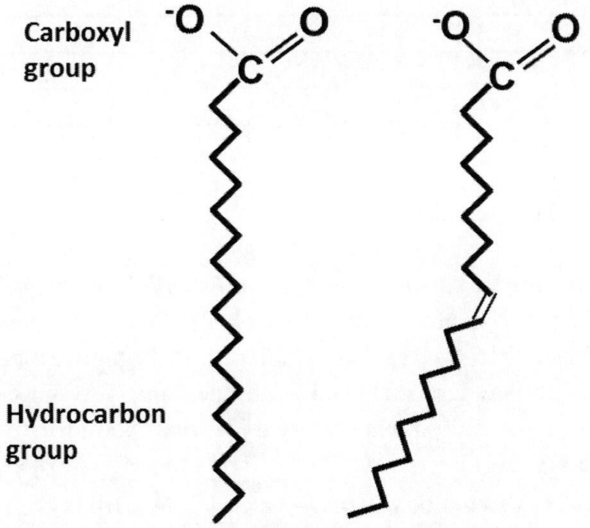

Figure 1. Fatty acids chains. On the left is presented a saturated fatty acid (C18:0 or stearic acid) and on the right is shown an unsaturated fatty acid (C18:1 or oleic acid).

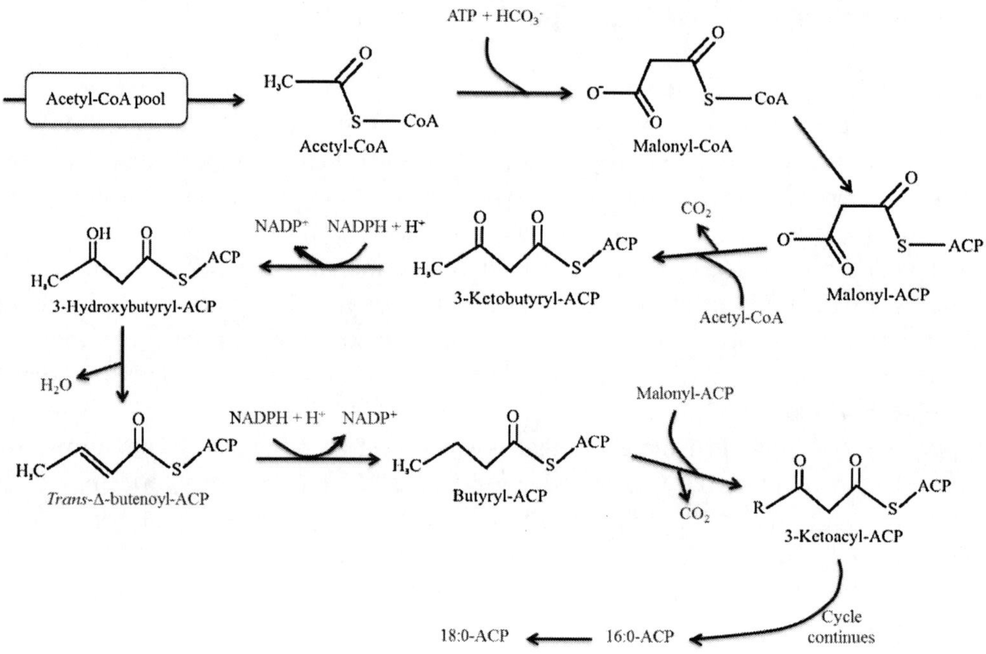

Figure 2. Synthesis pathway of fatty acids in chloroplast.

Table 1. Fatty acid composition of microalgal oil

Fatty acid	Chain length:n of double bonds	Oil composition (w/total lipid)	Cetane No.	Melting point (°C)
Palmitic acid	C16:0	12-21	74.5	62.9
Palmitoleic acid	C16:1	55-57	45	-0.1
Stearic acid	C18:0	1-2	86.9	69.6
Oleic acid	C18:1	58-60	55	14
Linoleic acid	C18:2	4-20	36	-5
Linolenic acid	C18:3	14-30	28	-11

(Schenk et al., 2008; Meng et al., 2009)

In microalgae, fatty acid FA and triacylglycerols (TAGs) biosynthesis pathways have been scarcely investigated, compared with higher plants. FA biosynthesis is performed in the chloroplast and produces a C16 or C18 fatty acids or both, which are then used to synthesize all lipid-containing cell molecules, such as membranes and TAGs as storage lipids, which primarily serve as a source of carbon and energy under deprived growth conditions (Hu et al., 2008; Schenk et al., 2008; Aguirre et al., 2013). The crucial step of FA biosynthesis is the carboxylation of acetyl CoA to malonyl CoA by acetyl CoA carboxylase (Figure 2) (Hu et al., 2008; Schenk et al., 2008).

Lipid production from microalgae can be performed by both, photoautotrophy (by photosynthesis) and heterotrophy. Table 1 shows the fatty acid composition of a typical oil from microalgae (Petkov et al., 2006; Meng et al., 2009), which is mainly composed of a mixture of unsaturated fatty acids such as palmitoleic acid (C16:1), oleic acid (C18:1), linoleic acid (C18:2), and the linolenic acid (C18:3) and saturated fatty acids such as palmitic (C16:0) and stearic (C18:0) are also present although to a small extent (Salis et al., 2010).

While *C. reinhardtii* is a model organism to study lipid biosynthesis in green algae, some unusual hydrocarbons and ether lipids from *Botryococcusbraunii* have also been described (e.g., n-alkadienes, trienes, triterpenoidbotryococcenes, methylated squalenes, tetraterpenoids and lycopadiene) (Schenk et al., 2008).

2.3. Microalgal Lipid for Biodiesel Production

In general, biodiesel is a renewable substitute for diesel fuel, this is, CO_2-neutral because the amount of CO_2 produced during their combustion is the same that the vegetable species absorbe during their growth. From the chemical point of view, biodiesel is a mixture of fatty acid alkyl esters of long chain fatty acids obtained by alcoholysis or transesterification of triacylglycerols (Figure 3) in the presence of a catalyst, either an acid (such as H_2SO_4 or HCl) or an alkali (such as NaOH or KOH) (Salis et al., 2010; Aguirre et al., 2013). Methanol is the most used alcohol, but ethanol and C3-C5 linear and branched alcohols might also be used (Anastopoulos et al., 2009; Ehimen et al,. 2010; Salis et al., 2010; Halim et al., 2012).

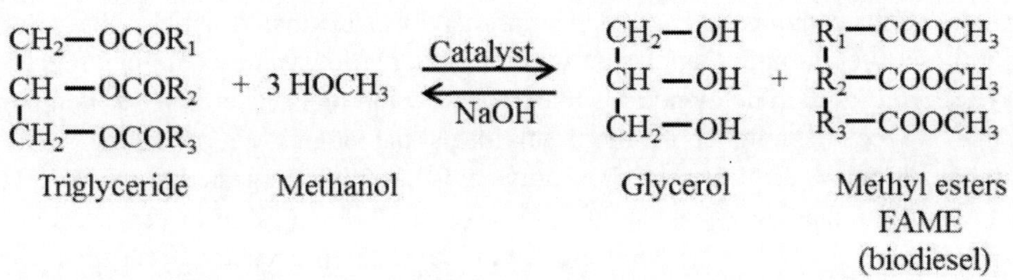

Figure 3. The chemical reaction for the production of Fatty Acid Alkyl Esters (FAAE).

Table 2. Lipid content of some microalgae

Microalga	Lipid content (% drywt)
Botryococcusbraunii	25-75
*Chlorella*sp.	28-32
Crypthecodiniumcohnii	20
*Cylindrotheca*sp.	16-37
Dunaliellaprimolecta	23
*Isochrysis*sp.	25-33
Monallanthus salina	>20
*Nannochloris*sp.	20-35
*Nannochloropsis*sp.	31-68
Neochlorisoleoabundans	35-54
*Nitzschia*sp.	45-47
Phaeodactylumtricornutum	20-30
*Schizochytrium*sp.	50-77
Tetraselmissueica	15-23

(Chisti, 2007)

Microalgal lipid productivity (the mass of lipid produced per unit volume of the microalgal broth per day) depends on the microalgae growth rate and the lipid content of the biomass (Chisti, 2007). Microalgae that present high lipid productivities are desired for producing biodiesel. The oil levels between 20 to 50% percent are commonly found (Table 2).

Lipid quality (chain length and saturation grade) is an important issue for biodiesel production, as the alkyl ester content dictates the stability and performance of the fuel, and this, in the end, is an important factor in meeting international fuel standards. Microalgal fatty acids range from 12 to 22 carbons in length and can be either saturated or unsaturated; however, they are predominantly polyunsaturated, then they have a low stability to oxidation which is a serious issue with biodiesel during the storage. It is therefore preferable that the level of polyunsaturated fatty acids in biodiesel is kept to a minimum. This drawback, however, can be corrected through partial catalytic hydrogenation of the oil although this would increase the costs (Chisti, 2007; Schenk et al., 2008; Salis et al., 2010; Halim et al., 2012). In contrast, higher levels of polyunsaurated fats lower the cold filter plugging point (CFPP), the temperature at which the fuel starts to form crystals/solidifies and blocks the fuel filters of an engine. Table 1 presents the melting point of the major fatty acids, it can be seen that the more unsaturated an oil is, the lower the melting point it presents. Therefore, colder climates require a higher unsaturated lipid content to enable the fuel to perform at low temperatures. Cetane number (Table 1) is another measure that describes the combustion quality of diesel fuel during compression ignition. In a particular diesel engine, higher cetane fuels have shorter ignition delay periods than lower cetane fuels. Therefore, it is important to ensure that the cetane number of biodiesel meets the engine cetane rating. With these considerations in mind, the "ideal mix" of fatty acids has been suggested to be 16:1, 18:1 and 14:0 – in regards of chain length and double bonds respectively (Table 1) – in a recommended ratio of 5:4:1, repectively. Such a biodiesel would have the properties of very low oxidative potential whilst retaining a good CFPP rating and cetane number (Schenk et al., 2008).

However, the composition and fatty acid profile of a lipid mixture is affected by the microalgal life cycle as well as the cultivation conditions, such as medium composition, temperature, illumination intensity, ratio of light/dark cycle, and aeration rate. Microalgal cells harvested during the stationary phase have lower polar lipid content than the same species obtained during the logarithmic phase (Sheng et al., 2011; Halim et al., 2012). In response to environmental stress like salinity, incident light intensity, extreme temperature or nitrogen deprivation, the yield of oil in several strains increases. Sometimes, in very few cases, phosphorus depletion also has the same effect, and this is characterized by the increase in synthesis of oil droplets (Salis et al., 2010; Halim et al., 2012; Aguirre et al., 2013; Campenni' et al., 2013). For diatoms, deprivation of silicon, which is a fundamental compound present in the cell wall, results in an increased yield of oil (Salis et al., 2010). It is important to highlight that the highest TAG formation rate is observed within the first hours or days of nutrient depletion and this formation rate gradually decreases thereafter (Klok et al., 2013).

Microalgae can produce about 10-300 times more lipid for biodiesel production than tradicional crops on an area basis (Schenk et al., 2008; Halim et al., 2012) and they have a short harvesting cycle depending on the process compared with convencional crop plants which are usually harvested once or twice a year. Greater light capture and conversion efficiencies lead to reduced fertilizer and nutrient inpunts and this in turn results in less waste and pollution. Microalgae are perceived to have the advantage of requiring less land area to produce an equivalent amount of fuel and the cultivation of microalgae can be performed on land that cannot be used for crop such as arid, drought or salinity-affected regions. Furthermore, microalgae require much less water than traditional crops and there are many species which can be grown using saline water and waste water (Chisti, 2007; Salis et al., 2010; Fon Sing et al., 2013; Mendoza et al., 2013 a).

3. Microalgae Cultivation

Several conditions affect simultaneously the microalgae cultivation, such as light avaiability and intensity, land topography, climatic conditions, water supply and access to the carbon source and other nutrients. Furthermore, biomass also depends on the mode of cultivation (photoautotrophic, mixotrophic and heterotrophic production), types of culture (open or closed systems), culture strategies (batch or continuous culture), inhibitors concentration, mixing, dilution rate, depth and harvests frequency (Aguirre et al., 2013). Therefore, the optimization of strain-specific cultivation conditions is required; however, there are many interrelated factors that can each be limiting. These include temperature, mixing, fluid dynamics and hydrodynamic stress, gas bubble size and distribution, gas exchange, mass transfer, light cycle and intensity, water quality, pH, salinity, mineral and carbon regulation/bioavailability, cell fragility, cell density and growth inhibition (Schenk et al., 2008). However, in this section only some topics will be described.

3.1. Nutrients Source

Microalgal cells consist mainly of carbohydrates, proteins and lipids, the actual composition of microalgae varies between species. Under optimum growth conditions, unicellular microalgae synthesize protein to maintain cell growth and carbohydrate and lipid are present in lesser quantities (Das et al., 2011).Optimal media formulation is also critical to ensure sufficient and stable supply of nutrients to attain maximal growth acceleration and cell density, and ultimately to produce biofuels at higher efficiencies (Schenk et al., 2008).

3.1.1. Carbon source

Most of microalgae use inorganic carbon as the carbon source for catabolism (photoautotrophic metabolism), but some species have the ability to grow in the presence of both, a fixed organic carbon and light, with a mixotrophic metabolism and even in the presence of organic carbon alone and darkness, with an heterotrophic metabolism (Barsanti and Gualtieri, 2006; Das et al., 2011). All three modes can be applied to closed bioreactor systems, but in open culture systems only photoautotrophic production is typically used (Das et al., 2011; Aguirre et al., 2013).

To achieve high biomass productivity, there is a need for addition of inorganic carbon to the culture of microalgae and most of the microalgae can import both, CO_2 and bicarbonate ion (HCO_3^-) through the cell membrane, therefore, pure CO_2 or that one from coal-fired power plants as well as HCO_3^- can be added to the medium because they have showed to improve biomass and lipid productivity (Prins et al., 1987; Kim et al., 2010; Chi et al., 2011; Kee Lam et al., 2013; Moheimani, 2013). It is important to highlight that microalgae growth in raceways is typically assumed to require 1.5-2.0 kg CO_2 kg^{-1} biomass with a range of CO_2 absorption efficiencies of 50-100% - assumend for different microalgal species and growth conditions (Handler et al., 2012).

According to the equilibrium reaction $H^+ + HCO_3^- <=> CO_2 + H_2O$, H^+ is consumed during the conversion of HCO_3^- to CO_2, and this CO_2 is ultimately fixed by Rubisco ezyme during photosynthesis. Thus, steady-state usage of HCO_3^- as the original carbon source for photosynthesis leaves OH^- in the cell, and this has to be neutralized by H+ uptake from the extracellular environment. The reduction of H^+ in the culture medium leads to an increased pH, which changes the equilibrium between different Ci species. The pKa of HCO_3^- in fresh water at 25 ºC and 1 atm is 10.33; thus, the acid/base bicarbonate/carbonate pair can act as a strong buffer around this pH (Kim et al., 2010; Chi et al., 2011; Moheimani, 2013).

On the other hand, microalgae such as *Chlorella* sp., *Chlorella protothecoides*, *Phaeodactylumtricornutum* and *Nannochloropsis* sp. have shown higher intracellular lipid when glucose and glycerol have been used as the carbon source (Das et al., 2011; Campenni' et al., 2013). Also, Acetic acid or carbonate have been used as carbon source in the actual commercial production of microalgal biomass (Hase et al., 2000).

3.1.2. Nitrogen source

Generally, all microalgae culture media contains nitrate ion as the nitrogen source, however, there are microalgae like *Chlorella vulgaris,* which is able to assimilate both, ammonium (NH_4^+) and nitrate (NO_3^-) as nitrogen source, with a preference for ammonium (Hadj-Romdhane et al., 2012). This affinity has been corroborated in the treatment of wastewater, in which only ammonium is utilized when the cultures contain both, nitrate and ammonium. Therefore, ammonium is depleted first in the culture, and then microalgal cell growth is carried out with nitrate as the nitrogen source until nitrate is depleted (Feng et al., 2011).

On the other hand, growth promotion of microalgae (*Chlorella vulgaris* and *Chlorella sorokiniana*) by plant growth-promoting bacteria, such as *Azospirillum* spp. and *Bacillus pumilus* (they are able to fix atmospheric nitrogen and release ammonium), has been reported (Hernandez et al., 2009). de-Bashan et al. (2002) reported an increase in pigments and lipid content, lipid variety, as well as the increased size of the cell and population of *Chlorella* spp. when it was co-immobilized in alginate beads with the microalgae-growth-promoting bacterium *Azospirillumbrasilenses.*

Microalgae production can also be a multi-phased process with each step set to have independent optimal conditions such as nitrogen limitation (low biomass productivity) in oil production or sulphur limitation in H_2 production (Schenk et al., 2008; Das et al., 2011; Santos et al., 2012; Klok et al., 2013). Particularly, in the absence of nitrogen, fixed carbon sources cannot be metabolized for protein synthesis and these results in a metabolic switch to intracellular lipid production. In this regard, there are reports about deprivation of essential nutrients to cyanobacteria, for which the intra-cellular lipid content showed a 2 - 3 fold increase (Das et al., 2011, Chu et al., 2013). *Chlorella vulgaris,* for example, when exposed to nitrogen starving conditions after normal nutrition, presented a 2 fold increase in the lipid content (Widjaja et al., 2009; Lv et al., 2010).

3.1.3. Phosphorus source

Phosphorus is another essential ingredient in the process of photosynthesis, and an important nutrient for algal growth. The assimilated phosphorus is in the form of polyphosphate (Poly-P). It was found that some microalgae can assimilate more phosphorus than required for growth under nutritional conditions unfavorable for growth. Therefore, more studies have been undertaken focusing on the Poly-P metabolism in algae and these studies have found that environmental factors such as light, osmotic shocks and nutrient availability all affect phosphorus accumulation (Eixler et al., 2006). There are reports on the accumulation of Poly-P by *Chlorella fusca* as energy storage under nitrogen starvation. On the other hand, under nitrogen deficient conditions but sufficient phosphate, *C. vulgaris* presented lipid productivity higher than a culture under complete nutrition requirements (Chu et al., 2013).

3.2. Light

Microalgae carry out the photosynthesis as a main metabolism for the acquisition of organic materials by using energy from light sources such as sunlight or artificial light. Therefore, multiple lines of evidences suggest that light is the most significant factor that governs the entire process of microalgal cultivation. Generally, microalgae use light at wavelengths from 400 to 700 nm for photosynthesis. The wavelengths absorbed by microalgae differ accordint to the microalgae species. For instance, green microalgae absorb light energy for photosynthesis through chlorophylls as a main pigment and they absorb light energy at wavelength ranges of 450–475 nm and 630–675 nm. Along the same lines, carotenoids function as an accessory pigment and they absorb light energy at a wavelength of 400–550 nm (Blair et al., 2014).

Based on the different approaches to obtain light source, the strategy for microalgal cultivation could be broadly categorized into open-pond or photobioreactor systems. While open-pond systems utilize free sunlight, microalgal cultivation in closed photobioreactors with artificial light source present the indisputable advantage regarding biomass production, which could be significantly increased particularly for value added products compared to those cultures from open pond systems. However, the supply of artificial light is the most expensive cost during the operation of the photobioreactors systems. Moreover, if light energy penetrates into the microalgal suspensions, it significantly decreases along with an increase of the light path-length at high microalgal cells concentrations. This fact is a bottleneck for microalgal cultivation. Therefore, it is crucial to develop efficient processes to maximize the utilization of light energy, thereby improving the economic feasibility in the microalgal cultivation process (Kim et al., 2014).

Recently, light-emitting diodes (LED) have emerged as a replacement of traditional artificial light source. Compared to the conventional tubular fluorescence lamps, the recent developed LEDs light makes significant advances in narrowing a specific wavelength with low power consumption. Due to these advantages, and also because LEDs could provide a particular wave-length to illuminate microalgal culture, it would be pertinent to select LEDs for the purpose of adequate manipulation of microalgal cultivation. Thus, a number of studies have addressed that there are optimum wavelengths for each one of the microalgal species, although contradictory results have been obtained about the influence of specific wavelength of LED on microalgal growth. For example, whereas a red light was the most effective for *Botryococcusbraunii* sp. growth, a blue light led to the best biomass productivity for *Nannochloropsis* sp. Additionally, an adequate manipulation of LED wavelengths at different growth stages could lead to a significant increase in biomass and lipid productivity for *Chlorella vulgaris* (Kim et al., 2014).

3.3. Temperature and pH

The optimal temperature range for microalgae cultivation is between 25 and 35 °C, but many microalgae can tolerate temperatures around 15 °C. Temperature affects the types of fatty acids produced. In general, at low temperatures, the amount of saturated fatty acids produced decreases (El-Sheekh et al., 1995; Aguirre et al., 2013).

The pH of the microalgae system affects the biomass regulation, photosyntesis rate, availability of phosphorous to microalgae and species competition. Hence, the maintenance of an acceptable pH range throughout culturing is very important as it impacts all aspects of the media biochemistry. Both, ionic absorption from the media and the metabolic biochemistry of the cell, exert significant pressure upon pH and in high performance cultures their effect is powerful enough to overcome the neutralizing capacity of exogenous buffering agents (Schenk et al., 2008; Aguirre et al., 2013). However, Moheimani and Borowitzska (2006) reported that the main reasons for the low contamination, especially by protozoa, of the *Pleurochrysiscarterae* (Haptophyta) culture in outdoor raceway ponds appear to be the increase in the pH of the culture to about 11 during the day, as well as the production of chemical defense compound(s) by this microalgae, because an increase in pH to11 inhibits the growth of some unwanted species of algae and cyanobacteria by restricting the available C_i source.

3.4. Salinity

High salt content can influence physiological processes of both, microalgae and plants. This salt stress affects through osmotic and ionic stress. Salt stress involves an excess of sodium ions, whereas osmotic stress is primarily due to a deficit of water without a direct role of sodium ions. Ionic imbalance occurs in the cells due to excessive accumulation of Na^+ and Cl^- and reduces the uptake of other mineral nutrients, such as K^+, Ca^{2+}, and Mn^{2+} (Sudhir and Murthy, 2004).

In *Synechocystis* sp., salt stress affects the photosyntesis. For this species, at a 1.026M NaCl concentration, the content of chlorophyll *a* and phycocyanin decreases, while the caroteniods content increases. This increased carotenoids content might diminish the amount of photons available for absorption by chlorophyll *a* by means of a shadowing effect. Also, in microalgae, salt stress inhibits Photosystem II and increases electron transport activity of Photosystem I (Alyabyev et al., 2007; Sudhir and Murthy, 2004).

In regards of salinity, Campenni' et al. (2013) studied the production of carotenoids and fatty acids by *Chlorella protothecoides* (grown in autotrophic conditions) and they found that the best production was attained with a 20 g L^{-1}NaCl solution, at which they reached 43.4%

of total lipids. On the other hand, open marine microalgae cultures suffer from evaporation and require addition of freshwater to maintain the salinity, a higher salinity would prevent or minimize the growth of microalgae and bacteria. Hence, any halotolerant microalgae grown for several days under phototrophic conditions without the addition of freshwater (for a higher salinity) would give an opportunity to maintain the culture under mixotrophic conditions for a second phase during a shorter culture time (Das et al., 2011).

3.5. Water and Recycling Culture Media

Particularly in open ponds, water can be lost by evaporation. Environmental factors that affect the evaporation rate include temperature, wind speed, net solar radiation and atmospheric water vapor pressure. Also, water is lost during the harvest of the diluted microalgae suspension. However, in outdoor cultures, rainfall can add water to the system (Handler et al. 2012).

Additionally, the economic production of microalgae also necessitates the continuous recycling of the nutrients after harvesting the biomass; also, recycling the culture media is a key issue for the development of large-scale cultures to minimize water and nutrients consumption, and to diminish the ecological impacts of the process (Moheimani and Borowitzska, 2006; Hadj-Romdhane et al., 2013). Hadj-Romdhane et al. (2013) refered to the work of Lívansky' et al. (1996), who recycled the culture medium during *Scenesdesmusobliquus* growth (in fed-batch mode) and this allowed saving 63% of water volumen and 16% of nutients. Hadj-Romdhane et al. (2012) estimated that using an optimized culture medium of *Chlorella vulgaris* with nutrient inputs limited to physiological requirements of microalgae would save about 75% of water and 62% of nutrients.

When culture medium is recycled, organic metabolites (lipids, polysaccharides, proteins or others) accumulate during growth when physiological stress such as nutrient limitation or cell lysis occur, and this lead to a decrease in the biomass productivity. This accumulation can also increase the contamination risks or, in some species such as *Spirulina*, can lead to autoinhibition of growth (Moheimani and Borowitzska, 2006; Hadj-Romdhane et al., 2013). Moreover, when culture medium supernatant is recycled, the counter ions of nutrients (e.g., Cl^-, Na^+, Ca^{2+}) will not or poorly be assimilated by microalgae and will accumulate in the supernatant, which leads to a change in the medium salinity a fact that may negatively affect the biomass growth. In order to avoid such mineral accumulation, Hadj-Romdhane et al. (2012) developed a highly assimilable minimal growth medium (HAMGM) in which counter ions were replaced by ammonium ion (NH_4^+).

4. Culture Systems for Microalgae

The basic culture systems for microalgal are: closed an open systems, however, a third type of system (hybrid systems) is briefly mentioned too.

4.1. Closed Systems

A very wide variety of closed photobioreactors has been developed, but fundamentally they can be classified as bags or tanks, towers, plate reactors, or tubular reactors. Large bag systems consist of clear plastic bags usually supported by a metal mesh frame or hung from supports and are widely used in aquaculture. Similarly the 'tower' systems, which are vertical cylinders usually made of fiberglass or acrylic, are also widely used in aquaculture. Plate reactors come in many forms ranging from alveolar panels with internal baffles, to vertical thin plate-like tanks made of glass, Perspex or metal mesh frames containing a plastic bag. The tubular photobioreactors are generally constructed of glass, clear Teflon tubing, or clear PVC tubing arranged either in parallel straight lines or helically wound around a central support tower. In the parallel tube photobioreactors the tubes may be arranged vertically or the tubes may be placed on the ground in one or more layers (Schenk et al., 2008).

To increase the process efficiencies, photobioreactors have to be designed to distribute light intensities for the cells. This is usually achieved by arranging tubular reactors in a fence-like construction. The fences are oriented in a north/south direction to prevent direct bright light hitting the surface, hence, sunlight is diminished in a horizontal and vertical direction (Fon Sing et al., 2013).

Moreover, the culturing of microalgae in closed systems, taking the biofuel production as the main objective is regarded as too energy intensive in addition to the high cost implied a fact that makes the process too expensive to be feasible. (Das et al., 2011).

4.2. Open Culture Systems

Open ponds can be built and operated at low cost, hence they offer many advantages, as long as the species selected for cultivation can be maintained. Open ponds have a variety of shapes ans sizes and can be classified as extensive shallow unmixed ponds, circular ponds mixed with a rotating arm, raceway ponds mixed with a paddle wheel and sloping thinlayer cascade systems (Fon Sing et al., 2013). The most commonly used design is the raceway pond, which generally is a pond with the shape of an oval with a central separation wall to obtain a closed loop recirculation channel, typically 0.25-0.30 m deep. Raceway channels are built of fiberglass, concrete or compacted earth and may be lined with white plastic. A paddle wheel is used to drive water flow continuously around the circuit and to prevent sedimentation. Flow can be guided around bends by baffles placed in the flow channel (Schenk et al., 2008; Salis et al., 2010; Sompech et al., 2012).

Raceway reactors present limitations, in terms of the type of microalgae that can be cultivated, the relatively large area required, the lower efficiency of light utilization, the poor gas/liquid mass transfer, the lack of temperature and evaporation control, the low final density of the microalgae, rainfalls that can dilute the available nutrients and the high risk of culture contamination (Pushparaj et al., 1997; Chisti, 2007; Salis et al., 2010; Fon Sing et al., 2013; Mendoza et al,. 2013 a and b). However, a greenhouse can be used to minimize such fluctuations and risks (Hase et al., 2010; Salis et al., 2010). Also, most of the species cultured in such systems are currently grown in selective environments (i.e., *Arthrospira*spp. under high alkalinity, *Dunaliellasalina*in under high salinity and *Chlorella*spp. under high nutrients concentration) and other species with normal growth requirements have also been grown successfully in open ponds, either in batch or continuous mode for very long periods of time without exhibiting significant contamination (Moheimani and Borowitzka, 2006; Das et al., 2011;Fon Sing et al., 2013).

The hydrodynamics of the system is an important factor in the behavior of any type of photobioreactor as it influences phenomena such as mass transfer and nutrient distribution (Mendoza et al., 2013 a and b). It is also important to notice the effect of the size of the system on the algal culture. For example, when the size of the raceway pond increases, the pattern of the water flow changes at the end of the pond and is very uneven. There are regions of very high flow and others with low flow and eddies in different parts of the pond, which may lead to a potential localized settling of the algal cells. These changes affect the productivity and culture stability. A method to minimize settling of the algal cells is the use of multiple (three) baffles and a small dumbbell-shaped island at the end of the central divider (Figure 4). This configuration eliminates the occurrence of dead zones and is energy efficient to operate (Sompech et al., 2012; Fon Sing et al., 2013).

4.3. Hybrid Systems

Hybrid culture systems are a combination of a closed photobioreactor and an open raceway. The closed photobioreactors produce a high density inoculum for the open ponds, thus optimizing the biomass production in the shortest time possible. This 2-step concept has been employed, for example, for the production of astaxanthin in *Haematococcuspluvialis* in Hawaii, but it is generally more expensive in both, capital and operating costs as it is a batch rather than a continuous culture process and thus is not suitable for a low cost product such as biofuels (Schenk et al., 2008; Fon Sing et al., 2013).

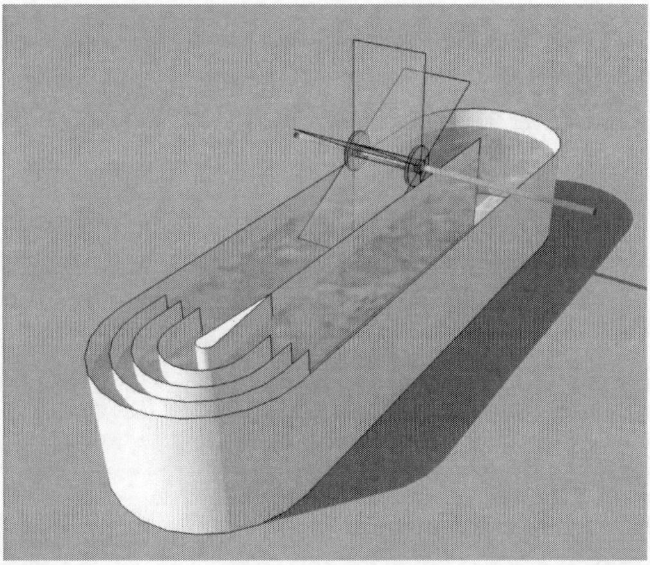

Figure. 4. A raceway with three baffles and a small dumbbell shaped island at the end of the central divider. Reproduced from Sompech et al. (2012).

5. HARVESTING

The harvesting of microalgae refers to the concentration of the microalgal suspensions until a thick paste or dry mass is obtained. There is no a single best method for harvesting microalgae and reducing their water content, but the main methods for harvesting are filtration, centrifugation, sedimentation, flocculation and flotation (Schenk et al., 2008; Parmar et al., 2011)

Firstly, filtration is a simple process to collect cells on membranes of various kinds with the aid of a suction pump, this technique could be the cheapest one but it requires backwashing to maintain the efficiency of the membrane, besides the electrical energy needed and the additional time that backwashing implies (Parmar et al., 2011; Suali and Sarbatly, 2012). Centrifugation involves the centrifugal force on the cells, but it needs additional energy too. Centrifugation technique may be expensive, but for commercial and industrial scales over a long term basis, it could be economically feasible (Parmar et al., 2011; Sualiand Sarbatly, 2012; Fon Sing et al., 2013). Sedimentation is an option because it does not generate a cost, but it is too slow. Thus, flocculation-sedimemtetion or flotation techniques are commonly considered. Flocculation is a process in which flocculants are added to the mixture to form big cell aggregates. Inorganic (alum, lime, salts, etc.) or organic (cellulose, chitosan, etc.) flocculants are chemical additives that have been studied. The manipulation of the pH in the microalgae suspension and bioflocculation (co-culturing with another organism) or the addition of a microbial flocculationg agent, which promotes the spontaneous microalgae flocculation, are other options to the chemical additives (Parmar et al., 2011; Suali and Sarbatly, 2012; Alam et al., 2014). Flocculation is followed by filtration, centrifugation, sedimentation or flotation. In flotation, air is bubbled through the microalgal culture causing cell clusters to float to the surface where these can be removed (Parmar, et al., 2011; Suali and Sarbatly, 2012).

6. CELL DISRUPTION AND EXTRACCTION OF LIPIDS

After the microalgal culture is harvested from the bioreactor, the concentrated microalgal culture is then processed in a pre-treatment step to prepare it for lipid extraction. During lipid extraction, lipids are extracted out of the cellular matrices with an extraction solvent. Microalgal lipid extraction generally uses either organic solvent or supercritical fluid (such as supercritical carbon dioxide) as an extraction solvent. The lipids are then separated from the cellular debris, isolated from the extraction solvent and any residual water, and finally converted to biodiesel in the transesterification step.

6.1. Cell Disruption or Pre-Treatment of Microalgal Biomass for Lipid Extraction

The efficiency of microalgal lipid extraction is known to increase with the degree of cell disruption. When intact cells are disintegrated during cell disruption, intracellular lipids are liberated from the cellular structures and released into the surrounding medium. During subsequent lipid extraction, the eluting extraction solvent can directly interact with these free lipids without penetrating into the cellular structures.

Laboratory-scale cell disruption methods are classified based on the manner in which they achieve microalgal cellular disintegration: mechanical or non-mechanical. Mechanical methods include bead mill, press, high-pressure homogenization, ultrasonication, autoclave, lyophilization, and microwave, while non-mechanical methods often involve lysing the microalgal cells with acids (HCl, H_2SO_4), alkalis, enzymes, or osmotic shocks. Bead mill, high-pressure homogenization, and ultrasonication are three of the most widely used methods for laboratory-scale microalgal cell disruption (Park et al., 2014; Halim et al., 2012).

Lee et al. (1998) assessed the effect of prior mechanical cell disruption on lipid extraction from the species *B. braunii*. They used a chloroform/methanol mixture (2/1 v/v) as an extraction solvent and found completely disrupted microalgal cells to yield almost twice the amount of crude lipids of intact microalgal cells.

Park et al. (2014) investigated the efficiency of acid-catalyzed hot-water extraction of lipids from *C. vulgaris*, with hexane as the extraction solvent. They extracted 337.4 mg lipids/g cell from *C. vulgaris* treated with a 1.0% sulfuric acid concentration heated at 120°C for 60 min.

6.2. Lipid Extraction from Microalgal Biomass

An ideal lipid extraction technology for microalgal biodiesel production needs to display a high level of specificity towards lipids in order to minimize the co-extraction of non-lipid contaminants, such as protein and carbohydrates. To reduce downstream fractionation /purification, the lipid extraction technology should also be more selective towards acylglycerols than other lipid fractions that are not as readily convertible to biodiesel, i.e., polar lipids and non-acylglycerol neutral lipids (free fatty acids, hydrocarbons, sterols, ketones, carotenes, and chlorophylls).

Thus, a sovent mixture of chloroform/methanol (1/2 v/v) is the most frequently used for lipid extraction from any living tissue (Folch et al., 1951). In using this organic solvent system, residual endogenous water in the microalgal cells acts as a ternary component that enables the complete extraction of both, neutral and polar lipids. It is noticed that this method does not require the complete drying of microalgal biomass. Once the cell debris is removed, more chloroform and water are added to induce biphasic partitioning. The lower organic phase (chloroform with some methanol) contains most of the lipids (both, neutral and polar) while the upper aqueous phase (water with some methanol) constitutes most of the non-lipids (pigments, proteins and carbohydrates). Extraction of lipids using the chloroform/methanol (1/2 v/v) mixture is fast and quantitative, however, chloroform is highly toxic and its usage is undesirable (Halim et al., 2012).

On the other hand, hexane/isopropanol (3/2 v/v) mixture has been suggested as a low-toxicity substitute to chloroform/methanol system (Halim et al., 2011). Upon biphasic separation, the upper organic phase (hexane with some isopropanol) contains most of the lipids (both, neutral and polar) while the lower aqueous phase (water with some isopropanol) contains most of the non-lipids (pigments, proteins and carbohydrates). The hexane/ isopropanol mixture has been shown to be more selective towards neutral lipids compared to chloroform/methanol mixtures. Guckert et al. (1988) attributed the neutral lipid selectivity of hexane/isopropanol mixture to the inability to extract the polar lipid constituents of microalgal membranes (chloroplast membranes contain glycolipids and cell membranes contain phospholipids).

Pure alcohol (such as butanol, isopropanol, and ethanol) is cheap, volatile, and has a strong affinity to membrane-associated lipid complex due to its ability to form hydrogen bonds. However, its polar nature is also a disadvantage as it limits interaction with free-standing neutral lipid globules. For this reason, when it is used as a microalgal lipid extraction solvent, alcohol is almost always combined with a non-polar organic solvent, such as hexane or chloroform, to ensure the total extraction of both forms of neutral lipids (free-standing globules and membrane-associated complexes) (Halim et al., 2011).

Supercritical fluid extraction (SFE) is an emerging green technology that has the potential to replace traditional organic solvent extraction. Supercritical fluid displays physical properties intermediate to a liquid and a gas. These transitional properties allow for a rapid penetration of the fluid through cellular matrices, thus resulting in a higher total lipid yield and a shorter extraction time (Halim et al., 2011).

Supercritical carbon dioxide ($SCCO_2$) is the primary solvent used in the majority of supercritical fluid extractions, its moderate critical pressure (72.9 atm) allows for a modest compression cost, while its low critical temperature (31.1°C) enables successful extraction of thermally sensitive lipid fractions without degradation. $SCCO_2$ facilitates a safe extraction due to its low toxicity, low flammability, and lack of reactivity. During the lipid extraction process, the microalgal biomass and diatoms are packed tightly inside the cylindrical extraction vessel. The supercritical carbon dioxide travels on the surface of the packed mixture and lipids are desorbed from the microalgal biomass. Immediately upon dissolution, the $SCCO_2$ encloses the lipids to form a $SCCO_2$–lipids complex. The complex, driven by concentration gradient, diffuses across the static $SCCO_2$ film and enters the bulk $SCCO_2$ flow (Halim et al., 2012).

7. Conversion to Biodiesel

The isolated lipids from microalgae can be gravimetrically quantified. The term 'total lipids' is primarily used for analytical purposes. As previously mentioned in Section 2, in addition to acylglycerols, crude lipids obtained from microalgal biomass frequently contain polar lipids and non-acylglycerol neutral lipids (such as free fatty acids, hydrocarbons, sterols, ketones, carotenes, and chlorophylls). From the perspective of biodiesel production, any non-acylglycerol biochemical fraction is a contaminant and will have to be removed from the crude lipids. As such, crude lipids arising from microalgal biomass are often subjected to a fractionation step before they are transesterified. Different purification methods, such as liquid chromatography, acid precipitation, and urea crystallizations, are used for lipid fractionation (Halim et al., 2012).

During transesterification, the fatty-acid-containing lipid fractions in the crude lipids are reacted with alcohol (methanol, ethanol, isopropanol, buthanol) and converted to fatty acid alkyl esters. When methanol is used, the reaction produces fatty acid methyl ester (FAME) or biodiesel. Either an acid (such as H_2SO_4) or an alkali (such as NaOH or KOH) can be used as a catalyst for transesterification. Since alkali catalysts have faster reaction rates and higher conversions than acid catalysts for the transesterification of acylglycerols, they are commercially used in the chemical industry for conversion of plant and animal oils to biodiesel. However, alkaline-catalyzed transesterification has limited efficiencies when applied to non-acylglycerol fatty-acid-containing lipid fractions, such as polar lipids and free fatty acids. During alkaline transesterification of acylglycerols, the catalyst cleaves the ester bonds holding the fatty acids to the glycerol backbone. The liberated fatty acids are then reacted with methanol to form FAME. In lab-scale experiments, where only small amounts of crude microalgal lipids are available, a large amount of methanol (substantially in excess of stoichiometric requirement) is often added to ensure quantitative transesterification (Halim et al., 2012).

Once transesterification is completed, the reaction mixture that contains biodiesel, glycerol, reformed alkali catalyst, excess of methanol, and un-transesterified lipids, then undergoes post- transesterification purification to remove by-product contaminants (glycerol, alkali catalyst, and excess of methanol). A laboratory-scale post-transesterification purification method typically consists of 2 steps. The reaction mixture is left to settle under gravity to induce bi-phasic partitioning (top biodiesel/un-transesterified lipids phase and bottom glycerol phase). Once the biodiesel/un-transesterified lipids phase is decanted off, it is washed repeatedly with water to eliminate any alkali catalyst and the excess of methanol (Halim et al., 2012).

CONCLUSION

The potential of microalgae to contribute to the world energy demand is high, however, it is necessary to solve the bottlenecks about microalgae production and lipid content. The biomass production cost must be also reduced to promote a successful commercialization. The novel technology then could be scaled-up several orders of magnitude to significantly contribute to the biofuels market. The main areas for development and innovation in biodiesel from microalgae are: the culture media, culture systems (raceways, photobioreactors), microalgae strains, culture regime (batch, continuous, semicontinuous) and control of cultures contamination in order to achieve high yields of biomass and lipids at low cost. Also, it is important to innovate the harvesting, drying, extraction and transesterification processes in order to minimize the costs and the energy requirements, which would make the biodiesel production from microalge a feasible investment and an environmentally friendly process.

REFERENCES

Acién, F. G., Fernández, J. M., Magán, J. J. & Molina, E. (2012). Production cost of a real microalgae production plant and strategies to reduce it. *Biotechnology Advances. 30*, 1344-1353.

Aguirre, A. M., Bassi, A. & Saxena, P. (2013). Engineering challenges in biodiesel production from microalgae. *Crit. Rev. Biotechnol. 33*, 293-308.

Alam, M. A., Wan, C., Guo, S. L., Zhao, X. Q., Huang, Z. Y., Yang, Y. L., Chang, J. S. & Bai, F. W. (2014). Characterization of the flocculating agent from the spontaneously flocculating microalga *Chlorella vulgaris* JSC-7. *Biosci.Bioeng*. http://dx.doi.org/10.1016/j.jbiosc.2013.12.021

Alyabyev, A. Ju., Loseva, N. L., Gordon, L. Kh., Andreyeva, I. N., Rachimova, G. G., Tribunskih, V. I., Ponomareva, A. A. & Kemp, R. B. (2007). The effect of changes in salinity on the energy yielding processes of *Chlorella vulgaris* and *Dunaliellamaritima* cells. *Thermochimica Acta. 458*, 65-70.

Anastopoulos, G., Zannikou, Y., Stournas, S. & Kalligeros, S. (2009). Transesterification of vegetable oils with ethanol and characterization of the key fuel property of ethyl esters.*Energies. 2*, 362-376.

Barsanti, L. & Gualtieri, P. (2006). Algae: Anatomy, Biochemistry, and Biotechnology. CRC Press Taylor & Francis Group. U. S. A.

Blair, M. F., Kokabian, B. & Gude, V. G. (2014). Light and growth medium effect on *Chlorella vulgaris* biomass production. *J. Environ. Chem. Eng. 2*, 665-674.

Breuer, G., Lamers, P. P., Martens, D. E., Draaisma, R. B. & Wijffels, R. H. (2012). The impacto f nitrógeno starvation on the dynamics of triacylglycerol accumulation in nine microalgae strains.*Bioresource Technology. 124*, 217-226.

Campenni', L., Nobre, B. P., Santos, C. A., Oliveira, A. C., Aires-Barros, M. R., Palabra, A. M. F. & Gouveia, L. (2013). Carotenoid and lipid production by the autotrophic microalga *Chlorella protothecoides* under nutricional, salinity, and luminosity stress conditions. *Appl. MicrobiolBiotechnol. 97*, 1383-1393.

Chen, F., Liu, Z., Li, D., Liu, C., Zheng, P. & Chen, S. (2012). Using ammonia for algae harvesting and as nutrient in subsequent cultures.*Bioresour. Technol. 121*, 298-303.

Chi, Z., O'Fallon, J. V. & Chen, S. (2011). Bicarbonate produced from carbon capture for algae culture. *Trends in Biotechnology. 29*, 537-541.

Chisti, Y. (2007). Biodiesel from microalgae.*Biotechnol Advances. 25*, 294-306.

Chu, F. F., Chu, P. N., Cai, P. J., Li, W. W., Lam, P. K. S. & Zeng, R. J. (2013). Phosphorus plays an important role in enhancing biodiesel productivity of *Chlorella vulgaris* under nitrogen deficiency. *Bioresour. Technol. 134*, 341-346.

Das, P., Aziz, S. S. & Obbard, J. P. (2011). Two phase microalgae growth in the open system for enhanced lipid productivity. *Renewable Energy. 36*, 2524-2528.

De-Bashan, L. E., Bashan, Y., Moreno, M., Lebsky, V. K. & Bustillos, J. J. (2002). Increased pigment and lipid content, lipid variety, and cell and population size of the microalgae *Chlorella* spp. when co-immobilized in alginate beads with the microalgae-growth-promoting bacterium *Azospirillumbrasilenses. Can. J. Microbiol. 48*, 514-521

Doucha, J., Straka, F. & Livansky, K. (2005). Utilization of flue gas for cultivation of microalgae (*Chlorella* sp.) in an outdoor open thin-layer photobioreactor. *J. Appl. Phycol. 17*, 403-412.

Ehimen, E. A., Sun, Z. F. & Carrington, C. G. (2010). Variables affecting the in situ transesterification of microalgae lipids. *Fuel. 89*, 677-684.

Eixler, S., Karsten, U. & Selig, U. (2006). Phosphorus storage in *Chlorella vulgaris* (Trebouxiophyceae, Chlorophyta) cells and its dependence on phosphate supply. *Phycol. 45*, 53-60.

El-Sheekh, M. M., Hammouda, O. H. & Kotkat, H. (1995). Photosynthesis, lipids and proteins in the cyanobacterium *Synechocystis* PCC 6803 as affected by temperature. *BiologiaPlantarum. 37*, 27-34.

Feng, Y., Li, C. & Zhang, D. (2011). Lipid production of *Chlorella vulgaris* cultured in artificial wastewater medium. *Bioresour. Technol. 102*, 101-105.

Folch, J., Ascoli, I., Lees, M. Meath, J. A. & Lebaron, F. N. (1951). Preparation of lipide extracts from brain tissue. *J. Biol. Chem. 1*, 833-841.

Fon Sing, S., Isdepsky, A., Borowitzka, M. A. & Moheimani, N. R. (2013). Production of biofuels from microalgae. *Mitig Adapt Strateg Glob Change. 18*, 47-72.

Grobbelaar, J. U. (2009). Factors governing algal growth in photobioreactors: the "open" versus "closed" debate. *J. Appl. Phycol. 21*, 489-492.

Guckert, J. B., Cooksey, K. E. & Jackson, L. L. (1988). Lipid solvent system are not équivalent for analysis of lipid classes in the microeukaryotic green alga, *Chlorella. J. Microbiol. Methods. 8*, 139-149.

Hadj-Romdhane, Jaouen, P., Pruvost, J., Grizeau, D., Van Vooren, G. & Bourseau, P. (2012). Development and validation of a minimal growth medium for recycling *Chlorella vulgaris* culture.*Bioresour. Technol. 123*, 366-374.

Hadj-Romdhane, F., Zheng, X., Jaouen, P., Pruvost, J., Grizeau, D., Croué, J. P. & Bourseau, P. (2013). The culture of *Chlorella vulgaris* in a recycled supernatant: Effects on biomass production and medium quality. *Bioresour. Technol. 132*, 285-292.

Halim, R., Gladman, B., Danquah, M. K. & Webley, P. A. (202). Oil extraction from microalgae for biodiesel production.*Bioresour. Technol. 102*, 178-185.

Halim, R., Danquah, M. K. & Webley, P. A. (2012). Extraction of oil from microalgae for biodiesel production: A review. *Biotechnology Advances. 30*, 709-732.

Han, F., Huang, J., Li, Y., Wang, W., Wan, M., Shen, G. & Wang, J. (2013). Enhanced lípido productivity of *Chlorella pyrenoidosa* through the culture strategy of semi-continuous cultivation with nitrógeno limitation and pH control by CO_2.*Bioresour. Technol. 136*, 418-424.

Handler, R. M., Canter, C. E., Kalnes, T. N., Lupton, F. S., Kholiqov, O., Shonnard, D. R. & Blowers, P. (2012). Evaluation of environmental impacts from microalgae cultivation in open-air raceway ponds: Analysis of the prior literature and investigation of wide variance in predicted impacts. *Algal Research. 1*, 83-92.

Hase, R., Oikawa, H., Sasao, C., Morita, M. & Watanabe, Y. (2000). Photosynthetic production of microalgal biomass in a raceway system under greenhouse conditions in Sendai City. *J. Bioscience and Bioengineering. 89*, 157-163.

Hernandez, J. P., de-Bashan, L. E., Rodriguez, D. J., Rodriguez, Y. & Bashan, Y. (2009). Growth promotion of the freshwater microalga *Chlorella vulgaris* by the nitrógeno-fixing, plant growth-promoting bacterium *Bacillus pumilus* from arid zone soils.*European Journal of Soil Biology. 45*, 88-93.

Hu, Q., Sommerfeld, M., Jarvis, E., Ghirardi, M., Posewitz, M., Seibert, M. & Darzins, A. (2008). Microalgaltriacylglycerols as feedstocks for biofuel production: perspectives and advances. *Plant. J. 54*, 621–639.

Kee Lam, M., Teong Lee, K. & Rahman-Mohamed, A. (2013). Current status and challenges on microalgae-based carbon capture. *International Journal of Greenhouse Gas Control. 10*, 456-469.

Kim, D. G., Lee, C., Park, S. M. & Choi, Y. E. (2014). Manipulation of light wavelength at appropriate growth stage to enhance biomass productivity and fatty acid methyl ester yield using *Chlorella vulgaris*. *Bioresour. Technol. 159*, 240-248.

Kim, J., Lingaraju, B. P., Rheaume, R., Lee, J. & Siddiqui, K. F. (2010). Removal of ammonia from wastewater effluent by *Chlorella vulgaris*. *Tsinghua Science and Technology*. *15*, 391-396.

Klok, A. J., Martens, D. E., Wijffels, R. H. & Lamers, P. P. (2013). Simultaneous growth and neutral lípido accumulation in microalgae.*Bioresour. Technol. 134*, 233-243.

Lakeman, M. B., von Dassow, P. & Cattolico, R. A. (2009). The strain concept in phytoplankton ecology. *Harmful Algae. 8*, 746-758.

Lee, S. J., Yoon, B. D. & Oh, H. M. (1998). Rapid method for the determination of lipid from the green algae *Botryococcusbraunii*.*Biotechnol. Tech. 7*, 553-556.

Lv, J. M., Cheng, L. H., Xu, X. H., Zhang, L. & Chen, H. L. (2010). Enhanced lipid production of *Chlorella vulgaris* by adjustment of cultivation conditions. *Bioresour. Technol. 101*, 6797-6804.

Mendoza, J. L., Granados, M. R., de Godos, I., Acién, F. G., Molina, E., Banks, C. & Heaven, S. (2013a). Fluid-dynamic characterization of real-scale raceway reactors for microalgae production. *Biomass and Bioenergy. 54*, 267-275.

Mendoza, J. L., Granados, M. R., de Godos, I., Acién, F. G., Molina, E., Heaven, S. & Banks, C. J. (2013b). Oxygen transfer and evolution in microalgal culture in open raceways. *Bioresorce Technology. 137*, 188-195.

Meng, X., Yang, J., Xu, X., Zhang, L., Nie, Q. & Xian, M. (2009). Biodiesel production from oleaginous microorganisms.*Renewable Energy. 34*, 1–5.

Moheimani, N. R. & Borowitzka, M. A. (2006). The long-term culture of the coccolithophore Pleurochrysiscarterae (Haptophyta) in outdoor raceway ponds. *J. Appl. Phycol. 18*, 703-712.

Moheimani, N. R. (2013). Inorganic carbon and pH effect on growth and lípido productivity of Tetraselmissuecica and *Chlorella*sp (Chlorophyta) grown outdoors in bag photobioreactors. *J. Appl. Phycol. 25*, 387-398.

Mutanda, T., Ramesh, D., Karthikeyan, S., Kumari, S., Anandraj, A. & Bux, F. (2011). Bioprospecting for hyper-lipid producing microalgal strains for sustainable biofuel production. *Bioresour. Technol. 102*, 57-70.

Oswald, W. J. & Golueke, C. G. (1968). Large scale production of microalgae. In: Mateless RI, Tannenbaum SR, editors. Single cell protein. Cambridge, MA: MIT Press; p. 271-305.

Park, J. Y., Oh, Y. K., Lee, J. S., Lee, K., Jeong, M. J. & Choi, S. A. (2014). Acid-catalyzed hot-water extraction of lipids from *Chlorella vulgaris*. *Bioresour. Technol. 153*, 408-412.

Parmar, A., Singh, N. K., Pandey, A., Gnansounou, E. & Madamwar, D. (2011). Cyanobacteria and microalgae: A positive prospect for biofuels. *Bioresour. Technol. 102*, 10163-10172.

Petkov, G. & Garcia, G. (2006). Wich are fatty acid of the green alga *Chlorella*?. *Biochemical Systematics and Ecology. 35*, 281-285.

Prins, H. B. A. & Elzenga, J. T. M. (1987). Bicarbonate utilization: Function and mechanism. *Aquatic Botany. 34*, 59-83.

Pushparaj, B., Pelosi, E., Tredici, M. R., Pinzani, E. & Materassi, R. (1997). An integrated clture system for outdoor production of microalgae and cyanobacteria. *J. App. Phycol. 9*, 113-119.

Ríos, S. D., Castañeda, J., Torras, C., Farriol, X. & Salvadó, J. (2013). Lipid extraction methods from microalgal biomass harvested by two different paths: Screening Studies Howard biodiesel production. *Bioresour. Technol. 133*, 378-388.

Salis, A., Nicolò, M., Guglielmino, S. & Solinas, V. (2010). Biodiesel from microalgae. In: *Handbook of hydrocarbon and lípid microbiology*. Timmis, K. N. Springer-Verlag. Berlin, Germany. 2827-2839 pp.

Santos, A. M., Janssen, M., Lamers, P. P., Evers, W. A. C. & Wijffels, R. H. (2012). Growth of oil accumulating microalga *Neochlorisoleoabundans*under alkaline-saline conditions.*Bioresour. Technol. 104*, 593-599.

Schenk, P. M., Thomas-Hall, S. R., Stephens, E., Marx, U. C., Mussgnug, J. H., Posten, C., Kruse, O. & Hankamer, B. (2008). Second Generation Biofuels: High-Efficiency for Biodiesel Production. *Bioenerg. Res. 1*, 20-43.

Sheng, J., Kim, H. W., Badalamenti, J. P., Zhou, C., Sridharakrishnan, S., Krajmalnik-Brown, R., Rittmann, B. E. & Vannela, R. (2011). Effects of temperature shifts on growth rate and lípido characteristics of *Synechocystis* sp. PCC6803 in a bench-top photobioreactor. *Bioresour. Technol. 102*, 11218-11225.

Sompech, K., Chisti, Y. & Srinophakun, T. (2012). Design of raceway ponds for producing microalgae.*Biofuels. 3*, 387-397.

Suali, E. & Sarbatly, R. (2012). Conversion of microalgae to biofuel.*Renewable and Sustainable Energy Reviews. 16*, 4316-4342.

Sudhir, P. & Murthy, S. D. S. (2004). Effects of salt stress on basic processes of photosynthesis.*Photosynthetica. 42*, 481-486.

Vonshak, A., Torzillo, G., Masojidek, J. & Boussiba, S. (2001). Sub-optimal morning temperature induces photoinhibition in dense outdoor cultures of the alga Monodussubterraneus (Eustigmatophyta). *Plant Cell Environ*, *24*, 1113–1118

Widjaja, A., Chien, C. C. & Ju, Y. H. (2009). Study of increasing lipid production from fresh water microalgae *Chlorella vulgaris. Journal of the Taiwan Institute of Chemical Engineers. 40*, 13-20.

In: Microalgae and Other Phototrophic Bacteria
Editor: Luis G. Torres Bustillos
ISBN: 978-1-63482-078-3

Chapter 11

DEWATERING AND DEHYDRATION OF MICROALGAE

Juan Rodríguez Ramírez** and *Lilia Méndez Lagunas
Instituto Politecnico Naciona, CIIDIR Unidad Oaxaca,
Cruz Xoxocotlan, Oaxaca, Mexico

ABSTRACT

Microalgae post-harvesting must include dewatering and drying processes in order to obtain a high concentration of solids. Post-harvest microalgae drying is necessary to process the material and to take advantage of microalgae as an energy product and in related uses. The dewatering, drying and storage of microalgae are studied so as to improve the yield and quality of biomass derived from microalgae as well as to establish a basis for further research.

This chapter presents the interaction of microalgae with water and their chemical equilibrium; evaluates microalgae drying methods; experimentally determines drying kinetics for process analyses; and calculates water diffusivity, physical changes and the effect on the lipid content as a result of the drying process. *Chlorella vulgaris* and *Chlorella protothecoides* are specifically analyzed in order to present water sorption isotherms, graphs depicting moisture loss in convection drying, the calculation of the diffusion coefficient and scanning electron and atomic force microscope images of these two dehydrated microalgae.

INTRODUCTION

In biodiesel production from microalgae, the drying process or dehydration is considered the main bottle neck due to the large amounts of energy required to reach high biomass concentrations. Drying is the final step in the harvesting process, which may include one or more thickening and dewatering stages, including sedimentation, filtration, flotation, centrifugation, microflotation or any combination of these processes [1, 2].

* Corresponding author: jrodrigr@ipn.mx.

The first separation of the biomass from the medium aims for a concentration of 2 to 7% of the total solid matter. In dewatering, the subsequent process, the concentration reaches 12 to 35% of the total solid matter [3].Intracellular water can only be removed by thermal processes, which results in a water content of less than 10%. The mass reduction is necessary for preservation, storage and for the reduction of transportation costs.

The separation of the biomass from the aqueous medium is influenced by the physicochemical properties of the microalgae such as shape, size, specific weight and surface charge [4]. The harvestable biomass is very low per unit area (<150 g/m^2) and usually contains between 0.2 and 2 g/L of solids, as the algae are very small (<25 μm) and because they are cultivated in large volumes of water; to obtain the biomass (dry weight basis), the diluted microalgae solutions must be concentrated up to 100 times [5, 6].

The final harvesting steps are dewatering and drying, which serve to reduce the humidity content to 10-15% and to ensure product stability during storage. For some types of lipid extraction, a high degree of concentrated substrates (>20 g/L of solids) is necessary and the presence of water at this stage can cause problems in the extraction units by promoting the formation of emulsions in the presence of broken cells and participating in secondary reactions. Intracellular water can also form a barrier between the solvent and the solute, reducing the efficiency of the process. Moreover, because biomass slurry is perishable, it must be processed rapidly following harvest so as to remain stable [3-4, 6-8].

Drying is a complicated process that involves a number of phenomena. In order to understand these phenomena, it is necessary to define the physical and chemical characteristics of the microalgae: the chemical bonds and the physical interactions of the material with water, the drying processes and methods, the changes that the material undergoes during drying and the final application of the dehydrated materials. To ensure the economic feasibility of biofuel production, a specific drying process must be designed for each microalgae species.

MICROALGAE AND WATER

Microalgae Biology

Microphytes or microalgae are microscopic algae, typically found in freshwater and marine systems. They are unicellular species which exist individually, in chains or in groups. Depending on the species, their size can range from a few micrometers (μm) to a few hundreds of micrometers. *Chlorella* is a genus of single-cell green algae, belonging to the Phylum *Chlorophyta*; it is spherical in shape with a diameter of 2 to 10 μm. The preference for microalgae as an energy product, is largely due to their less complex structure, fast growth rates, and high oil-content (for some species).

The cytoplasm of eukaryotic microalgae contains the nucleus and different kinds of organelles and compartments formed by the invagination of the plasma membrane and the endoplasmic reticulum (figure 1). Among the organelles, chloroplast, lipid globules and cytoplasmic lipids represent an important source of polyunsaturated fatty acids [9].

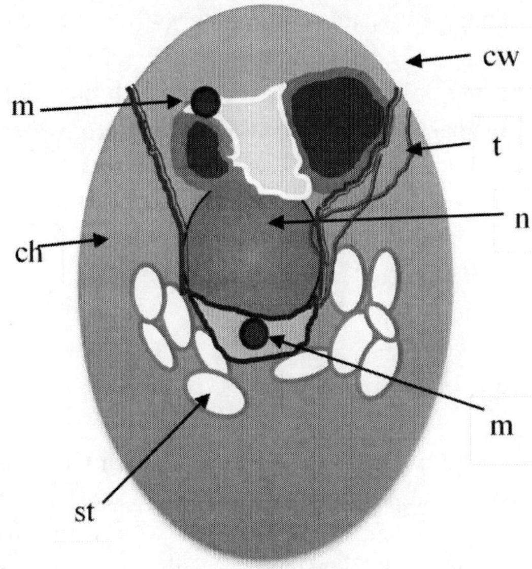

Figure 1. Diagram of the structures of a *Chlorella vulgaris* cell. Abbreviations: Cw cell wall, ch chloroplast, t thylakoids, st starch grain (leucoplasts), n nucleus, m mitochondria. (Adapted from Tomaselli [9]).

Microalga and Water Bonds

Water, with the chemical formula H_2O (one molecule of water has two hydrogen atoms covalently bonded to a single oxygen atom) is primarily a liquid, under standard conditions. The presence of a charge on each of these atoms (oxygen positive charge and hydrogen negative charge) gives each water molecule a net dipole moment. Electrical attraction between water molecules due to this dipole pulls individual molecules closer together, making it more difficult to separate the molecules and thus raising the boiling point. This attraction is known as hydrogen bonding and it is responsible for a number of water's physical properties such as high melting and boiling point temperatures. More energy is required to break the hydrogen bonds between molecules, which is responsible for liquid water's high specific heat capacity. This high heat capacity makes water a good heat storage medium (coolant) and heat shield, as water molecules tend to stay close to each other (cohesion). Water also has high adhesion properties because of its polar nature.

Water has a high surface tension of 72.8 mN/m at room temperature, caused by the strong cohesion between water molecules. Pure water containing no exogenous ions is an excellent insulator but not even "deionized" water is completely free of ions. One of water's most important properties is its high dielectric constant. This constant represents its ability to form electrostatic bonds with other molecules, meaning that it can eliminate the attraction of the opposite charges of surrounding ions.

When an ionic or polar compound enters water, it is surrounded by water molecules (hydration). The relatively small size of water molecules (~ 3 Angstroms) allows many water molecules to surround one molecule of solute. The partially negative dipole ends of the water molecules are attracted to the positively charged components of the solute, and vice versa for the positive dipole ends. Chemically, water is amphoteric, because the oxygen atom in water has two ion pairs.

In biological cells and organelles, water is in contact with the hydrophilic membrane and protein surfaces; that is, surfaces with a strong attraction to water. The dehydration of hydrophilic surfaces requires doing substantial work against hydration forces. These forces are very large but decrease rapidly over a nanometer or less. They are of great importance in biology, particularly when cells are dehydrated as a result of exposure to dry atmospheres or extracellular freezing.

Equilibrium of Microalgae-Water

Water is the dominant component in concentrated microalgae slurry. The water concentration greatly influences sub-sequential processing steps, such as the extraction of organic compounds. In biological systems, water exists as free water, like liquid water, and as bond water. Bond water is generally defined as sorbent associate water that differs thermodynamically from pure water. It has been suggested that this water is bound to hydrogen bond acceptors. Bound water is not available to other compounds and requires a great deal of energy to separate it from the solid. An understanding of thermodynamic properties, particularly of water-solid equilibria, such as water sorption behavior in microalgae, is of fundamental importance to drying and energy considerations.

In water-solid equilibria the activity of the water, or aw, is a thermodynamic property and, in a simplified way, is expressed as the relationship of the partial vapor pressure of water (p) in a substance, divided by the reference state partial vapor pressure of water (p_0); the reference state is most often defined as the partial vapor pressure of pure water at the same temperature. Using this particular definition, distilled water has a water activity of one.

$$aw = p/p_0 \tag{1}$$

The water activity can be related to relative humidity of air in equilibrium with a sample and is called Equilibrium Relative Humidity (ERH) expressed in percent.

$$ERH = aw \cdot 100 \tag{2}$$

The graphically expressed relationship between the total moisture content of microalgal material and the corresponding aw over a range of values at a constant temperature is referred as the Moisture Sorption Isotherm (MSI). MSIs are nonlinear with varying shapes that depend on the structure and solid compounds. The sigmoid or S-shaped adsorption isotherms are found in most food products.

The isotherm can be divided into three intervals based on the aw to interpret the degree of interaction of the water with the material (figure 2). Water in the A region is strongly linked to hydrophilic groups and polar components of material and presents an enthalpy of vaporization much higher than pure water. In region B, water molecules are linked with less force than in region A and the enthalpy of vaporization is slightly greater than pure water. The properties of water in area C, are similar to those of free water. The water is retained in the capillaries and cracks of the material and is able to act as a solvent [10].

Various equations exist to describe sorption isotherms. The Guggenheim-Anderson-de Boer (GAB) model is widely used (Equation 3). This model is based on the assumption of multilayer physical adsorption without lateral interactions. The first layer of water covers the surface evenly and is strongly linked in a monolayer. Subsequent water layers have less interaction with the surface, these layers are called multilayers. Successive layers have properties similar to those of free water [11]. Xeq is the moisture content in equilibrium in kg w/kg dm kgdm (w and dm refer respectively to water and dry mater), Xm is the moisture content of the monolayer in kg w/kg dm, C is a constant related to the heat of sorption of the monolayer, K is a constant related to the heat of sorption of the multilayer and aw is the activity of water.

$$X_{eq} = \frac{X_m \cdot C \cdot K \cdot a_w}{(1 - K \cdot a_w) \cdot (1 - K \cdot a_w + C \cdot K \cdot a_w)} \quad (3)$$

The parameters of this model have physical meanings and the values that can be established are $C \geq 1$, $0 < K < 1$ and $X_m > 0$. The parameter C is a measure of the strength of the water molecule bonds in the primary active sites. The greater the value of C, the more strongly linked the water is in the monolayer and the greater the difference in enthalpy between the molecules of the monolayer and the multilayer. The parameter K is a factor that corrects the properties of the water molecules in the multilayer, in respect to free water. When K is close to 1, there is no distinction between the molecules of the multilayer and free water. When the value of K is low, the adsorbed water molecules are structured in a multilayer. X_m is the monolayer value and is a measure of the availability of the active sites for adsorption of water of a material [11]. The solid surface area is the total area available for hydrophilic bonds and can be calculated using equation 4 [12]. Its value will give us an indication of the degree of difficulty of water removal: the more area available, the greater the number of bonds between the water and the material. This interaction is defined by the chemical composition of microalgae, for example, compounds with a polar or hydrophilic group (-OH, -COOH) or nonpolar or hydrophobic (hydrocarbon chains). In microalgae both types of groups exist, in sugars and lipids, with the proportions varying according to the species.

$$S_A = X_m \cdot \left(\frac{1}{PM_{H_2O}} \right) \cdot N_A \cdot A_{H_2O} = 3.5 \times 10^3 X_m \quad (4)$$

Where S_A is the surface area of the solid in m^2/g of solid, X_m is the moisture content of the monolayer in kg w/kg dm, PM_{H2O} is the molecular weight of water (18 kg/kmol), N_A is the number of Avogadro (6×10^{23} molecules/mole) and A_{H2O} is the area of a molecule of water (10.6×10^{-20} m^2).

There are different methods to obtain the isotherms, related to the measurement of change in weight and water activity. The samples are kept in a controlled environment until equilibria is reached. Water activity determinations based on psychrometrics measurements are widely used, with dew-point temperature measurement instruments, wet-bulb temperature and conductance sensor based hygrometers.

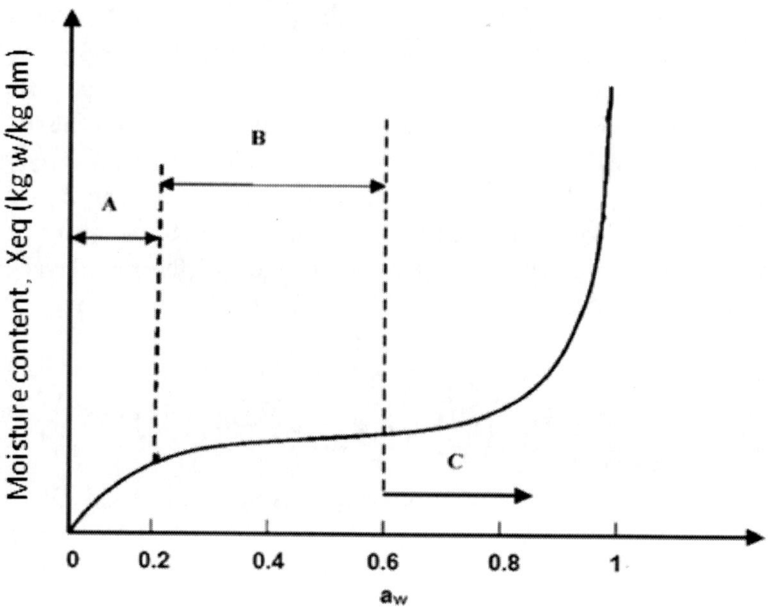

Figure 2. Moisture adsorption isotherm graph (adapted from Basu et al.[10]).

Isotherms of Chlorella vulgaris and Chlorella protothecoides. The Dynamic Vapor Sorption method (DVS) [13] which consists of measuring the differential mass variation for each level of relative air humidity, was used to determine the isotherms of *Chlorella vulgaris* (*cv*) and *Chlorella protothecoides* (*cp*). Previously, the microalga were dehydrated using a co-current pilot-scale spray dryer (Niro, Copenhagen, Denmark) equipped with a rotary atomizer (TS-Minor, M02/A). The operating parameters were: inlet air temperature, Ti (150 °C); feed flow rate, Fe (2.0 L/h), atomization speed (28, 000 rpm). Adsorption isotherms were determined using a dynamic vapor sorption analyzer VTI-SA+ (TA Instruments) with an aw range of 0.20 to 0.80 at 25°C. Samples of 10 mg were placed on the VTI pan and a 25°C air flow was passed over the sample with a controlled relative humidity, until the sample weight ratio (dw/dt) became lower than 0.005% in 2 minutes. The relative air humidity was increased by increments of 10% for each equilibria step.

Experimental data was fitted to the GAB model [14]. The parameters of this model were estimated with the MS Excel Solver algorithm (Microsoft, Redmond, USA). The initial values of the constants introduced into the software were delimited for their physical significance. The solid surface area was determined with the equation used by Cassini et. al.[12].

The adsorption isotherms of spray dried *Chlorella vulgaris* and *Chlorella protothecoides* (figure 3) were sigmoid shaped or type II adsorption isotherms, in accordance with Brunauer's classification.

The moisture content at equilibria is low in a wide aw range (0<aw<0.6). *Spirulina* isotherm was included as a comparison. *Chlorella protothecoides* adsorbs slightly more water than *Chlorella vulgaris*. *Spirulina* adsorbs more water than *cp* and *cv*.

The parameters of the GAB model for microalgae powders are shown in Table I. The model presented a good fit to the experimental data (R^2> 0.99). The value of the monolayer

(X_m) indicates the amount of water that is strongly absorbed to specific sites at the powder surface and this can be related to the stability of dried microalgae. Once dehydrated, microalgae are stable at ambient conditions and therefore do not require particular storage conditions.

The solid surface area is associated with the quantity of pores present in the materials. The solid surface area values (less to 160 m^2/g dm) of the microalgae powders were similar to those commonly obtained for non-hygroscopic food products 100-250 m^2/g dm.

The values of K (>0.9) demonstrate that the properties of the multilayer water molecules are similar to those of liquid water. The energy needed to separate the water from the multilayer, is comparable to the energy of the sorption heat of pure water.

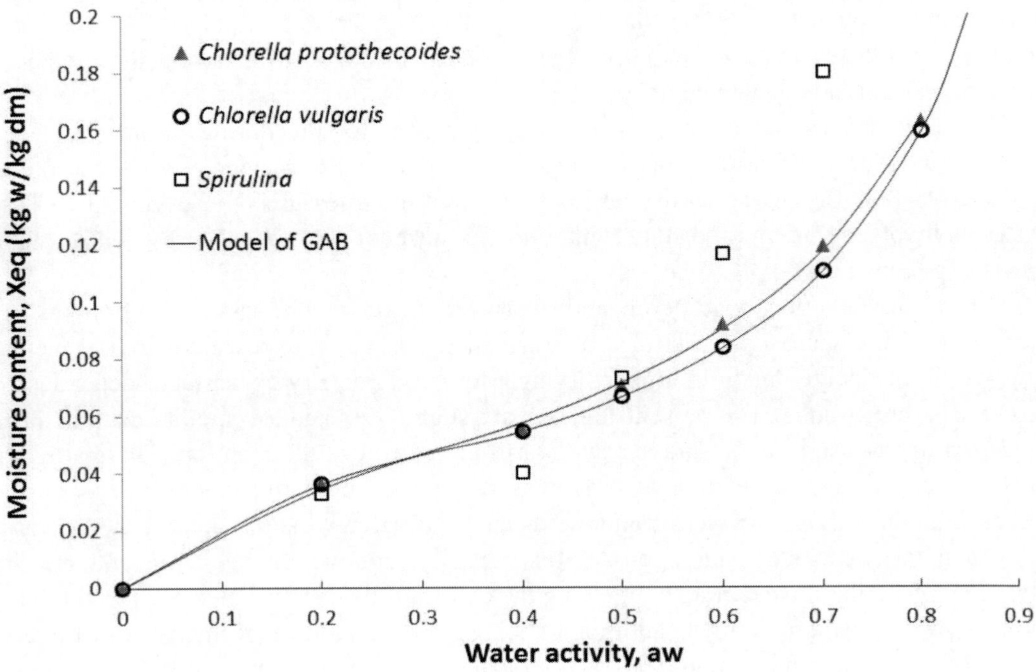

Figure 3. Experimental adsorption isotherm of *Chlorella vulgaris* and *Chlorella protothecoides* at 25°C, obtained by the DVS method. Data of *Spirulina* from Dexmorieux and Decaen [15].

Table I. Adsorption parameters for the GAB model

Parameter	*Chlorella vulgaris*	*Chlorella protothecoides*
Xm	0.0366	0.0443
C	16.4783	7.678
K	0.9682	0.9260
Correlation coefficient R^2	0.9994	0.9990
Residual of sum of squares RSS	9.21E-06	1.72E-05
Standard error SE	0.0018	0.0024
Surface area S_A (m^2/g)	129.34	156.58

The values of C (>7) show that the water is more strongly linked in the monolayer and that the enthalapy values of the monolayer and multilayer water molecules are significantly different. Removing the water to a moisture content of less than 0.045 kg w/kg dm could be a poor energy strategy.

DRYING ALGAE

Algae Drying Methods

Microalgae drying methods are either in batches or continuous, depending on energy and mass transfer. Convection dryers transfer heat to the biomass and eliminate water vapor by circulating hot gases over and between the particles. In conduction drying the biomass is placed in an externally heated container.

The following drying methods have been evaluated for the drying of microalgae:sun drying, drum drying, fluidized bed drying, freeze drying, Refractance Window, rotatory dryers, cross-flow drying, spray drying [3, 15-18] and low-pressure shelf drying [19]. These processes involve removing moisture with a hot air current (convection) or by contact with a hot surface (conduction).

Direct solar radiation, solar dryers and solar water heaters with glass panels or tubes have been used to dry microalgae. The use of solar energy is the cheapest dehydration method, however disadvantages include longer drying times, a large drying surface, dependent on weather conditions and it also presents the risk of material loss and low quality material [19].

Drum dryers use rotation, gravity and warm gasses to dry the material. Thermal energy can be either electrical or steam-heated. However, drum drying with an electric dryer appears to reduce the microbial load, yeasts and molds more effectively than a steam-heated dryer [3, 20]. Advantages include a large production capacity, continuous operation and the dual advantage of sterilizing and quickly breaking the cell wall. Its disadvantages include high heat loss and low thermal efficiency in addition to being a longer and energy intensive method.

In spray drying, the liquid feed is atomized within a chamber containing hot gas, generally at temperatures from 150 to 220°C [21], to cause the evaporation of the liquid and the formation of particles that are recuperated using a cyclone. The atomization system includes a centrifugal rotor, a nozzle and pneumatic and sonic atomization. The droplet size and distribution depends on the type of atomizer, while the droplet shape depends on the rheological properties of the liquid and the interaction with the hot drying medium. Droplet characteristics and drying conditions influence the physical and chemical properties of the resulting powders. This process has an extremely short drying time and requires minimal exposure to heat and oxygen, which reduces compound degradation. However, the high drying temperatures can partially evaporate the lipids and modify fatty acid composition. This process is relatively expensive and can deteriorate some algal pigments [15, 22-23].

Freeze drying is a dehydration process in which the material is frozen below its eutectic temperature and placed in a vacuum chamber. The partial pressure of the vapor surrounding the product is maintained lower than the pressure of the vapor from the ice. In order to sublime the water in the material, heat is transferred to the frozen water.

Freeze drying is the gentlest drying method and also facilitates oil extraction. However, it is too costly for small entrepreneurs and results in the deterioration of some of the algal pigments.

The Refractance Window method uses circulating water at atmospheric pressure to carry thermal energy to the material to be dehydrated. The products are spread out on a transparent plastic conveyor belt and any unused heat is recycled. This method is relatively inexpensive and the equipment is easy to operate.

Drying Mechanisms

Understanding the mechanisms of heat and mass transfer that occur in the drying process is necessary to calculate the size of the dryer as well as to improve product quality.

Heat transfer occurs from the surrounding towards the material, heat conduction from inside the material and evaporation at the surface. Mass transfer occurs from the interior towards the surface and then towards surrounding. Water transport mechanisms within the material have been explained by water migration theories such as: water movement under capillary forces, diffusion of liquid due to concentration gradients, surface diffusion, water vapor diffusion in air-filled pores, flow due to pressure gradient, and flow due to vaporization-condensation sequence [24].

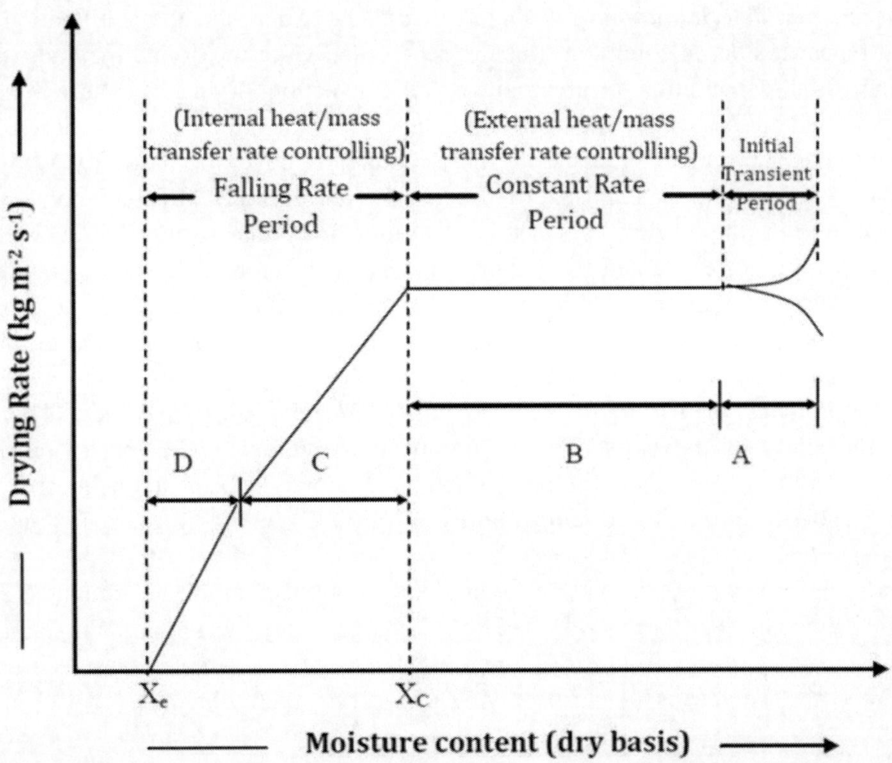

Figure 4. Drying Curve.

The drying processes of materials are extremely complex. Experimental data from drying kinetics of microalgae is indispensable in determining the operating conditions, the time required to reach the desired degree of drying, to understand mechanism of moisture movement and to design the dryers.

In convection drying, external parameters (drying temperature and drying air rate) and internal parameters (structural properties, the chemical composition of the material and the types of bonds with which the water interacts with the compounds) both influence the internal transfer of water. The influence of these parameters is represented in the drying curves (dX/dt vs X) where the constant and falling rate periods can be identified (figure 4).The constant rate period occurs when the free water evaporates, the surface is saturated with water and the rate at which the interior water molecules of the material travel to the surface is equal to the rate of evaporation. The evaporation phenomenon is predominant and is largely dependent on external air conditions.

The falling rate period begins when the interior water molecules are not able to travel to the surface quickly enough and the surface is no longer wet. At this point, referred to as critical moisture (Xc), the process is primarily dependent on the properties of the material and the drying phenomenon is a function of water diffusion or capillarity.

Drying Kinetics

Experimental determination of drying kinetics. The experimental quantification of water loss must be carried out in conditions that are representative of the drying methods to be used in the industrial dehydration of microalgae. In convection drying, moisture kinetics are generally carried out by placing the sample in a tray onto which the drying air flows. In pastes or concentrated suspensions, the material must completely cover the base of the tray, so that the surface of the solid is exposed to the drying air current. Moisture content loss in microalgae during the drying process is determined by measuring the weight of the microalgae with a scale adapted to support the tray at different time intervals. The drying conditions (velocity, humidity, temperature and air direction) must be similar to those of the production processes, the sample must not be too small and the form and depth of the tray must be similar to that used in production.

Drying kinetics of Chlorella vulgaris and Chlorella protothecoides. This chapter presents the study of the convection drying of two microalgae, *Chlorella vulgaris* and *Chlorella protothecoides,* grown in 100L photobioreactors. Two layer thicknesses of algae slurry (1.5 and 2.5 mm) and two temperatures (40 y 60°C) with a constant airflow of 1 m/s were evaluated.

The drying process was carried out with a convection tray dryer, denominated as *drying tunnel for pilot tests* (Pat. No. 304462). The microalgae slurry was placed in stainless steel trays 15.5 cm in diameter (figure 5). The weight loss of the samples was recorded every minute. With the weight loss data, the drying rate curves were obtained for each microalga strain.

Figure 5. Convective drying equipment and trays.

Data analysis. The initial moisture content of the microalgae was determined in triplicate by drying 3.0 g of slurry in an oven at 100°C for 48 h and was calculated with the following equation:

$$X_t = \frac{(W_o - W_s)}{W_s} \quad (5)$$

Where: X_t is the initial water content of the microalgae in a dry base, w_o is the initial sample mass and w_s is the dry weight of the sample.

The drying rate was calculated from the moisture content data measured at different times with the following equation:

$$R = \frac{L_S}{A}\frac{\Delta X}{\Delta t} \quad (6)$$

Where R is the drying rate in kg w/s m^2, L_s is the dry weight of microalgae in kg and A is the surface area of the trays exposed to drying in m^2.

The initial water content of *C.vulgaris* was 12.42 kg w/kg dm, which is equivalent to 92.6% wet basis (kg w/ kg algae slurry).

The drying curves of *Chlorella vulgaris* are presented in figure 6. Experimental repetitions show the reproducibility of the microalgae drying curves for the same conditions: thickness, temperature and drying surface. The drying times at 40°C were less for the 1.5 mm thick layer than for the 2.5 mm thick layer. When the layer is thinner and the temperature greater, the drying time is less.

The constant and falling rate periods were identified in the drying curves (figure 7). During the first period, the surface of the solid is wet, continuously covered with a film of water. This free water layer acts as though there was no solid present. The evaporation rate, with the conditions established for the process, is independent of the solid and is essentially equal to the evaporation rate at free liquid surface. The drying rate of *Chlorella vulgaris* increases as the temperature and air velocity increase and the air moisture decreases, due to its relationship with water evaporation.

The duration of the constant rate period is greater when the layer is thicker, as shown in figure 7 for *Chlorella protothecoides.*

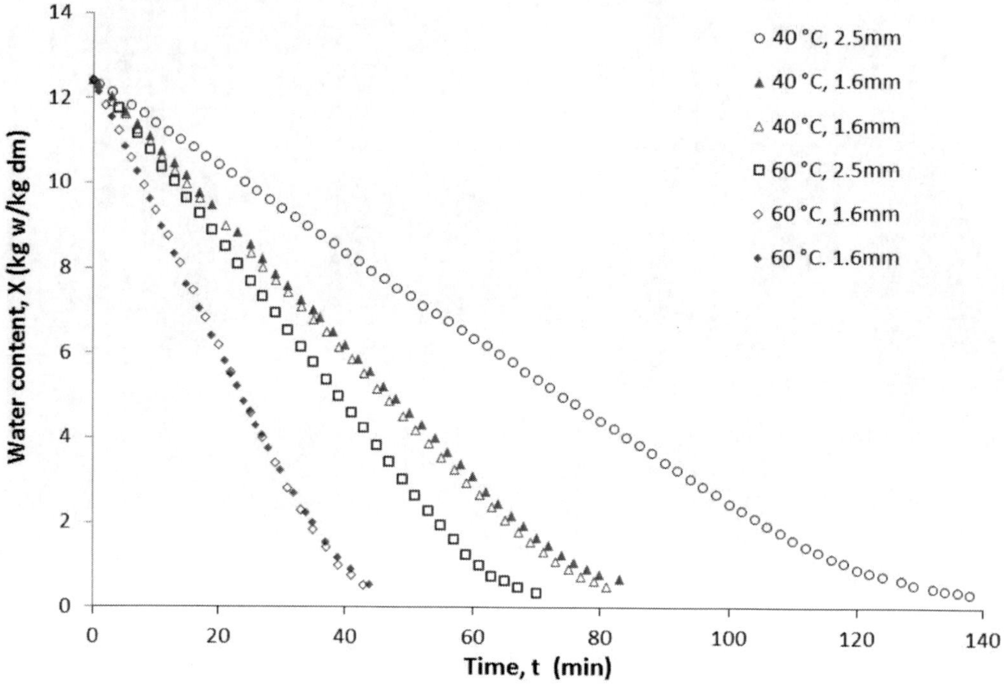

Figure 6. Drying curves of *Chlorella vulgaris.*

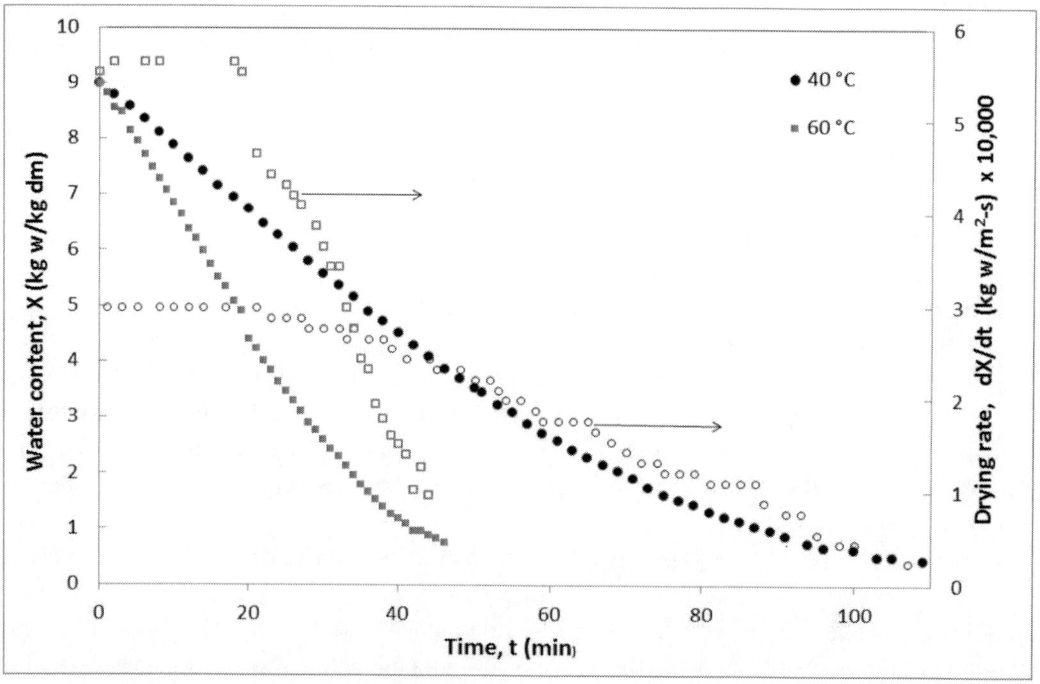

Figure 7. Drying curves of *Chlorella protothecoides*.

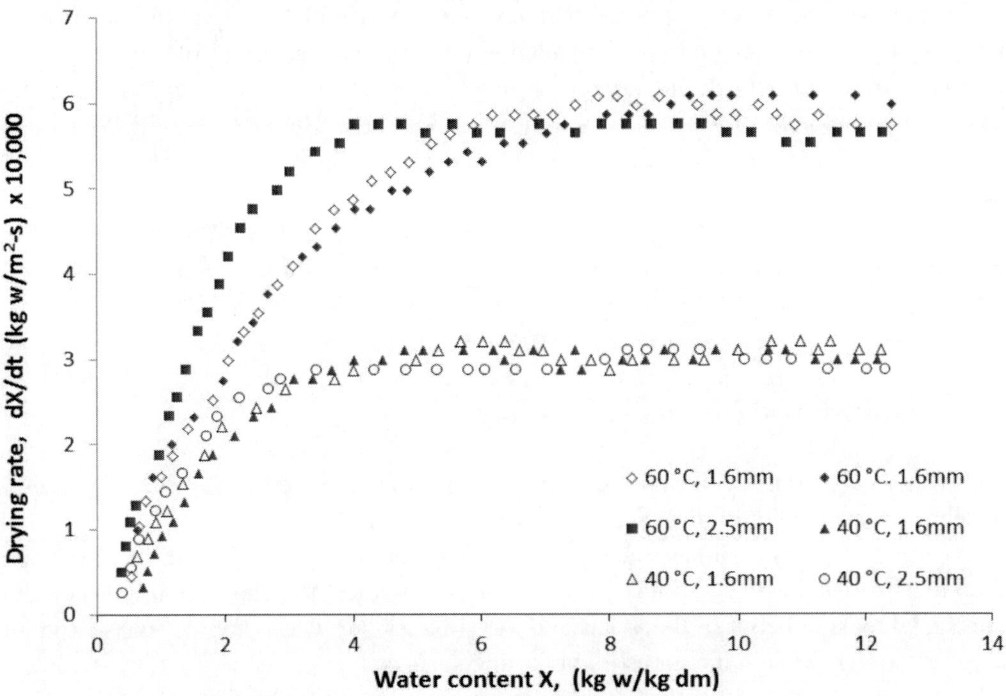

Figure 8. Drying rate vs water content curve for *Chlorella vulgaris.*

The drying rate of *Chlorella vulgaris* is similar to that of *Chlorella protothecoides* at the same drying temperature (figure 8). The constant rate period was more evident in *Chlorella vulgaris,* as its initial humidity content was greater than that of *Chlorella protothecoides.*

In the falling rate period, the surface is no longer completely wet; as the surface begins to dry, the water is transferred from the interior to the surface until the material is completely dry. In the falling rate period the same tendency was observed for both microalga strains and both of the layer thicknesses evaluated.

Drying kinetics of other species. Little data on microalga drying kinetics is available in the bibliography and all of it has been obtained using convective dryers. The species evaluated are *Spirulina, Aphanothece microscopica* Nageli, *Chlorella protothecoides and Chlorella vulgaris.*

A microalga mixture of *Chlorella minutissima*, *Chlamydomonas globosa* and *Scenedesmus bijuga* was evaluated in a thin layer dryer. Aliquots of a microalga slurry mixture with a thickness of 3.15 mm, dried in a convective dryer at a constant parallel air flow of 0.25 m/s a 30, 50, 70 and 90°C were studied by Viswanathan et al.[25]. Drying times of 1200 and 100 min were found at temperatures of 30 and 90°C, respectively.

Cylinders of *Spirulina* biomass were dried in a natural convection dryer at 50°C. Samples 4 mm in height and with diameters of 3, 4 and 6 mm were evaluated. Drying time and drying rate were dependent on the diameter of the cylinder. Drying times of 266, 350 and 475 min and drying rates between 0.2 and 0.005 kg w/s-m^2 were found [26]. In contrast, drying rates between 0 and 0.002 kg w/kg dm-s were reported for *Spirulina* in a thin layer subjected to drying in a convective dryer with a temperature range of 40-60°C and an air speed of 1.9 to 3.8 m/s. Drying rate was found to be dependent on drying temperature [15].

Biomass of *Aphanothece microscopica* Nageli was dried in a tray dryer, using an air speed of 5 m/s in parallel flow at 40, 50 and 60°C with layer thicknesses of 3, 5 and 7 mm. Drying time was dependent on layer thickness and drying temperature and varied from 100 to 550 min [27].

All kinetics show both drying periods, constant and falling. In drying conditions at low temperatures the constant rate period was identified, while at high temperatures only the falling rate period was found. In this period the principal mechanism is diffusion.

Moisture Diffusion in Microalgae Drying

The water mass transfer during the falling rate period is related to the rate of the internal movement of water by liquid diffusion or capillary movement.

The capillary theory assumes that a non-porous sphere-packed bed contains empty spaces between the spheres, known as pores. As the water evaporates, the capillary forces act, due to the interfacial tension between the water and the solid, giving way to an impulsive force that displaces the water through the pores to the drying surface.

The diffusion theory states that during the drying of a solid material, the water moves to of the center to the surface due to an impulsive force that is the difference of the concentration of moisture content. This theory is widely used because of the diffusivity is a parameter of proportionality of the migration of the material in a medium. The term of effective diffusivity considers the different mechanisms by which the water moves inside the material (liquid diffusivity, vapor diffusivity, Knudsen, etc.).

Identifying the mass transfer mechanism requires the analysis of the experimental drying data [28]. The drying mechanisms in the falling rate period are analyzed with logarithm graphs of the kinetic drying experimental data. Diffusivity is determined using the analytical solution to Fick's equation.

Figure 9, shows the diffusion of water from the base of the trays towards the surface of the material, where x=0 is the base of the tray when the microalgae slurry is placed in trays in thin layers for drying.

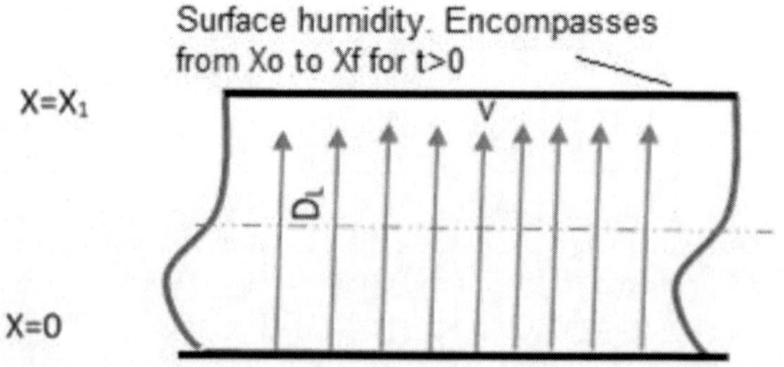

Figure 9. Moisture diffusion in the tray-drying of microalgae.

When the liquid diffusion of the moisture controls the drying rate, in the falling rate period, Fick's equation of mass transfer can be applied.

$$\frac{\partial X}{\partial t} = D_L \frac{\partial^2 X}{\partial x^2} \tag{7}$$

Where:
X: Moisture content (kg w/kg dm)
D_L: Coefficient of liquid diffusion (m^2/s)
x: Coordinate of the microalgae layer thickness (m)

The solution of the equation assumes an initial and uniform moisture distribution over the solid, as well as a surface equilibrium moisture content (X^*). For long drying times the equilibrium moisture content X^* can be substituted by the final humidity content. The solution of Fick's equation simplified to one term in the series is:

$$W = \frac{X_t - X_f}{X_0 - X_f} = \frac{8}{\pi^2} e^{-\frac{D_L \pi^2 t}{x_1{}^{2 \cdot 4}}} \tag{8}$$

Where:
x_1: Total layer thickness (algae) (m).
X_0: Initial moisture content at the time t =0 (kg w/kg dm)
X_f: Final moisture content (kg w/kg dm)
X_t: Moisture content at time t (kg w/kg dm)
t: Drying time (s)

Diffusivity is calculated using the slope of the data (ln W vs t) in a linear form Y=aX+b with the equation (9):

$$D_L = \frac{a \cdot x_1{}^{2 \cdot 4}}{\pi^2} \tag{9}$$

Table 2. Water diffusion coefficient in *Chlorella vulgaris* and *Chlorella protothecoides*

Algae strain	Drying temperature °C	Thickness mm	Diffusion coefficient constant, D_L m^2/s
Chlorella vulgaris	60	2.5	8.1182×10^{-10}
Chlorella vulgaris	40	2.5	4.7746×10^{-10}
Chlorella vulgaris	40	1.6	3.0693×10^{-10}
Chlorella vulgaris	60	1.6	5.6595×10^{-10}
Chlorella protothecoides	40	1.6	3.3254×10^{-10}
Chlorella protothecoides	60	1.6	6.1563×10^{-10}

This method assumes that D_L is constant, but in reality the diffusion coefficient varies according to the moisture content, temperature, etc. The analysis of diffusion during the drying of microalgae should take into consideration the variations in layer thickness over the course of the drying due to shrinking and the decrease in the thickness of the microalgae layer.

Diffusion coefficients in Chlorella vulgaris and Chlorella protothecoides. The mass diffusion coefficients were determined for both strains dried at two temperatures and two layer thicknesses (table 2).

The data for *cp* and *cv* (table 2) reveals that, for both strains, with the same layer thickness, if the drying temperature increases, the diffusion increases. For the same type of algae dried at the same temperature, if the thickness of the layer increases, the diffusion increases. The type of strain does not significantly influence the order of magnitude of the diffusion coefficient.

The results are of the same order of magnitude as the data found in the literature. The diffusivity in *Aphanothece microscópica* Nageli was 9.30 x 10^{-8} , 15.3 10^{-8} and 17.7 x 10^{-8} m^2/s for biomass dried at 40, 50 and 60°C y 3mm thick layers [27]. In *Spirulina* dried at 50°C, the diffusivity reported was 1.06 x 10^{-10} and 1.67 x 10^{-10} m^2/s. In this work layer thickness was not measured, however, a decrease in thickness has been reported in the drying of *Spirulina* and has been incorporated into the expression of diffusivity [26].

STRUCTURAL CHANGES PRODUCED BY THE DRYING PROCESS

Collapse, shrinking and other structural changes produced by the stress of the material, as well as other chemical changes on the surface, occur during drying. These changes can lead to the formation of a crust or case-hardening that impedes mass transfer.

Structural changes in Chlorella vulgaris and Chlorella protothecoides. The macroscopic and microscopic physical characteristics of both microalgae changed during and after drying. Shrinking and reduction of the spaces between cells resulted in thin sheets of microalgae (fig 10). *C. protothecoides* formed thin sheets, opaque on top and shiny on the bottom with a dark, almost black, as can be seen in fig. 10a. *C. vulgaris* produced fractured leaves, forming opaque pieces, dark green on top and light green on the bottom (Figure10b). The final characteristics of the material must be considered to determine the appropriate recuperation method.

Microstructure Photographs from a scanning electron microscope (SEM) and atomic force microscopy(AFM). To compare the cell changes after the thermic treatments, SEM images were obtained using a scanning electron microscope focused ion beam (SEM-FIB), Jeol (JIB 4500). The samples were placed on a stand with carbon conductive double-sided adhesive tape, with no cover. Samples of *Chlorella vulgaris* dried at 40°C in 2.5 mm thick-layers and *Chlorella protothecoides* dried at 40°C in 1.5 mm thick-layers were examined. Samples of spray dried C*hlorella vulgaris* were also analyzed for comparison.

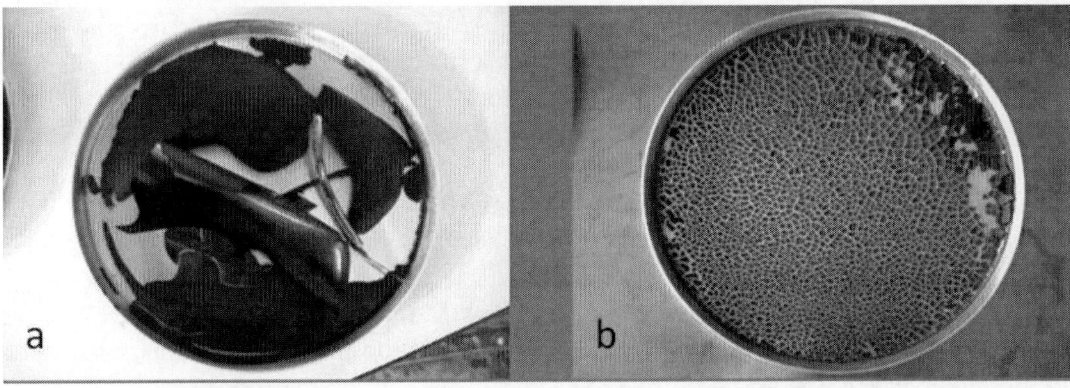

Figure 10. Physical aspect of a) *C. protothecoides* and b) *C. vulgaris* post- tray drying.

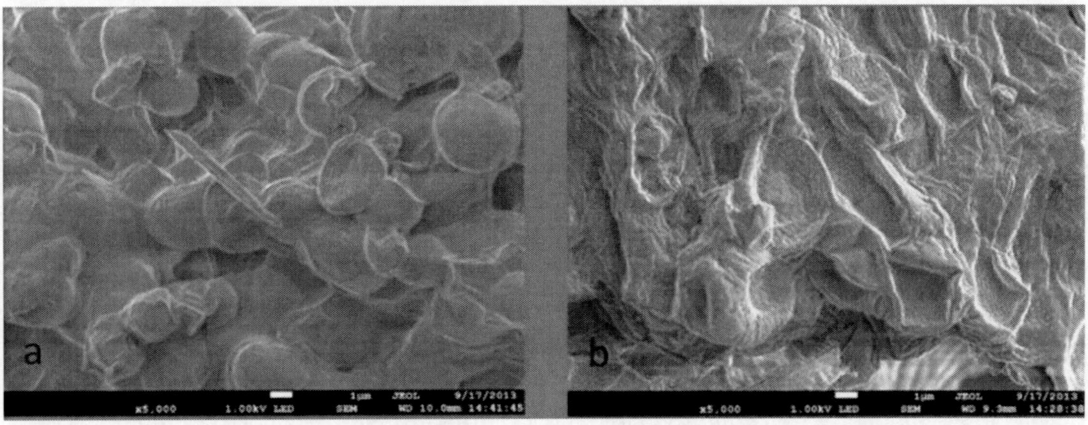

Figure. 11. Tray-dryer samples from a*)* *C. protothecoides* and b) *C. vulgaris* analyzed with a scanning electron microscope (SEM) magnified 5,000X.

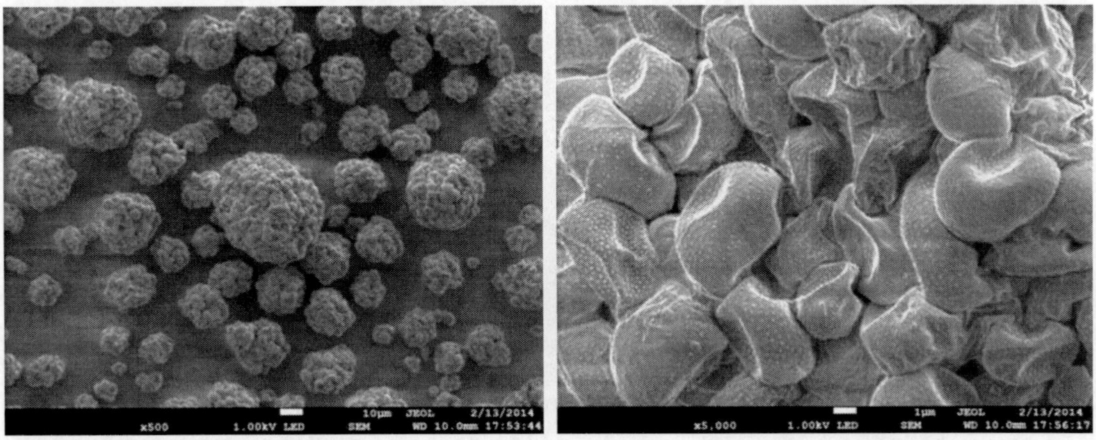

Figure 12. Spray-dried samples of *C. vulgaris* analyzed with a scanning electron microscope (SEM) a) magnified 500X and b) 5,000X.

In the dehydrated *C. protothecoides* material (Figure 11a), structured cells with shrinking can be observed. On the edges of the dry sheets are collapsed cells. The dehydrated microalgae are between 2 and 4 μm. The *C. vulgaris* microalga (Figure11b) has a wrinkled appearance with bumps on the cell surface. The majority of the cells appear to have collapsed. The cellular contraction is due to the large quantity of liquid lost during drying. The SEM images allowed for the identification of the two microalgae strains, based on differences in size and surface structure.

Spray-drying produces spherical powder from 10-50 μm of agglomerated cells (figure 12a). A close-up view reveals a partial collapse. The cell size are among 3 to 6 μm (figure 12b).

The external structure provides the cell walls with some characteristics that explain the tendency of microalgae to group together, to sediment and to interact with other strains of a consortium of algae.

Images of biomass from *Chorella vulgaris* and C*hlorella protothecoides* dried at room temperature were obtained with an atomic force microscope (Multimode V Veeco, USA) using the software Nanoscope 7.3. The samples were held in place with double sided adhesive tape.

Figure 13a shows the structure of a round *C. protothecoides* cell. Figure 13b shows the structure of a *C. vulgaris* cell with its identifying wrinkled cell wall.

The protuberances on the surface of *C. vulgaris* influence interactions with other microalgae and with the culture medium in a way that affects microalgal association and separation from the culture medium. The external morphology of *C. vulgaris* explains why the sedimentation rate of this alga is greater than that of *C. protothecoides*.

Structural changes in other species. It has been suggested that when *Spirulina* is dried, a crust is formed that limits internal heat and mass transfer [15]. Shrinking, porosity and density were studied in *Spirulina* cylinders [26]. Shrinking was linear in the initial stages of drying, but showed non-linear behavior in the final drying stage. Density decreased from an initial 850 kg/m^3 to 300 kg/m^3 at the end of drying, while porosity behaved exponentially. Microscopic images revealed the formation of a crust and cracks.

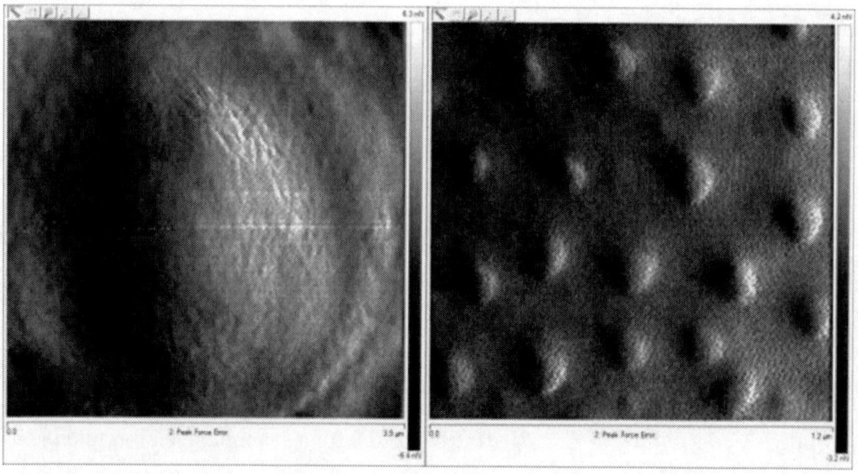

Figure13. Samples dried at room temperature of a) *C. protothecoides* and b) *C. vulgaris* analyzed with an atomic force microscope (AFM).

EFFECT OF THE DRYING ON MICROALGAE LIPIDS

Expressions used in drying. The term total suspended solids (TSS), at times referred to simply as suspended solids (SS), encompasses both inorganic and organic solids. Inorganic solids include clay, silt and sand; organic solids include algae and detritus. Suspended solids are a measurement of the dry weight of the suspended solid per unit of volume of water and are reported in milligrams of solids per liter of water (mg/L).

Dry mass factor (% DMF) is the percentage of dry biomass in relation to fresh biomass. For example, if the dry mass is 5%, 20 kg (44lb) of algae slurry (algae in the medium) are needed to obtain 1 kg (2.2 lb) of dry algae cells. This can be related to the water content in wet base (kg w/kg algae slurry).

$$\%DMF = (1 - X) \cdot 100 \tag{10}$$

Lipid content (% LC) is the percentage of oil in relation to the dry biomass needed to obtain it. For example, if the lipid content in the algae is 40%, 2.5 kg (5.5 lb) of dry biomass are needed to obtain 1 kg of oil. The% of oil in relation to the algae slurry (algae in the medium) is calculated with the following equation.

$$\%\, LC_{wb} = [\%\text{LC}]\left[\frac{\%DMF}{100}\right] \tag{11}$$

Dewatered algae is dried to improve the efficiency of the extraction of lipids, pigments and other products with polar solvents. Drying, prior to extraction is necessary to prevent the humidity from interfering with the solvents. The water blocks the access to the solvents forming a layer around the liquids and making it more difficult for the polar solvents to contact, solubilize and extract lipids [29-31]. Although drying alone is not enough to improve the lipid extraction yield, a 7% yield increase was found for dried, powdered biomass, in comparison with the yield of wet, powdered biomass. In contrast, the lipid yield of dried biomass in comparison with the yield of wet, unground biomass was less than 9% [32]. This has been attributed to the irreversible closure of pores in the cell wall, slowing down or inhibiting solvent access [32].

Drying temperature has been found to affect lipid yield and composition. A greater lipid yield was found when higher drying temperatures were used on a mixture of *Chlorella minutissima, Chlamydomonas globosa* and *Scenedesmus bijuga* [25] and for *Aphanothece microscopic* [33], while an inverse relation was found for *Chlorelle vulgaris* [34].The lipid yields of biomass from *Scenedesmus* sp. And *Nannochloropsis* sp. Freeze dried, oven dried or solar dried were not significantly different [35]. The drying method did affect phytochemical retention in *spirulina*; a phycocyanin retention of less than 54% was reported in *Spirulina* biomass subjected to a crossed flow in a spray dryer [36].

Discontinuous convective drying was used to dry *Spirulina platensis*. The composition of many fatty acids was not significantly affected by drying the biomass at 55°C with a layer thickness of 3.7 mm [37]. High drying temperatures can produce polymerization, geometrical isomerization and intramolecular tryacyglyceride cyclisation reactions, the main precursors to biodiesel production [38-40].

Energy Consumption in Microalgae Drying

Biomass drying is highly dependent on energy. The energy consumed in this process depends on the initial moisture content and the final moisture content desired. The energy required for drying is greater than the latent heat of water. However, few references contain complete information. In the evaluation of dewatering models, it was shown that the production of algal biodiesel requires 3.4 kWh/kg algae dm of which 1.9 kWh/kg dm are necessary to take the biomass from 30 to 90% dry basis [41]. This represents 55% of the energy used in the drying process alone. Other analyses have shown that the drying process represents between 69% (natural gas drying) to 75% of the total harvesting cost because the energy required for drying is greater than the latent heat of water (0.54 kWh/kg) [42-43].

References

[1] Chen, C. Y., Yeh, K. L., Aisyah, R., Lee, D. J. & Chang, J. S. (2011). Cultivation, photobioreactor design and harvesting of microalgae for biodiesel production: A critical review. *Bioresource Technology, 102*, 71-81.

[2] Petrick, I., Dombrowski, L., Kröger, M., Beckert, T., Kuchling, T. & Kureti, S. (2013). *Algae biorefinery material and energy use of algae*: Report No. 16 Deutsches Biomasseforschungszentrum GmbH, Leipzig, Germany: DBFZ

[3] Show, K. Y., Lee, D. J. & Chang, J. S. (2013). Algal biomass dehydration, *Bioresource Technology, 135*, 720–729

[4] Fon Sing, S., Isdepsky, A., Borowitzka, M. A. & Moheimani, N. R. (2013). Production of biofuels from microalgae, *Mitigation and Adaptation Strategies for Global Change 18(1)*, 47-72

[5] Benemann, J. R., Goebel, R. P., Weissman, J. C. & Augenstein, D. C. (1982). Microalgae as a source of liquid fuels, final technical report, U.S. Departament of Energy, Office of Energy Research

[6] Ferrell, J. & Sarisky-Reed, V. (2010). National Algal Biofuels Technology Roadmap, National Algal Biofuels Technology Roadmap. U.S. Department of Energy, Office of Energy Efficiency and Renewable Energy, Biomass Program

[7] Brennan, L. & Owende, P. (2010). Biofuels from microalgae—A review of technologies for production, processing, and extractions of biofuels and co-products, *Renewable and Sustainable Energy Reviews, 14*, 557–577

[8] Molina, G. E., Belarbi, E. H., Acien, F. F. G., Robles, M. A. & Chisti, Y. (2003). Recovery of microalgal biomass and metabolites: process options and economics, *Biotechnology Advances, 20*, 491 – 515

[9] Tomaselli, L. (2004). *In Handbook of Microalgal Culture: Biotechnology and Applied Phycology:The Microalgal Cell.* Negev, Israel.Wiley-Blackwell.

[10] Basu, S., Shivhare, U. S. & Mujumdar, A. S. (2006). Models for sorption isotherms for foods: a review. *Drying Technology, 24(8)*, 917-930.

[11] Quirijns, E. J., Van Boxtel, A. J., Van Loon, W. K. & Van Straten, G. (2005). An improved experimental and regression methodology for sorption isotherms. *Journal of the Science of Food and Agriculture, 85(2)*, 175-185.

[12] Cassini, A. S., Marczak, L. D. F. & Noreña, C. P. Z. (2006). Water adsorption isotherms of texturized soy protein. *Journal of Food Engineering, 77(1)*, 194-199.

[13] Levoguer, C. L. & Williams, D. R. (1997). Moisture sorption properties of foods products and packaging materials studied by dynamic vapor sorption. *Food Technology Europe, 6*, 28–30.

[14] Van den Berg, C. & Bruin, S. (1981). *Water Activity: Influences on Food Quality: Water activity and its estimation in food systems: theoretical aspects*, Rockland, L. B., Stewart, G. T., Eds., London/New York: Academic Press

[15] Desmorieux, H. & Decaen, N. (2006). Convective drying of spirulina in thin layer. *Journal of Food Engineering, 66(4)*, 497–503.

[16] Millamena, O. M., Aujero, E. J. & Borlongan, I. G. (1990). Techniques on algae harvesting and preservation for use in culture and as larval food. *Aquacultural Engineering, 9*, 295–304

[17] Leach, G., Oliveira, G. & Morais, R. (1998). Spray-drying of *Dunaliella salina* to produce a b-carotene rich powder. *Journal of Industrial Microbiology and Biotechnology, 20*, 82-85

[18] Nindo, C. I. & Tang, J. (2007). Refractance window dehydration technology: a novel contact drying method. *Drying Technology,* 25, 37–48

[19] Prakash, J., Pushparaj, B., Carlozzi, P., Torzillo, G., Montaini, E. & Materassi, R. (1997). Microalgal biomass drying by a simple solar device. *International Journal of Solar Energy*, *18*(4), 303–311

[20] Mahadevaswamy, M. & Venkataraman, L. V. (1981). Microbial load in mass cultures of green algae *Scenedesmus acutus* and its processed powder. *Journal of Biosciences, 3*(4), 439-448

[21] Gharsallaoui, A., Roudaut, G., Chambin, O., Voilley, A. & Saurel, R. (2007). Applications of spray-drying in microencapsulation of food ingredients: An overview, *Food Research International, 40*, 1107–1121

[22] Tiburcio, P. C., Galvez, F. C. F., Cruz, L. J. & Gavino, V. C. (2007). Optimization of low-cost drying methods to minimize lipid peroxidation in *Spirulina platensis* grown in the Philippines, *Journal of Applied Phycology, 19*, 719–726

[23] Becker, M. (1994). *Microalgae: biotechnology and microbiology.* Cambridge: University Press

[24] Barbosa-Cánovas, G. V. & Vega-Mercado, H. (1996). *Dehydration of Foods*. International Thomson Publishing, Springer-Science+Business Media, B.V.

[25] Viswanathan, T., Mani, S., Das, K. C., Chinnasamy, S., Bhatnagar, A., Singh, R. K. & Sing, M. (2012). Effect of cell rupting methods on the drying characteristics and lipid compositions of microalgae, *Bioresource Technology, 126,* 131-136.

[26] Dissa, A. O., Desmorieux, H., Savadogo, P. W., Segda, B. G. & Koulidiati, J. (2010). Shrinkage, porosity and density behaviour during convective drying of S*pirulina*. *Journal of Food Engineering, 97*(3), 410–418.

[27] Jacob-Lopes, E., Zepka, L. Q., Pinto, L. A. A. & Queiroz, M. I. (2007). Characteristics of thin-layer drying of the cyanobacterium *Aphanothece microscopica* Nageli, *Chemical Engineering and Processing, 46*, 63-69.

[28] Geankoplis, C. J. (2006). *Transport processes and separation process principles.* Pearson Education Inc.

[29] Ventaka, M. S., Prathima, D. M., Ventaka, S. G. & Chandra, R. (2014). *Biofuels from Algae*: *Algae Oils as Fuels*. A. Pandey, D.-J Lee, Y. Chisti, C. Soccol (Eds.), (1ª. Edition). Burlington, MA, USA:Elsevier

[30] Lee, J. Y., Yoo, C., Jun, S. Y., Ahn, C. Y. & Oh, H. M. (2010). Comparison of several methods for effective lipid extraction from microalgae, *Bioresource Technology, 101*, 75-77

[31] U.S. DOE (2010). *National Algal Biofuels Technology Roadmaop. Office of Energy Efficiency and Renewable Energy. Biomass Program.* Washington: D.C. U.S. Department of Energy (DOE).

[32] Cooney, M., Young, G. & Nagle, N. (2009). Extraction of Bio-oils from Microalgae, *Separation & Purification Reviews, 38*, 291–325

[33] Zepka, L. Q., Jacob-Lopes, E., Goldbeck, R. & Queiroz, M. I. (2008). Production and biochemical profile of the microalgae *Aphanothece* microscopica Nägeli submitted to different drying conditions, *Chemical Engineering and Processing, 47(8)*, 1311-1316.

[34] Widjajaa, A., Chien, C. C. & Ju, Y. H. (2009). Study of increasing lipid production from fresh water microalgae *Chlorella vulgaris*, *Journal of the Taiwan Institute of Chemical Engineers, 40*, 13–20

[35] Guldhe, A., Singh, B., Rawat, I., Ramluckan, K. & Bux, F. (2014). Efficacy of drying and cell disruption techniques on lipid recovery from microalgae for biodiesel production, *Fuel, 128*, 46-52.

[36] Sarada, R., Pillai, M. G. & Ravishankar, G. A. (1999). Phycocyanin from *Spirulina* sp: influence of processing of biomass on phycocyanin yield, analysis of efficacy of extraction methods and stability studies on phycocyanin, *Process Biochemistry, 34*, 795–801

[37] Oliveira, E. G, Duarte, J. H., Moraes, K., Crexi, V. T. & Pinto, L. A. A. (2010). Optimisation of *Spirulina platensis* convective drying: evaluation of phycocyanin loss and lipid oxidation, *International Journal of Food Science & Technology, 45*(8), 1572-1578

[38] Berdeaux, O., Fournier, V., Lambelet, P., Dionisi, F., Sebedio, J. L. & Destaillats, F. (2007). Isolation and structural analysis of the cyclic fatty acid monomers formed from eicosapantaenoic and docosahexaenoic acids during fish oil deodorization. *Journal of Chromatograph A., 1138*, 216-224

[39] Ceriane, R. & Meirelless, J. A. (2007). Formation of *trans* PUFA during deodorization of canola iol: a study through computational simulation, *Chemical Engineering and Processing: Process Intensification, 46*(5), 375-385

[40] Kirrolia, A., Bishnoi, N. R. & Singh, R. (2013). Microalgae as a boon for sustainable energy production and its future research & development aspects. *Renewable & Sustainable Energy Reviews, 20*, 642-656

[41] Lohrey, C. & Kochergin, V. (2012). Biodiesel production from microalgae: Co-location with sugar mills, *Bioresource Technology, 108*, 76–82

[42] Sander, K. & Murthy, G. S. (2010). Life cycle analysis of algae biodiesel, *The International Journal of Life Cycle Assessment, 15*, 704-714.

[43] Mohn, F. H. & Soeder, C. J. (1978). Improvement technologies for harvesting and processing of microalgae and their impact on production costs. *Archiv für Hydrobiologie–Beiheft Ergebnisse der Limnologie, 11*, 228-253

In: Microalgae and Other Phototrophic Bacteria
Editor: Luis G. Torres Bustillos
ISBN: 978-1-63482-078-3

Chapter 12

MICROALGAE AS SOURCE OF HIGH ADDED-VALUE COMPOUNDS: CAROTENOIDS

León Sánchez[1] and Marcia Morales[2]
[1]Biotecnología, Universidad Autónoma Metropolitana-Iztapalapa, México
[2]Departamento de Procesos y Tecnología, Universidad Autónoma Metropolitana-Cuajimalpa, Colonia Santa Fe Cuajimalpa, México

ABSTRACT

Microalgae can be used as cell factories (biofactories) for the industrial production of carotenoids. Carotenoids are high-value chemicals with health benefits. An economically feasible production of carotenoids with microalgae is possible because microalgae produce these compounds directly from the solar energy at high photosynthetic efficiencies. In addition, microalgal production systems have high volumetric and areal productivities. Research contribution in biosynthesis, metabolic regulation, and metabolic engineering of carotenoids will support the future exploration of novel carotenoid production strategies. However, it is necessary to improve culture methods and downstream processing because they represent the major challenge in the massive production due to the high cost and energy requirement.

INTRODUCTION

Over the last years microalgae have received a lot of interest in order to obtain a variety of marketable chemicals products because of their potential uses as a feedstock (Wijffels et al., 2010). Microalgae have the ability to accumulate a variety of biomolecules such as polyunsaturated ω-3 fatty acids (PUFA's), sterols, phenolic compounds, terpenes, enzymes, polysaccharides, alkaloids and pigments, all of them enhances the nutritional value of human food and animal feed and some are used as supplements, cosmetics, biomaterials, nanostructures or in pharmaceutical industry. In addition, microalgae are used to produce renewable biofuels (e.g., biodiesel, biohydrogen and bioethanol) based on the carbon dioxide (CO_2) capture and providing an alternative to reduce greenhouse gas emissions (Ferreira et al., 2013; Marques et al., 2011; Wang et al., 2008). Despite of being more popular for biofuel, microalgae have gained a lot of attention as a source of biomolecules and biomass for food and feed purposes.

The main limitation on the use of microalgae is the high cost of industrial production of biomass (Becker, 1994). Nowadays, large-scale production of chemical products, obtained from microalgae biomass, involves the use of raceway ponds or tubular photobioreactors to produce the algal broth, where the biomass productivity is the principal bottleneck (Hejazi and Wijffels, 2004). In addition, the biomass must be recovered and the metabolite of interest must be isolated and purified, these processes increase the cost and the energy requirements (Mercer and Armenta, 2011).

This chapter will provide a brief overview of the steps involved in the production of carotenoids from microalgae focusing in the technologies of production and separation of the value added chemical.

CAROTENOIDS

Carotenoids are the most common pigments in nature; they are synthesized by all photosynthetic organisms such as microalgae or plants but can also be produced by bacteria, yeast and fungi. Microalgae are prokaryotic or eukaryotic photosynthetic microorganisms that produce carbohydrates, proteins, pigments and lipids as a result of photosynthesis (Travieso et al., 2006). In microalgae, carotenoids have important biological function as accessory light-harvesting pigments, structural components for photosystem assembly, moderators of non-photochemical quenching, and scavengers of reactive oxygen species (ROS) (Britton et al., 1999). At least 700 carotenoids have been characterized from the two known C30 and C40 carotenoid biosynthetic pathways. Most of them are distributed in the C40 pathway, which is found in thousands of plant and microbial species (Naik et al., 2003; Hornero-Mendez and Britton, 2002).

Carotenoids represent a large group of biological chromophores with an absorption range between 400 and 550 nm. The basic structural elements of carotenoids are two hexacarbon rings joined by an 18-carbon, conjugated double-bond chain. They are usually either hydrocarbons, named carotenes with a linear chain terminated by cyclic groups (rings), the most abundantare α-carotene and β-carotene; or oxygenated carotenoids globally named xanthophylls that can be classified as those containing the hydroxyl (as in lutein) or keto (as in cantaxanthin) groups, or the combination of both groups (as in astaxanthin) (Figure 1). Carotenoids have also been classified as primary and secondary compounds. Primary carotenoids, such as lutein, act as structural and functional components of the photosynthetic apparatus, so they are essential for survival in photoautotrophic culture conditions. Secondary carotenoids, such as astaxanthin, accumulate in large quantities in oil bodies within or outside cells when chloroplasts exposed to various stress conditions (Baroli and Niyogi 2000; Demming-Adams et al., 1996). As a member of carotenoids, β-carotene is found in various configuration, the double bonds enable the formation of cis or trans isomers. The all-*trans*, 9-*cis*, 13-*cis* and 15-*cis* stereoisomers of β-carotene have been identified in natural sources (Patrick, 2000). Stereoisomers of β-carotene display various biokinetics and biological activities. The 9-*cis* to all-*trans* ratio is proportional to the integral light intensity during cell-division cycle (Ben-Amotz et al., 1996).

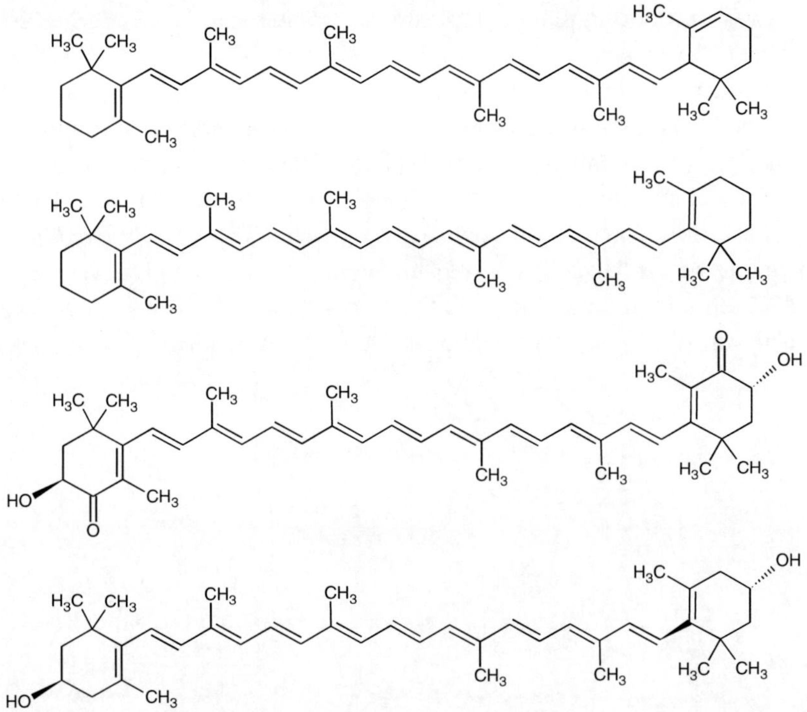

α-carotene

β-carotene

Astaxanthin

Lutein

Figure 1. Chemical structure of α-carotene, β-carotene and of the xanthophylls astaxanthin and lutein, main carotenoids from microalgae with commercial interest.

CAROTENOID BIOSYNTHESIS

The carotenogenesis pathways and their enzymes have been investigated in cyanobacteria and land plants (Britton, 1998; Takaichi and Mochimaru, 2007). Findings show that algae have common pathways with land plants but there are also additional specific pathways for algae. These pathways are solely proposed based on the chemical structures of carotenoids. Some common carotenogenesis genes in algae are suggested from homology of the known genes, but most genes and enzymes for algae-specific pathways are still unknown. Although the biosynthetic pathways of chloroplast pigments have been extensively studied in higher plants and green algae, the research on carotenoid biosynthesis in diatoms is still in its infancy (Bertrand, 2010; Frommolt et al., 2008).

Isopentenyl pyrophosphate (IPP) is the precursor for carotenoid synthesis (Lichtenthaler, 1999), is a C5-compound, and is also the source of isoprenoids, terpenes, quinones, sterols and phytol of chlorophylls. Two distinct pathways for IPP biosynthesis have been found: the mevalonate (MVA) pathway in the cytosol and the non-mevalonate 1-deoxy-D-xylulose-5-phosphate pathway in the chloroplast (DOXP pathway or MEP pathway) (Eisenreich et al., 2004; Lichtenthaler et al., 1997; Ramos-Valdivia et al., 1997). The main differences between the two pathways are in: 1) its location and, 2) in the initial step condensation. In the MVA pathway, three molecules of acetyl-CoA condense, and six enzymes are required, two NADPH, being regulated, and three ATPs to finally yield IPP, being regulated by the HMG-CoA reductase. The pathway is found in plant cytoplasm and some bacteria (Miziorko, 2011; Lichtenthaler, 1999). The DOXP pathway was found in the 1990s, and in this pathway, condensation of pyruvate and D-glyceraldehyde-3-phosphate is produced, resulting in the last step IPP, subsequently catalyzes the isomerization of IPP by isopentenyl pyrophosphate isomerase (IPI) to dimethylallyl diphosphate (DMAPP) (**Figure 2**) (Lichtenthaler, 1999). The MEP pathway is found in cyanobacteria, the plastids in algae, land plants and some bacteria (Miziorko, 2011). In unicellular green microalgae *H. pluvialis* and *Chlamydomonas reinhardtii*, IPP is believed to be synthesized solely from the non-mevalonate DOXP pathway (Disch et al., 1998). Carotenoids are normally synthesized in plastids. But exceptionally, *Euglenophyceae* and *Chlorophyceae* have only the MVA or MEP pathways, respectively (Miziorko, 2011).

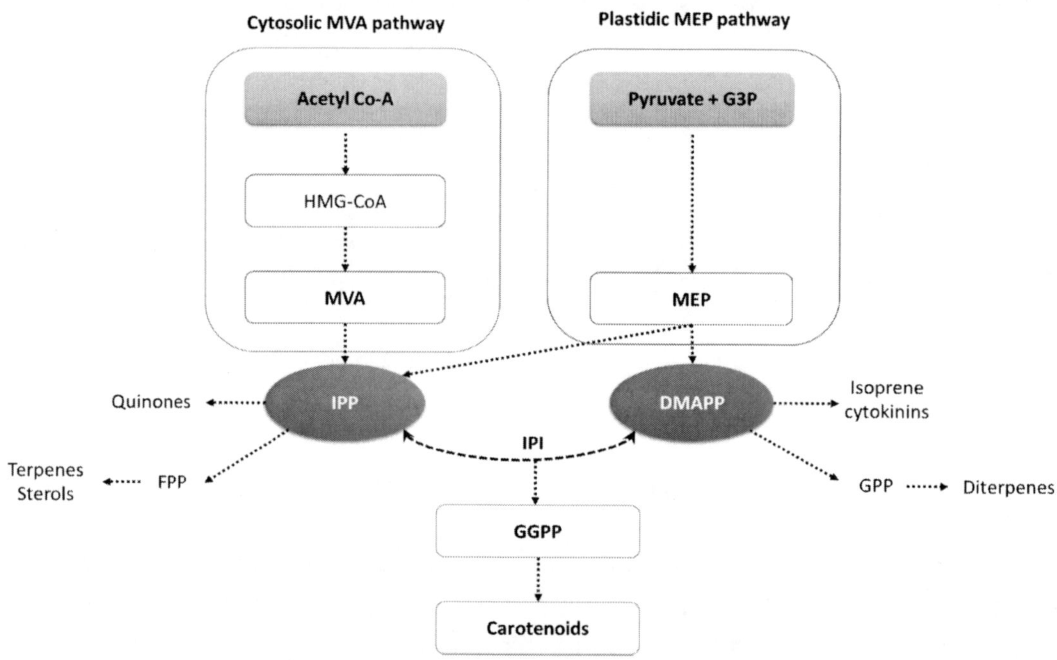

Figure 2. Scheme for the synthesis of IPP by pathways MVA and MEP, also DMAPP by pathway MEP. G3P, glyceraldehyde-3-phosphate, GPP, geranyl pyrophosphate, HMG-CoA hydroxyl-methylglutaryl-CoA, IPI, isopentenyl pyrophosphate isomerase, GGPP, geranylgeranyl pyrophosphate, FPP farnesyl pyrophosphate.

There is evidence that green algae have lost cytosolic MVA pathway for the formation of IPP and MEP pathway provides the building blocks for the biosynthesis of all cellular isoprenoid (Lohr et al., 2012; Rohmer, 1999). The next step in the biosynthesis of isoprenoids continues by adding several IPP molecules synthesized by either of the two pathways, a head-to-tail arrangement depending of the final product to be formed. The prenyltransferases and their different chain-length specificities develop chain elongation, and finally the short-chain prenyl transferases synthesize geranyl diphosphate (GPP, C10), farnesyl diphosphate (FPP,

C15), or geranylgeranyl diphosphate (GGPP, C20), which are the direct precursors of mono-, di-, and tri-terpenes and carotenoid (Lee and Schimdt-Dannert, 2002).

BIOSYNTHESIS OF CAROTENES IN MICROALGAE

In cyanobacteria, microalgae, and photosynthetic bacteria, the biosynthesis of isoprenoids follows the non-mevalonate, plastidic DOXP/MEP pathway (Miziorko, 2011; Lichtenthaler, 2004). The absence of the MVA pathway in Chlorophyta is supported in the fact, that, could not find inhibitor effects, when applied mevinolin and cervastin, highly specific inhibitors of HMG-CoA reductase (Bach and Licchtenthaker, 1987) on the growth of *Scenedesmus obliquus* (Schwender et al., 1996) or *Chlamydomonas reinhardtii*. Also, it was not detected gene sequences homologous to HMG-CoA reductase in several Chloropyta species such as *Chlamydomonas* (Schwender et al., 2001). After IPP has been synthesized by MEP phatway, the condensation of two molecules of GGPP, with elimination of two diphosphate groups, is catalyzed by phytoene synthase (PSY, crtB, al-2). It produces the first colorless carotene phytoene (C_{40}) and represents the first-committed step in the carotenoid biosynthetic pathway. PSY has been considered to be a key enzyme in carotenoids biosynthesis, and it may be the regulatory point determining the flux of carbon source towards carotenoids (Cordero et al., 2011; Huang et al., 2006; Schwender et al., 1996). The gene PSY have been studied and associated with some green algae such as: *Chlamydomonas reinhardtii, Chlorella vulgaris, Volvox carteri*, also with diatoms *Thalassiosira pseudonana* and *Phaeodactylum tricornutum* (Tran et al., 2009). It seems that other gene copies were lost in these species. Actually, gene duplication of PSY has been reported in the green algae, *Dunaliella salina* and *Dunaliella bardawil*; the existence of two classes of PSY gene families in some algae and suggests that carotenoid biosynthesis in these algae is regulated in response to development and environmental stress as well (Ye et al., 2008).

Once the pythoene is biosynthesized, the next step in carotenogenesis is the desaturation reaction (or dehydrogenation reaction) resulting a conversion from phytoene to lycopene. Phytoene is desaturated four times by phytoene desaturase (CrtP, PDS) and ζ-carotene desaturase (CrtQ, ZDS) and isomerized by 15-cis-ζ-carotene isomerase (CrtH, Z-ISO) (Chen et al., 2010) and carotene isomerase (CrtI) to form the linear all trans-lycopene. CrtP catalyzes the first two-desaturation steps, from phytoene to δ-carotene through phytofluene, and CrtQ catalyzes two additional desaturation steps, from δ-carotene to lycopene through neurosporene. During desaturation by CrtQ, neurosporene and lycopene are isomerized to poly-cis forms, and then CrtH isomerizes to all-trans forms. In bacteria and fungi, these four dehydrogenation reactions are performed by a single enzyme, the CrtI type PDS, which is coded by CrtI (Albrecht et al., 1999). Light is also effective for their photoisomerization to all-trans forms (Masamoto et al., 2001). The functions of these enzymes have been mainly confirmed in cyanobacteria, green algae and land plants such as in CrtP from *Synechocystis* sp., PCC 6803 (Martínez-Férez and Vioque, 1992), *Synechococcus elongatus* PCC 7942 (Chamovitz et al., 1992), *Chlamydomonas reinhardtii* (Vila et al., 2008) and *Chlorella zofingiensis* (Huang et al., 2008).

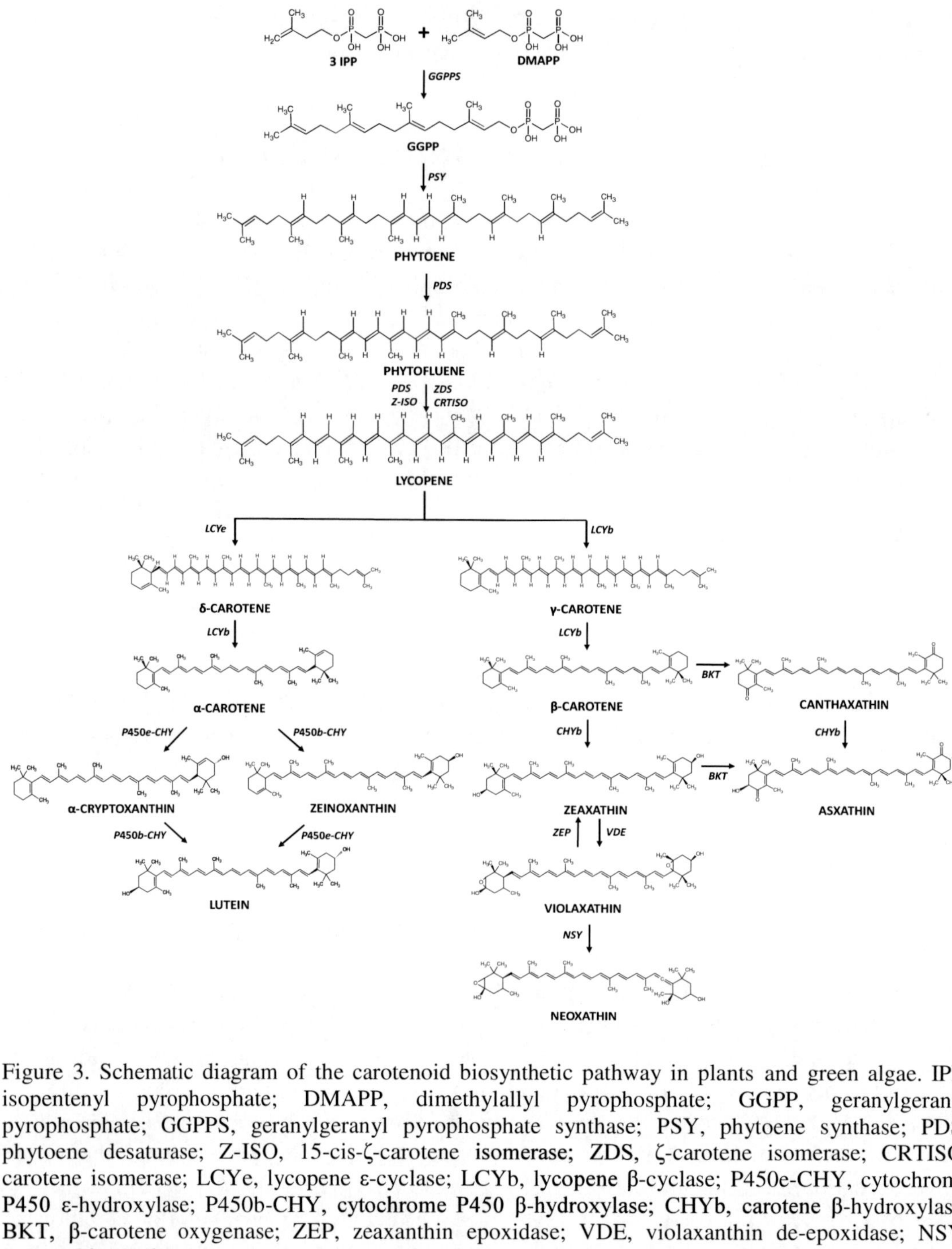

Figure 3. Schematic diagram of the carotenoid biosynthetic pathway in plants and green algae. IPP, isopentenyl pyrophosphate; DMAPP, dimethylallyl pyrophosphate; GGPP, geranylgeranyl pyrophosphate; GGPPS, geranylgeranyl pyrophosphate synthase; PSY, phytoene synthase; PDS, phytoene desaturase; Z-ISO, 15-cis-ζ-carotene isomerase; ZDS, ζ-carotene isomerase; CRTISO, carotene isomerase; LCYe, lycopene ε-cyclase; LCYb, lycopene β-cyclase; P450e-CHY, cytochrome P450 ε-hydroxylase; P450b-CHY, cytochrome P450 β-hydroxylase; CHYb, carotene β-hydroxylase; BKT, β-carotene oxygenase; ZEP, zeaxanthin epoxidase; VDE, violaxanthin de-epoxidase; NSY, neoxanthin synthase.

The carotenoid biosynthetic pathway is divided again into two diverging branches to level lycopene in plants, some types of algae, such as green algae, and some cyanobacteria. In one branch, the lycopene is converted into β-carotene by the action of lycopene β-cyclase (LCYb). LCYb introduces a β ring at both ends of the linear lycopene molecule to form β-carotene, which is converted to zeaxanthin, violaxanthin and only in some green microalgae, astaxanthin. In the second into α-carotene by the action of lycopene ε-cyclase (LCYe) and

LCYb that introduces a ring at one end of lycopene to form δ-carotene, which is transformed by the LCYb in α-carotene, which is then hydroxylated to lutein (Cunningham et al., 1996). Cyclization of α-carotene lycopene (with ε ring ring and β) or β-carotene (β two rings) is therefore an important branch point in carotenoid biosynthesis and has been proposed as a point regulation; the relative activities of LCYb and LCYe determine the proportion of carotenoids (Cordero et al., 2011; Sandmann et al., 2006). α-carotene is modified into lutein by the hydroxylases P450b-CHY and P450e-CHY, and β-carotene is hydroxylated by CHYb to zeaxanthin. Zeaxanthin epoxidase (ZEP) and violaxanthin de-epoxidase (VDE) catalyze the interconversion of zeaxanthin and violaxanthin (Cordero et al., 2011; Kim et al., 2009), and neoxanthin is formed from violaxanthin by the action of neoxanthin synthase (NSY). A limited number of organisms including some green algae such as *Haematococcus pluvialis* and *C. zofingiensis* can synthetize astaxanthin from β-carotene by the action of a ketolase/oxygenase (BKT) and the hydroxylase (CHYb) (Vidhyavathi et al., 2009; Ojima et al., 2006; Fan et al., 1995). Currently, lutein and astaxanthin are widely used as feed additives in poultry farming and aquaculture. They have also important applications in food, nutraceutical and pharmaceutical industries because of their antioxidant activity and beneficial effects on human health (Guedes et al., 2011).

β -CAROTENE

Only a few microbial organisms accumulate β-carotene to a high extent. The fungi *Phycomyces blakesleanus* and *Brakeslea trispora*, the yeast *Rhodotorula*, and the alga *Dunaliella* (Johnson and Schroeder, 1996). The most common carotenoid, β-carotene, is present in green leafy plants (parsley, spinach, broccoli), fruits (mandarin, peach), and several vegetables (carrot, pumpkin). Industrial production methods using the mucoral fungus *Brakeskea* (30 mg of β-carotene per gram mycelium and 3 gL^{-1}) have been developed (Ninet and Renault, 1979). Nevertheless, the most important process for natural production of β-carotene is the culture of the green, unicellular alga *Dunaliela.* It is a cell wall-less green flagellate belonging to the order Volvocales (Chlorophyceae, Chlorophyta) (Polle et al. 2009; Oren 2005). There are about 26 saltwater species described for the genus of *Dunaliella* (Polle et al. 2009; Borowitzka and Silva 2007; Oren, 2005). *D. salina* teodoresco is, a model organism, characterized for the high of β-carotene accumulation in oil globules in the interthylakoid spaces of their chloroplast, depending of environmental conditions (Polle et al., 2009; Lamers et al., 2010). Under conditions of high salinity, a stress temperature, high light intensity, and nitrogen limitation *Dunaliella* accumulates up to 12% of the algal dry weight as β-carotene (Ben-Amotz, 1999). *D. salina* represents an interesting model of carotenogenic pathway. Carotenoid biosynthesis in *D. salina* depends on the supply of the MEP-derived precursors (Capa-Robles et al., 2009; Paniagua-Mitchel et al., 2009). Actually only the sequences of phytoene synthase (PSY) and phytoene desaturase (PDS) of some species of *Dunaliella* have been analyzed and published basis on the data of National Center for Biotechnology Information (NCBI, http:/ /www.ncbi.nlm.nih.gov). Recently, Ye et al. (2008) cloned and sequenced the lycopene clyclase (LYC) of *D. salina* and *D. bardawil*. Globally, the carotenogenic pathway in this microalga could be divided in three steps: 1) the biosynthesis of GGPP, 2) the lycopene biosynthesis and 3) the generation of carotenoids with cyclohexenes rings that added to either or both ends of lycopene and are catalyzed by lycopene cyclase (LYC) leading to the formation of δ-, α-, γ-, and β –carotene (Ye et al., 2008). In other words, the metabolic pathway for cyclic carotenoids is dominant in carotenogenesis in *Dunaliella.* Two types of LCY have been found in plants, lycopene β–cyclase (LCY-B) and lycopene ε-cyclase (LCY-E). It has been postulated that these two cyclases migth exist in *Dunaliella* based on the analysis of *Dunaliella* carotenoids (Ye et al., 2008).

Considerable advances in the carotenogenesis have been addressed, but there are some mechanisms affecting the transmission of signals between the genes precursors, therefore, the identification of mechanisms and processes by which carotenoid biosynthesis is regulated consititute one of the great challenges from studies of carotenoid metabolism. Commonly, oxidative stress occurs by high light exposure that induces ROS generation and causes oxidative damage to cellular components such as DNA, proteins, lipids, and pigments (Skórzynska-Polit and Krupa, 2006; Foyer et al., 1994). However, ROS are also a molecular messenger involved in signaling pathways in biological systems under physiological conditions or stress conditions (Hideg, 2006). It has been suggested that ROS are involved in trigger of β-carotene accumulation in *Dunaliella.* Commonly, internal antioxidants systems in photosynthetic organisms consist of two protection mechanisms, enzymatic and non-enzymatic (Bouvier et al., 1998). In the enzymatic mechanism, ROS act as cytotoxic compounds and also mediate the induction of stress tolerance. To protect cellular membranes and organelles of ROS damaging effects, complex antioxidant systems are activated, including antioxidants and antioxidant enzymes such as superoxide dismutase (SOD), catalase (CAT), glutathione peroxidase (GPX), ascorbate peroxidase (APX), and glutathione reductase (GR). Changes in antioxidant enzyme activities and isozyme patterns under stress conditions may serve as indicators of biotic and abiotic stresses (Shin and Lee, 2003; Lee, 2002). SOD comprises a group of metaloenzymes that catalyzes the disproportionation of superoxide to H_2O_2 and O_2 and plays an important role in protection against superoxide-derived oxidative stress in the cells (Woo et al., 2007; Asada et al., 1974). The detoxification of cellular H_2O_2 through the activity of the Asada-Halliwell scavenging cycle is an important step in the defense mechanism against active oxygen species. The cycle, which occurs in the chloroplast and cytosol, involves the oxidation and re-reduction of ascorbate and glutathione through APX and GR (Mittler and Zilinskas, 1993). APX catalyzes the reaction of ascorbic acid with H_2O_2, and GR catalyzes the regeneration of ascorbic acid (Kang et al., 1999). Catalase can also reduce H_2O_2 to water, but it has a very low affinity for H_2O_2 compared to APX (Kuk et al., 2003). To overcome the damage of ROS under stress conditions, *Dunaliella* increases the activity of SOD and also catalyse the accumulation of large amounts of β-carotene (Rabinowitch et al., 1987). Nokookar et al. (2005) found that the accumulation of β-carotene and the increase in APX activity occurs when *D. salina* and *D. teriolecta* is under copper stress. Bouvier et al. (1998) and Rabbani et al. (1998) found that the amount of either PSY or PDS of *D. brardawil* changed, as well as the transcript levels for PSY, during the high light treatment. They also revealed that intensively increased carotenogenesis typically occurred after acyl lipids biosynthesis. Sánchez-Estudillo et al. (2006) studied the carotenogenesis under nitrogen-sufficient and nitrogen-limited conditions and established that the dxs gene, whose produce 1-deoxy-d-xylulose-5-phophate (DXS), diminished under nitrogen-limited conditions while PSY is constantly expressed in both conditions. Moreover, acyl lipid biosynthesis may affect the carotenogenesis in *Dunaliella* to some extent because of the relation between acyl lipids biosynthesis and carotenogenesis (Ye et al., 2008). Further, it seems that under adverse conditions, enhanced carotenogenesis in *Dunaliella* may undergo a regulatory mechanism closely related to ROS that triggers expression of carotenogenic genes. Also, due to insufficient study, relation between the expression on carotenogenic genes and massive carotenoids accumulation in *Dunaliella* is still poor. However, further work must been carried out to clarify the regulation of β–carotene accumulation under carotenogenic conditions, in *Dunliella.*

ASTAXANTHIN

Astaxanthin is a ketocarotenoid synthesized in by microalgae (*Haematococcus pluvialis* and *Chlorella zofingiensis*), plants, bacteria (*Agrobacterium aurantiacum*), yeast (*Phaffia rhodozym, Xanthophyllomyces dendrorhous*), and fungi. In microalgae, primary carotenoids (e.g., β-carotene, zeaxanthin, and lutein) constitute structural and functional components of the photosynthetic apparatus and astaxanthin is a secondary carotenoid accumulated in cytosolic lipid bodies (LBs) only under environmental stress or adverse culture conditions, such as high light intensity, high salinity, or nutrient deprivation (Johnson and Schroeder, 1996; Johnson and An, 1991). Astaxanthin has thirteen conjugated double bonds arranged as alternating single-double bonds. This configuration confers astaxanthin strong antioxidant properties thereby scavenging ROS and neutralizing free radicals (Vershinin, 1999; Miki, 1991). Astaxanthin can be produced in various amounts by a number of microalgae showed in Table 1.

Table 1. Production of astaxanthin by microalgae

Microalgae	Astaxanthin (% in dwt)	Reference
Botryococcus braunii	up to 0.01	Grung et al., 1994
Chlamydocapsa spp.	0.04	Leya et al., 2009
Chlorella zofingiensis	0.7	Orosa et al., 2000
Chlorococcum sp.	0.7	Liu and Lee, 2000 Ma and Chen, 2001
Chloromonas nivalis	0.004	Leya et al., 2009; Remias et al., 2005
Haematococcus pluvialis	4	Droop, 1954; Lee and Ding, 1994
Neochloris wimmeri	1.9	Orosa et al., 2000
Protosiphon botryoides	1.4	Orosa et al., 2000
Scenedesmus sp.	0.3	Orosa et al., 2000; Qin et al., 2008
Scotiellopsis oocystiformis	1.1	Orosa et al., 2000

Only two of these species have been commercially atractive to produce astaxanthin, the green microalga *H. pluvialis*is recognized as one of the most promising producer in nature due to its exceptional ability to accumulate large amounts of astaxanthin under stress conditions (Lemoine and Schoefs, 2010; Boussiba, 2000), and *C. zofingiensis* has attracted some interest due to its ability to grow rapidly (with a maximum specific growth rate of 1.03 d^{-1} and biomass concentration of 53 g L^{-1} when grown on glucose) and adopt a photoautotrophic, mixotrophic, or heterotropic culture mode (Sun et al., 2008; Del Campo et al., 2004; Ip et al., 2004; Orosa et al., 2000).

The biosynthesis of astaxathin by *H. pluvialis* follows the MEP pathway where IPP is isomerized by IPI and produced GGPP. In *H. pluvialis* has been cloned and characterized two cDNAs of IPI, IPI1 and IPI2 (Wang et al., 1999; Kajiwara et al. 1997). Transcripts of both IPI genes increased in response to oxidative stress, but only IPI2 was up-regulated at the translational level. Moreover, only IPI2 protein was detected in mature red cysts in which astaxanthin was accumulated, suggesting that IPI2 is responsible for synthesis of the secondary carotenoids, whereas the IPI1 is responsible for primary carotenoid synthesis in the chloroplast of *H. pluvialis* (Sun et al., 2008). Then GGPP is catalyzed by PSY to synthetize phytoene. Two classes of PSY were found in certain green algae like *Ostrecoccus* and *Micromonas*, while some other green algae like *C. reinhardtii* and *C. vulgaris* only possess one class of PSY (Tran et al., 2009). One copy of PSY gene has been cloned and characterized from a number of microorganisms including *H. pluvialis* (Steinbrenner and Linden, 2001) and *C. zofingiensis* (Cordero et al., 2011). The *Haematococcus* PSY has an N-terminal extension similar to a chloroplast targeting sequence, indicating that PSY is likely to be targeted to the chloroplast in *H. pluvialis* (Steinbrenner and Linden, 2001). On the next step, a series of dehydrogenation reactions by PDS and ZDS convert phytoene to lycopene. These dehydrogenation reactions extend the conjugated carbon-carbon double bonds to form the chromophore of carotenoids. These FAD-containing enzymes require PTOX and plastoquinone (PQ) as electron acceptors (Carol et al., 1999; Wu et al., 1999). The PDS and two PTOX genes such as PTOX1 and PTOX2, and have been cloned and characterized in *H. pluvialis* (Li et al., 2010; Wang et al. 2009; Grunewald et al. 2000). High light illumination and nitrogen deprivation increase the transcripts of PDS and PTOX simultaneously in *H. pluvialis*, suggesting that PDS and PTOX may act in concert to dehydrogenate phytoene and remove excess electrons under stress, thereby preventing over-reduction of the photosynthetic electron transport chain and the formation of excess ROS (Li et al., 2010; Wang et al., 2009) Grunewald et al., 2000). Then the cyclization of lycopene by LCY-b and LCY-e end in the case of LCY-b catalyzes two β-cyclization reactions, at each end of lycopene to form β-carotene. A LCY-b gene has been cloned in *H. pluvialis* and was observed that its transcripts increased concomitantly with the formation of β-carotene and astaxanthin at increased intensities of red or blue light (Steinbrenner and Linden, 2003). Oxygenation of β-carotene at the 4rd position by β-carotene ketolase (BKT) gives rise to echinenone and canthaxanthin. In *H. pluvialis*, three different BKT genes have been cloned, including BKT1 (Lotan and Hirschberg, 1995), BKT2 (Kajiwara et al., 1995), and BKT3 (Huang et al., 2006). Under stress, the multiple BKT genes were up-regulated and when total BKT mRNAs reached a certain threshold, *H. pluvialis* began to synthetize astaxanthin (Huang et al., 2006). Heterologously expressed *Haematococcus* BKT1 was not able to use dihydroxy carotenoid zeaxanthin as a substrate, indicating that the oxygenation steps likely precede hydroxylation steps (Lotan and Hirschberg, 1995). In *H. pluvialis*, it has been proposed that astaxanthin is synthesized from the hydroxylation of canthaxanthin (Han et al., 2013).

The microalgae *C. zofingiensis* has the identical β-carotene synthesis pathway found in *H. pluvialis*, but may take a different route to make astaxanthin (Li et al., 2010). Recently the genes involved in biosynthesis of astaxanthin in *C. zofingiensis* have been cloned and characterized, including PSY (Cordero et al., 2011), PDS (Huang et al., 2008) LCY-b (Cordero et al., 2010), LCY-e (Cordero et al., 2012) BKT (CrtO) (Huang et al. 2006), CrtR-b (Chy-b) (Li et al. 2008). Under high light intensity conditions, PSY, PDS, BKT CrtR-b genes were up-regulated, whereas the mRNA levels of LCY-b and LCY-e remained constant, leading to formation of secondary carotenoids (Cordero et al., 2012; Li et al., 2009). Functional analysis of *C. zofingiensis* BKT demonstrated that this enzyme not only converted β-carotene to canthaxanthin via echinenone, but also exhibited high enzymatic activity of converting zeaxanthin to astaxanthin via adonixanthin (Huang et al., 2006). When the BKT inhibitor diphenylamine was applied, *C. zofingiensis* accumulated zeaxanthin while *H. pluvialis* accumulated β-carotene (Wang and Chen, 2008), suggesting that *C. zofingiensis* synthesizes astaxanthin through zeaxanthin (Han et al., 2013). Two possible types of signals from the mitochondrial alternative pathway may regulate carotenogenesis in *C. zofingiensis*, ROS and organic acids from the tricarboxylic acid cycle (TCA) cycle. Chemically generated ROS induce carotenogenesis and astaxanthin formation in *C. zofingiensis* (Li et al., 2009; Ip and Chen, 2005,). On the other hand, TCA cycle acids such as citrate may act as a signal at the gene expression level to induce mitochondrial alternative pathway respiration to facilitate carotenogenesis (Van lerberghe and McIntosh, 1996). The accumulation of astaxanthin in microalgae is a strategy to cope with oxidative stress. Astaxanthin synthesis pathways interact with multiple metabolic pathways such as the photosynthetic electron transport, fatty acid synthesis, and ROS generation. The new knowledge obtained from the studies of astaxanthin biosynthesis and stress response will be applied for enhanced astaxanthin production (Han et al., 2013).

LUTEIN

Lutein ((3R, 3′R, 6′R)-β, ε-carotene-3,3′-diol) is a xanthophyll that, together with zeaxanthin, has gathered increasing attention to prevent or ameliorate the effects of degenerative human diseases, such as macular degeneration (AMD) or cataract and also in skin health (Fernandez-Sevilla et al., 2010; Arnal et al., 2009; Carpentier et al., 2009; Roberts et al., 2009; Chiu and Taylor, 2007; Granado et al., 2003). Lutein is present in dark, leafy green vegetables, such as spinach and kale, as well as in corn, egg yolk, and some other foods with yellow color. Currently, the major source of marketable lutein is the petals from marigold (*Tagetes erecta* and *Tagetes patula*) flowers that can content 0.03% dwt of lutein (Piccaglia et al., 1998). More than 95% of the lutein in these plant sources is esterified, with about half of the weight corresponding to fatty acids. Thus, the conventional plant pigment workup process comprises chemical saponification (Del Campo et al., 2007; Zorn et al., 2003). Microalgae, where lutein usually appears in the free non-esterified form, represent a most interesting alternative to the plant system. In addition marigold flowers have low content of lutein to difference with microalgae (Table 2). Several microalgae have been proposed as potential producer of lutein, as *Muriellopsis* sp. (Del Campo et al., 2001; 2007), *Chlorella zofingensis* (Del Campo et al., 2000), *Chlorella protothecoides* (Shi et al., 2000) and recently *Scenedesmus almeriensis* (Sánchez et al., 2008a).

Table 2. Lutein content of different microalgal strain and marigold flower (*Tagetes erecta*)

Microalage	Lutein content (mg g^{-1} dwt)	Lutein productivity	Conditions	Reference
Tagetes erecta	17-570 mg $100g^{-1}$ in the petals	--	Grown in an experimental field in Northern Italy and harvested at full flowering stage.	Piccaglia et al., 1998.
Scenedesmus almeriensis	5.5	4.9 mg $L^{-1}d^{-1}$	Laboratory, continuous culture, solar cycle illumination, 2 L	Sánchez et al., 2008.
	4.5	290 mg m^{-2} d^{-1}	Outdoors, tubular system, continuous culture, 4,000 L, year average	Sánchez et al., 2008.
Scenedesmus obluquus FSP-3	4.52	4.15 mg L^{-1} d^{-1}	Laboratory, batch culture, 1L, continuous light, 6-10 days	Ho et al., 2014
Chlorella sorokiniama	3.0	2.2 mg L^{-1} d^{-1}	Laboratory, batch 9-10 days, continuous illumination, 1 L	Cordero et al., 2011
Chlorella zofingensis	3.5	1.2 mg L^{-1} d^{-1}	Laboratory, batch, continuous, illumination, 0.2 L	Del Campo et al., 2000.
Chlorella protothecoides	4.6	10 mg L^{-1} d^{-1}	Laboratory, batch, heterotrophic, 16 L	Wei et al., 2008
Chorococcum citriforme	7.2	1.05 mg L^{-1} h^{-1}	Laboratory, batch, continuous illumination, 0. 2 L	Del Campo et al. 2000.
Muriellopsis sp.	5.5	1.4-0.8 mg L^{-1} d^{-1}	Laboratory, batch 4-7 days, continuous illumination, 0.2 L	Del Campo et al. 2000.
	4.3	180 mg m^{-2} d^{-1} 7.2 mg L^{-1} d^{-1}	Outdoors, tubular systems, continuous cultures 55L	Del Campo et al. 2001.
	4-6	100 mg m^{-2} d^{-1} 1.0 mg m^{-2} d^{-1}	Outdoors, open tank, semicontinuous culture, 100L	Blanco et al. 2007.

The most significant factors affecting the lutein content in microalgae are irradiance, pH, temperature, availability and source of nitrogen, salinity (i.e., ionic strength), and presence of oxidizing substances (Sánchez et al., 2008a). The specific growth rate also plays a crucial role. In conditions of high temperature the accumulation of lutein is favored, as in the case for other carotenoids like β-carotene (Garcia-Gonzalez et al., 2005). The temperatures close to the limit cause environmental stress and trigger the lutein accumulation; but, the operational range is narrow because higher temperature could be harmful and eventually cause decreases in biomass productivity. A high irradiance level is beneficial, but its effect also depends on the culturing mode (indoor or outdoor). Cultures of *Muriellopsis* sp. and *Scenedesmus almeriensis* produced diference reponses and make belive that exist an interaction between irradiance and temperature. Thus, it might be more useful to study these factors in combination rather than separately (Fernandez-Sevilla et al., 2010; Garcia-Gonzalez et al., 2005) Moreover, the O_2 concentration in outdoors cannot be manipulated, but it may also interact with illumination and temperature levels. Similarly, the effect of pH is not consistent between batch and continuous cultivation. In the former case, lutein content increased at extreme pH values, whereas the best results under continuous operation were obtained at the optimum pH for growth rate. The pH is particularly relevant in microalgal cultures because it also determines the CO_2 availability afecting equilibrium between CO_2, H_2CO_3, HCO_3^-, and

CO_3^{2-}. The nitrogen concentration in the culture medium, supplied in the form of nitrate, does not apparently cause any significant effect on the lutein content. However, nitrogen limitation decreases biomass productivity, thus leading to low lutein synthesis. Consequently, nitrate should be supplied to a moderate excess so that growth rate is not hampered, while avoiding saline stress caused by nutrient excess that may severely affect the culture performance (Sánchez et al., 2008 a,b) . Finally, an effect of specific growth rate was also observed in both continuous and semicontinuous cultures. Lutein tends to accumulate at low dilution rates but at this condition the biomass productivity could decrease. Therefore, the maximum lutein productivity was again attained at the dilution rate that is optimal for biomass production (Fernandez-Sevilla et al., 2010; Sánchez et al., 2008b).

GENETIC MODIFICATION

Genetic modification as a tool to improve algal performance is now considered a necessity to achieve new and economical viable productions systems (Wijffels et al., 2010). Transgenesis in algae is a complex and fast-growing technology. Selectable marker genes, promoters, reporter genes, transformation techniques, and other genetic tools and methods are already available for various species and currently ~25 species are accessible to genetic transformation (Hallmann, 2007). Fortunately, large-scale sequencing projects are also planned, in progress, or completed for several of these species; the most advanced genome projects are those for the red alga *Cyanidioschyzon merolae* (Minoda et al., 2004), the diatom *Thalassiosira pseudonana* (Barbier et al. 2005; Armbrust et al., 2004; Matsuzaki et al., 2004), and the three green algae *Chlamydomonas reinhardtii* (Grossman, 2005; Grossman et al. 2003; Shrager et al., 2003), *Volvox carteri* (Barbier et al. 2005; Matsuzaki et al., 2004) and *Ostreococcus tauri* (Derelle et al., 2006; Derelle et al., 2002). Also transcriptome analyses in the form of expressed sequence tags (EST) projects have been performed and several strains have been genetically modified. The vast amount of genomic and EST data coming from these and a number of other algae has the potential to dramatically enlarge not only the algae's molecular toolbox. A powerful driving force in algal transgenics is the prospect of using genetically modified algae as bioreactors. Algal genome research is needed as the basis for a new level of efficiency and success in the application of biotechnology and gene technology to algae and their products. Table 3 shows an overview of genome projects on algae situation. All commercially used algae are non-transgenic, but this could change quickly. In several applications using wild-type species, commercial companies, could quickly replace the currently used wild-type organisms by optimized transgenic organisms. These transgenics could result in new or modified products or show a reduced content of components that interfere with current production. Transgenics could also allow growth to higher densities or could permit the use of atypical, particularly cheaper, growth conditions (Hallmann, 2007).

Table 3. An overview of genome projects on algae situation (Adapted from Wijffels et al., 2012)

Organism	Status	Data available at
Chlorophyta		
Chlamydomonas reinhardtii	Completed	http://genome.jgi-psf.org/Chlre4/Chlre4.home.html
Chlorella sp NC64A	Draft	http://genome.jgi-psf.org/cgi-
Chlorella vulgaris C-169	Completed	http://genome.jgi-psf.org/Chlvu1/Chlvu1.info.html
Coccomyxa sp. C-169	Completed	http://www.jgi.doe.gov/genome-projects/
Dunaliella salina CCAP 19/18	In progress	http://www.jgi.doe.gov/genome-projects/
Micromonas pusilla CCMP1545	Completed	http://genome.jgi-psf.org/MicpuN3/MicpuN3.home.html
Micromonas pusilla. RCC299	Completed	http://genome.jgi-psf.org/MicpuN3/MicpuN3.home.html
Ostreococcus lucimarinus CCE9901	Completed	http://genome.jgi-psf.org/Ost9901_3/Ost9901_3.home.html
Ostreococcus tauri OTH95	Draft	http://genome.jgi-psf.org/Ostta4/Ostta4.info.html
Ostreococcus sp. RCC809	Completed	http://genome.jgipsf.org/OstRCC809_2/OstRCC809_2.home.
Volvox carteri f. nagariensis	In progress	http://www.jgi.doe.gov/genome-projects/
Botryococcus braunii	In progress	http://www.jgi.doe.gov/sequencing/why/bbraunii.html
Asterochloris sp	Draft	http://genome.jgi-psf.org/Astpho1/Astpho1.home.html
Rhodophyta		
Cyanidioschyzon merolae	Completed	http://www.biomedcentral.com/1741-7007/5/28/
Galdieria sulphuraria	In progress	http://genomics.msu.edu/galdieria/
Heterokontophyta		
Thalassiosira pseudonana	Completed	http://genome.jgi-psf.org/thaps1/thaps1.home.html
Phaeodactylum tricornutum	Completed	http://genome.jgi-psf.org/Phatr2/Phatr2.info.html
Fraqilariopsicylindrus	Complete	http://genome.jgi-psf.org/Fracy1/Fracy1.home.html
Haptophyta		
Emiliania huxleyi	Completed	http://genome.jgi-psf.org/Emihu1/Emihu1.home.html
Cryptophyta		
Guillardia theta	Draft	http://genome.jgi-psf.org/Guith1/Guith1.home.html
Chlorarachniophytes		
Bigelowiella natans	Draft	http://genome.jgi-psf.org/Bigna1/Bigna1.home.html
Cyanobacteria		
Nostoc azolla	Completed	http://www.ncbi.nlm.nih.gov/genomes/lproks.cgi
Acaryochloris mmarina	Completed	http://www.ncbi.nlm.nih.gov/genomes/lproks.cgi
*Synechococcus**	Completed	http://www.ncbi.nlm.nih.gov/genomes/lproks.cgi
*Prochlorococcus marinus**	Completed	http://www.ncbi.nlm.nih.gov/genomes/lproks.cgi
*Cyanothece**	Completed	http://www.ncbi.nlm.nih.gov/genomes/lproks.cgi
Microcystis aeruginosa	Completed	http://www.ncbi.nlm.nih.gov/genomes/lproks.cgi
Arthrospira platensis	Completed	http://www.ncbi.nlm.nih.gov/genomes/lproks.cgi

The promising opportunity to use transgenic algae derived from a well-investigated, fast-growing species, like *Chlamydomonas*, as a bioreactor, has already resulted in several business start-ups in this field during the last few years and also some established biotechnology companies consider the use of transgenics (Hallmann, 2007). Three types of targets for genetic modification of algae can be recognized:

1) *Improvement of photosynthetic efficiency.* Biofuel production efficiency with algae is directly dependent on the solar photon capture and conversion efficiency of the system. However, daylight intensity is most of the time above the maximum photosynthetic efficiency of algae and therefore growth is reduced, a phenomenon known as photoinhibition. Research in this area focuses on the light harvesting antenna complex (LHC) (Anastasios, 2009; Mussgnug et al., 2007).
2) *Improvement in productivity of selected products.* The rising market demand for pigments from natural sources has promoted large-scale cultivation of microalgae for synthesis of such compounds. Genes encoding enzymes that are directly involved in specific carotenoid syntheses have been investigated and further development of transformation techniques will permit considerable increase of carotenoid cellular contents, and accordingly contribute to increase the volumetric productivities of the associated processes (Guedes et al., 2011). One example of such a gene (PSY) has already been published (Steinbrenner and Sandmann, 2006).
3) *Development of new products.* An emerging field in the biotechnology of algae is the introduction of genes or metabolic pathways in order to produce components of economic interest and that are not yet present in the wild type. Two major groups of new products can be distinguished: energy products (like ethanol, hydrogen and fatty acids) and recombinant proteins. There is clearly more basic research that needs to be performed before algal transgenics and algal biotechnology reach a capacity to compete with other systems. But since many physiological, morphological, biochemical, or molecular characteristics of algae are quite different from higher plants or animals, algae can meet several requirements than other systems cannot sufficiently accomplish. This is one reason why algal systems gain more and more influence in the production of substances of economic, industrial, and pharmaceutical importance (Hallmann, 2007).

COMMERCIAL PRODUCTION IN THE WORLD

Carotenoids belong to the most important components in foods. They are natural colorants, as yellow to red colors, so they have great influence on the acceptability of many foods. Moreover, some carotenoids are precursors of vitamin A; in terms of human health, they are among the bioactive phytochemicals credited that reduce risks for degenerative diseases such as cancer, cardiovascular diseases, macular degeneration and cataract (Guedes et al., 2011). Algae culturing has been adopted since the late of 1800s and early 1900s, while microalgae mass culturing begin to develop since the achievement of Allen and Nelson in 1910 by cultivating *Chlorella* sp. for aquaculture purposes in Berlin, Germany (Preisig and Andersen, 2005). To date, it seems that microalgae have the potential to play pivotal roles in order to remedy the energy, environment and food crisis prevailing in the world. According to a recently published report, the global carotenoids market is projected to reach US $1.2 billion by the year 2015, because of the rising consumer awareness about health benefits offered by various carotenoids, and the shift towards healthy and natural food products (Global Industry Analysts, Inc. http://www.strategyr.com/Carotenoids_Market_Report.asp/), and tends to reach US$1.3 billion by 2017 with a compound annual growth rate of 2.3

(http://www.prweb.com/pdfdownload/8849957.pdf). β-carotene and the xanthophylls astaxanthin, cantaxanthin, and lutein are the major carotenoids with commercial interest. The market value of β -carotene was estimated about US$253 million in 2009 (According to a report on the global market for carotenoids) and this market increased around to US $261 million in 2010 and is expected to grow to US $334 million by 2018 at an annual growth rate of 3% (http://www.companiesandmarkets.com/News/Chemicals/Beta-carotene-the-Largest-Carotenoid-Sector-is-Forecast-to-be-Worth-US-334-Mn-by-2018/NI2439), and the price per Kg varied from US $300 to US$3000 (Spolaore, 2006). In the USA, several Universities and algae-biotech Companies, have been engaged for more than a decade in comprehensive screening for bioactive metabolites produced by marine organisms mainly microalgae and cyanobacteria. Major contributions to bioactive compound libraries have accordingly been made, as well as assessments on their potential for cultivation at industrial scales (Guedes et al., 2011). However, the development of innovative and efficient bioprocesses are neccesary to achieve economical commercialization at a large scale. Despite this obstacle, the commercial potential of microalgae is widely recognized. The production of astaxhantin-rich biomass from *Haematoccocus*, for example, has been already pursued by Cyanotech (Hawaii), Mera Pharmaceutical (Hawaii), and Fuji Health Science (Japan). The global astaxanthin market is estimated at about $257 million, most of which is used in fish coloration (2009 data; estimates by BCC Research for astaxanthin market size are however lower). The human uses market is growing and estimated at about US $ 27-40 million. The salmon feed industry is the principal astaxanthin consume by. Commercial production of astaxanthin is being carried out in USA, India, Japan and Israel (at: http://www.oilgae. com /non_fuel_products/astaxanthin.html#sthash.yVrh9E1h.dpuf). Likewise, β-carotene production from *Dunaliella* is the focus of a number of companies (Table 4), including Betatene, Western Biotechnology, and AquaCarotene, all in Australia. Cyanotech (Hawaii) and Inner Mongolia, Biological Engineering (China) also pursues β-carotene production. (Del Campo et al., 2007). Global lutein market in 2004 accounted for US$139 million and is expected to increase by 6.1% per year (the fastest projected growth in individual carotenoid sales), to reach US$187 million in 2009 (http://www.bccresearch.com). The microalga *Muriellopsis* sp. and *Scendesmus*, species are able to accumulate lutein as a part of their biomass (Del Campo et al., 2004, 2000; Sansawa and Endo 2004). An established commercial system for the production of lutein from microalgae does not exist yet, although the basis for outdoor production of lutein-rich cells of strains of *Muriellopsis* (Blanco et al., 2007; Del Campo et al. 2001) and *Scenedesmus almeriensis* (Fernandez-Sevilla et al., 2010) at a pilot scale have already been set up. On the other hand, the global production of *Dunaliella*is estimated at about 1,200 t year^{-1} (Pulz and Gross, 2004). Open ponds, with no or scarce process control, represent the conventional method used in commercial production plants for *Dunaliella* (Borowitzka and Borowitzka 1988). Plants producing β–carotene operate nowadays in Israel, China, USA, and Australia (Table 4). The production of specific metabolites by microalgae requires not only culturing for biomass build-up followed by secondary metabolism, but also entails recovery of the biomass and further downstream processing aimed at obtaining the desired metabolites in pure form. Unfortunately, the downstream processing is often more expensive than the bio- reaction itself.

Table 4. Commercial producer of β-carotene (Fernandez-Sevilla et al., 2010; Ho et al., 2014)

Company	Location	Culture area (ha)	β-carotene production (tonnes per year)	Culture system
Betatene http://www.betatene.com.au	Australia	400	13–14	Extensive unmixed ponds
Western technology http://www.cognis.com	Australia	240	4–6	Extensive unmixed ponds
AquaCarotene http://www.aquacarotene.com	Australia	na	na	Extensive unmixed ponds
Cognis nutrition and health http://www.foodproductdesign.com	Australia	na	na	Lagoons
Cyanotech http://www.cyanotech.com	Hawaii (USA)	na	na	Raceways ponds
Inner Mongolia Biological Eng.	China	na	na	Raceways ponds
Nature beta technologies http://chlostanin.co.jp	Israel	5	3–4	Raceways ponds
Tianjin Lantai Biotechnology	China	na	na	Raceways ponds
Parry Agro Industries http://www.murugappa.com/0_our_companies/parry_agro.htm	India	na	na	Raceways ponds

Processing for Metabolite Recovery

Pigments extraction processes applied to microalgae are mainly derived from phytochemical techniques developed on superior plants and macroalgae. The main parameters driving selection of an extraction technology are biochemical characteristics of extracted molecules, rapidity, limitation of solvent use, reproducibility, extraction yield, selectivity, and protection of extracted molecules against chemical transformation, dimension, cost and easiness (Rodriguez- Bernardo et al., 2006; Wang et al., 2006). It is also necessary to maintenance the bioability and integrity of the biomolecule. Microalgae represent a challenge due to the composition and the presence of a thick hard cell wall. So that the first step in the pigment recovery is the cell wall disruption. The most common methods are: soaking, cryogrinding, mortar, bead mill, ultrasound and autoclave (Ming-Chang et al., 2013; Ceròn et al., 2008). Ceròn et al. (2008) demonstrated that cell disruption is necessary and found that bead mill disruption was the best option among the treatments tested for lutein extraction from *Scenedesnus almeriensis*. Other cell disruption techniques based on mechanical or osmotic shock, such as bead beating and ultrasound assisted extraction have been proposed (Rostagno et al., 2007). Mechanical treatments are very drastic, and usually induce thermal denaturation of molecules of interest, especially when using an ultrasonic sharp probe (Jeffrey et al., 1997). Osmotic shock might not be efficient when working with frustulated species or armoured dinoflagellates.

The next step is organic solvent extraction techniques, should be scalable, this includes maceration (soaking), percolation, counter current extraction, pressurized liquid extraction, and soxhlet are widespread technologies described to extract lipids and pigments. These processes are reproducible, allow the rapid extraction of chemicals, but usually imply the use of large amounts of solvents, and the risk of thermal denaturation or transformation of molecules of interest. Coupling steam distillation or hot water extraction with maceration in solvent increases extraction yields for plant essential oils and bioactive compounds (Manzan et al., 2003), but thermolabile molecules are damaged using this technology. The use of enzymes, such as xylanases, pectinases or cellulases, to enhance pigments extractability rates was proposed and validated for superior plants tissues (Choudhari and Ananthanarayan, 2007; Kim et al., 2005) and macroalgae (Deniaud et al., 2003), and could be of interest for unfrustulated microalgae. Various techniques limiting chemical transformation of pigments have been proposed. Samples can be frozen (−80 °C, liquid nitrogen), freeze-dried, dessicated, or stored in water vapour saturated atmospheres (Esteban et al., 2009) to avoid oxidation and thermal denaturation. Maceration in liquid nitrogen followed by buffer extraction, consisting in pigment precipitation in 50% ammonium sulphate, was described for the cyanobacterial blue pigment phycocyanin. However, subsequent dialysis and gel filtration chromatography steps are needed to desalt pigments solutions (Sachindra et al., 2005). Extraction in aqueous solutions implies a strict control of the pH because many porphyrin pigments can undergo chemical transformations in acidic or alkaline conditions. For instance, chlorophylls can undergo epimerization, dephytylation and demetallation (pheophytination) (Jeffrey et al., 1997). The presence of chlorophyllases in microalgae extracts also explains the rapid degradation of porphyrin pigments (Karboune et al., 2005; Barrett and Jeffrey, 1964). Changing water to selected solvents, such as sec-butanol, allows a rapid and efficient extraction from plants and microalgae with a good stability of pigments (Martinson and Plumley, 1995). This solvent can also be favourably used to concentrate pigments extracted with acetone (Martinson and Plumley, 1995). Hejazi et al. (2002) proposed a biocompatible strategy to extract carotenoids from *Dunaliella salina* without killing the cells, using solvents having a *Log P* (octanol) > 6. The use of supercritical carbon dioxide (SCF-CO_2) or combination of solid phase extraction with SCF-CO_2 to extract lipids, pigments or bioactive compounds from plants (Kim et al., 2008; Jarent-Galan et al., 1999) or microalgae (Macias-Sanchez et al., 2005) was also described as a very efficient technology, in spite of a high cost. For instance, paprika and *Lithospermum* pigments are easily and efficiently extracted and purified using SCF-CO_2, in optimized conditions favoring differential solubilization in CO_2 (Jarent-Galan et al., 1999). Another extraction technique involves microwaves. MAE and VMAE have been proposed as efficient and rapid processes to extract antioxidants or pigments from plants or spices, oils from vegetables, allowing reduced solvent consumption and shorter extraction times, with equivalent or higher extraction yields. MAE and VMAE have been applied to marine microalgae to extract lipids, and allowed the highest recovery for all tested species compared to other extractions techniques (Pasquet et al., 2010).

MICROALGAE AS PHARMACEUTICAL AND NUTRITIONAL SOURCES

Several researchers have actively focused on carotenoids from microalgal sources; the major areas, in terms of actual or potential industrial applications, are food and health and the antioxidant properties constitute at present its core interest (Guedes et al., 2011). Traditionally, carotenoids have been used in the feed, food and nutraceutical industries. The recent discoveries of health-related beneficial properties attributed to carotenoids have spurred great interest in the production of structurally diverse carotenoids for pharmaceutical applications. In humans, the most relevant biological functions of carotenoids are linked to their antioxidant properties, which directly emerge from their molecular structure. In recent years, the understanding of ROS-induced oxidative stress mechanisms and the search for suitable strategies to fight oxidative stress has become one the major goals of medical research efforts. A number of studies have been reported which implicate oxidative stress involvement in degenerative pathogenesis, e.g., Alzheimer and Parkinson (Murthy et al., 2005; Guerin et al., 2003). In parallel, a carotenoid-enriched diet has been found to diminish the risk of suffering from degenerative diseases (Guerin et al., 2003).

Human Health

Cardiovascular diseases are the leading cause of death in developed countries, and have become the main health problem also in developing countries (American Heart Association, 2013). These include acute myocardial infarction and disorder of high morbidity and mortality (Lloyd-Jhones et al., 2009). Oxidative stress and inflammation are the main factors contributing to the pathophysiology of these disorders (Riccioni, 2009; Heller et al., 1998). In particular, the oxidative stress induced by ROS can cause low-density lipoproteins oxidation (LDL) that plays a key role in the pathogenesis of atherosclerosis (Yla-Herttuala et al., 1991). Numerous epidemiological studies suggest that diets rich in carotenoids could protect the human body from certain cardiovascular diseases due to the involvement of oxidizing substances and oxidative stress in the development and clinical expression of coronary heart disease (Kohlmeier and Hastings, 1995). In fact, high lycopene levels in plasma and tissues have been inversely linked to coronary heart disease (Kristenson et al., 1997), myocardial infarction (Kohlmeier et al., 1997) and risk to suffer from arteriosclerosis (Klipstein-Grobush et al., 2000). Low lutein levels in plasma have also been associated with an increased tendency to suffer from myocardial infarction (Street et al., 1994), while a high intake of lutein has been inversely related with the risk of stroke (Ascherio et al., 1999). Likewise, low α-carotene levels in serum have been shown to inversely correlate prevalence of coronary artery disease and formation of arterial plaque, by which α-carotene has been proposed as a potential marker for human atherosclerosis. In addition, carotenoids displaying high levels of provitamin A activity, including α-carotene, β-carotene and β-cryptoxanthin, have been associated with educed risk of angina pectoris disease (Kritchewsky, 1991). Vitamin A is well recognized as a factor of great importance for child health and survival, its deficiency causes disturbances in vision and various related lung, trachea and oral cavity pathologies. Other epidemiological studies have also found low levels of oxygenated carotenoids (xanthophylls: lutein, zeaxanthin, lycopene, β-cryptoxanthin, β-carotene and α-carotene) in plasma of patients with acute and chronic coronarysyndromes (Lidebjer et al., 2007; Sesso et al., 2005). Particularly, in the recent study by (Koh et al., 2011) high levels of β-cryptoxanthin and lutein in plasma have been shown to decrease risk for suffering from myocardial infarction, but no statistically significant associations with other carotenoids were found.

In eye health, many research studies showed that lutein and zeaxanthin are the main responsible pigments for both the yellowing and the maintenance of normal visual function of the human eye macula (Ma and Lin, 2010; Khachik et al., 2002), while other major carotenoids in serum (α-carotene, β-carotene, lycopene and β-cryptoxanthin), are absent or found in trace amounts in the human macula (Bates et al., 1996). In the eye macula, lutein and zeaxanthin absorb blue light and also attenuate pernicious photooxidative effects caused by the excess blue light, while reducing eye chromatic aberration. Due to their antioxidant properties, carotenoids protect the eye macula from adverse photochemical reactions (Landrum and Bohne, 2001). In people over the age of 64, visual sensitivity directly depends on lutein and zeaxanthin concentrations in retina (Hammond et al., 1998).

It is well known that cellular proliferation is controlled by the communication stablished between the cells in a tissue. Cell communications reset or stimulation becomes essential if abnormal cell proliferation occurs. In that respect, it has been mentioned that carotenoids might stimulate expression of genes directly involved in regulation of cell communication processes. In more detail, carotenoids would directly act on DNA in order to regulate the production of RNA that is responsible for gap-junctions communications, which could successfully explain some anti-tumor activities of carotenoids (Blunt et al., 2003; Muller-Feuga et al., 2003). Immune system cells also require intercellular communication to conduct their activity efficiently, so the previous action mechanism of carotenoids could also apply for supporting the immune system activity. As an example, high doses of β-carotene increase the CD4 to CD8 lymphocyte ratio, which is very low in patients suffering from HIV disease (Plaza et al., 2009). Recent clinical trials have demonstrated that phytomedicine is effective to treat pathologies related to vascularization and cell proliferation in prostate hyperplasia (Gerber et al., 2002; Klippel et al., 1997) Several marine microalgae can synthesize bioactive compounds with antitumoral effect. For example, campesterol, a sterol found in *Tetraselmis suecica*, has been reported to be antiangiogenic (Fabregas et al., 1997). Extracts of *Chlorella vulgaris* have also been found to be active against liver cancer in vitro and in vivo, where the inhibition of proliferation and increased apoptosis were reported (Saad et al., 2006).

CONCLUSION

The rising market demand for pigments from natural sources has promoted large-scale cultivation of microalgae for synthesis of such compounds, so significant decreases in production costs are expected in coming years. The development of microalgae-based biotechnology has been constrained by their limited growth rates in industrial photobioreactors. So that is necessary to selected microalgal strains containing the desired high levels of bioactive metabolites will require use of systems capable of maintaining defined growth conditions. The use of genetic and metabolic engineering tools to improve the selective pathways will lead towards highly productive systems for carotenoids.

REFERENCES

Albrecht M., Klein A., Hugueney P. Sandmann G. Kuntz M. (1999). Molecular cloning and functional expression in *E. coli* of a novel plant enzyme mediating ζ –carotene desaturation. *Biotechnol. Lett.*21: 791-795.

American Heart Association. *Heart Disease and Stroke Statistics* (2013) Update; American Heart Association: Dallas, TX, USA, 2013.

Anastasios M. (2009). Solar energy conversion efficiencies in photosynthesis: Minimizing the chlorophyll antennae to maximize efficiency. *Plant Science.* 177(4): 272-280.

Armbrust E.V., Berges J.A., Bowler C., Green B.R., Martinez D., Putnam N.H., Zhou S.G., Allen A.E., Apt K.E., Bechner M., Brzezinski M.A., Chaal B.K., Chiovitti A., Davis A.K., Demarest M.S., Detter J.C., Glavina T., Goodstein D., Hadi M.Z., Hellsten U., Hildebrand M., Jenkins B.D., Jurka J., Kapitonov V.V., Kroger N., Lau W.W.Y., Lane T.W., Larimer F.W., Lippmeier J.C., Lucas S., Medina M., Montsant A., Obornik M., Parker M.S., Palenik B., Pazour G.J., Richardson P.M., Rynearson T.A., Saito M.A., Schwartz D.C., Thamatrakoln K., Valentin K., Vardi A., Wilkerson F.P. and Rokhsar D.S. (2004). The genome of the diatom *Thalassiosira pseudonana*: Ecology, evolution, and metabolism. Science 306: 79–86.

Arnal E., Miranda M., Almansa I., Muriach M., Barcia J. M., Romero F. J., Diaz-Llopis M., Bosch-Morell F. (2009). Lutein prevents cataract development and progression in diabetic rats. *Graefes Arch Clin Exp Ophthalmol* 247(1):115–120.

Asada K., Takahashi M., Nagate, M. (1974) Assay and inhibitors of spinach superoxide dismutase. *AgricBiol Chem* 38:471-473.

Ascherio A.; Rimm E. B.; Hernán M.; Giovannucci E.; Kawachi I.; Stampfer M.; Willet W. (1999), Relation of consumption of vitamin E, vitamin C and carotenoids to risk for stroke among men in the United States. *Ann. Intern. Med.* 130, 963–970.

Bach T. J. Lichtenthaler H. K. (1987) Plant growth regulation by mevinolin and other sterol biosynthesisi inhibitors. *In*: Fuller G. Nes W. D. (Eds) Ecology and metabolism of plant lipid, American Chemical Society, Washington, D.C. pp 109.139.

Barbier G., Oesterhelt C., Larson M.D., Halgren R.G., Wilkerson C., Garavito R.M., Benning C., Weber A.P. (2005) Comparative genomics of two closely related unicellular thermo-acidophilic red algae, *Galdieria sulphuraria* and *Cyanidioschyzon merolae*, reveals the molecular basis of the metabolic flexibility of Galdieria sulphuraria and significant differences in carbohydrate metabolism of both algae. Plant Physiology 137, 460-474.

Baroli I., Niyogi K. K. (2000). Molecuar genetics of xanthophyll-dependent photoprotection in green algae plants. *Philos Trens R Soc London B Biol. Sci.* 355 (1402) 1385-94.

Barrett J., Jeffrey S. W. (1964). Chlorophyllase and formation of an atypical Chlorophyllide in marine algae. *Plant Physiol.* 33:911–7.

Bates C. J., Chen S. J., MacDonald A., Holden R. (1996). Quantitation of vitamin E and a carotenoid pigment in cataractous human lenses and the effect of a dietary supplement. *Internat.* K 66, 316–321.

Becker E.W. (1994). *Microalgae: biotechnology and microbiology.* Cambridge, United Kingdom: Cambridge University *Press.*

Ben-Amotz A. (1996). Effect of low temperature on the stereoisomer composition of β-carotene in the halotolerant alga, *Dunaliella bardawil* (Chlorophyta). *J Phycol.* 32:272–275.

Ben-Amotz A. (1999). *Dunaliella* β-carotene: from science to commerce. *In*: Seckbach J (Ed) Enigmatic microorganisms and life in extreme environments. Kluwer, Netherlands, pp 401–410.

Bertrand, M. (2010) Carotenoid biosynthesis in diatoms. *Photosynth. Res.*, 106, 89–102.

Blanco A. M., Moreno J., Del Campo J. A., Rivas J., Guerrero M. G. (2007). Outdoor cultivation of lutein-rich cells of *Muriellopsis*sp. In open ponds. *Appl Microbiol Biotechnol 73:1259–1266.*

Blunt J. W., Copp B. R., Hu W. P., Munro M. H., Northcote P. T., Prinsep M. R. (2003). Marine natural products. *Nat Prod Rep.* 25:35–94.

Borowitzka M. A., Borowitzka L. J. (1988) *Dunaliella. In*: Borowitzka M. A., Borowitzka L. J. (eds) Microalgal biotechnology. Cambridge University Press, Cambridge, UK, pp 27–58,

Borowitzka M. A., Siva C. J. (2007). The taxonomy of the genus *Dunaliella* (Chlorophyta, Dunaliellales) with emphasis on the marine and halophilic species. *J Appl Phycol* 19:567–590.

Boussiba S. (2000). Carotenogenesis in the green alga *Hematococcus pluvialis:* cellular physiology and stress response. *Physiol Plant.* 108:111-117.

Bouvier F., Backhaus R. A. Camara B. (1998). Induction and control of chromoplast-specific carotenoid genes by oxidative stress.*J. Biol. Chem.* 273: 343-349.

Britton, G. (1998) Overview of carotenoid biosynthesis. In Carotenoids: Biosynthesis and Metabolism; Britton, G., Liaaen-Jensen, S., Pfander, H., Eds.; Birkhäuser: Basel, Switzerland, Volume 3, pp. 13–147.

Britton G., Frank H. A., Young A., Cogdell R. J. (1999) The Photochemistry of Carotenoids. *In* Advances in Photosynthesis and Respiration. The Photochemistry of carotenoids, Kluwer Academic Publishers.

Capa-Robles W., Paniagua-Michel J., Olmos Soto J. (2009). The biosynthesis and accumulation of beta-carotene in Dunaliella salina proceed via the glyceraldehyde 3-phosphate/pyruvate path-way. *Nat Prod Res* 23:1021–1028.

Carol P., Stevenson D., Bisanz C., Breitenbach J., Sandmann G., Mache R., Coupland G., Kuntz M. (1999). Mutations in the Arabidopsis gene immutans cause a variegated phenotype by inactivating a chloroplast terminal oxidase associated with phytoene desaturation. *Plant Cell* 11:57-68.

Carpentier S., Knaus M., Suh M. (2009). Associations between lutein zeaxanthin, and age-related macular degeneration: an overview *Crit Rev Food Sci Nutr.* 49(4):313–326.

Cerón M.C., Campos I., Sánchez J.F., Acién F.G., Molina E., Fernández- Sevilla J.M. (2008). Recovery of lutein from microalgae biomass: development of a process for *Scenedesmus* almeriensis biomass. *J Agric Food Chem* 56: 11761–11766.

Chamovitz D., Misawa N., Sandmann G., Hirschberg J. (1992). Molecular cloning and expression in *Escherichia coli* of a cyanobacterial gene coding for phytoene synthase, a carotenoid biosynthesis enzyme. *FEBS Lett.* 296: 305–310.

Chen Y., Li F., Wurtzel E. T. (2010). Isolation and characterization of the Z-ISO gene encodimg a missing component of carotenoid biosynthesis in plants. *Plant. Physiol.* 153: 66-79.

Chiu C. J., Taylor A. (2007). Nutritional antioxidants and age-relate cataract and macular degeneration. *Exp. Eye. Res* 84:229–245.

Choudhari S. M., Ananthanarayan L. (2007). Enzyme aided extraction of lycopene from tomato tissues. *Food Chem.* 102:77–81.

Cordero B. F., Obraztsova I., Martín L., Couso I., León R., Vargas M. A., Rodriguez H. (2010). Isolation and characterization of a lycopene β-cyclase gene from the astaxanthin-producing green alga *Chlorella zofingiensis* (Chlorophyta). *J. Phycol.* 46:1229-1238.

Cordero B. F., Couso I., Leon R., Rodrguez H., Vargas M. A. (2011) Enhancement of carotenoids biosynthesis in *Chlamydomonas reinhardtii* by nuleas transformation using phytoene synthase gene isolated from *Chlamydomonas zoofigensis. Appl Microbiol. Biotech.*91: 341-351.

Cordero B. F., Couso I., León R., Rodriguez H., Vargas M. A. (2012). Isolation and characterization of a lycopene epsilon-cyclase gene of *Chlorella* (Chromochloris) *zofingiensis*: regulation of the carotenogenic pathway by nitrogen and light. *Mar. Drugs* 10:2069-2088.

Cunningham F. X, Pogson B., Sun Z., McDonald K. A., DellaPenna D., Gantt E. (1996). Functional analysis of the β and ε lycopene cyclase enzymes of *Arabidopsis* reveals a mechanism for control of cyclic carotenoid formation. *Plant Cell.* 8:1613–26.

Del Campo J. A., Moreno J., Rodríguez H., Vargas M. A., Rivas J., Guerrero M. G. (2000) Carotenoid content of chlorophycean microalgae. Factors determining lutein accumulation in *Muriellopsis* sp. (Chlorophyta).*J Biotechnol* 76:51–59.

Del Campo J. A., Rodríguez H., Moreno J., Vargas M. A., Rivas J., Guerrero M. G. (2001). Lutein production by *Muriellopsis* sp. in an outdoor tubular photobioreactor.*J Biotechnol* 81:289–295.

Del Campo J. A., Rodríguez H., Moreno J., Vargas M. Á., Rivas J., Guerrero, M. G. (2004). Accumulation of astaxanthin and lutein in *Chlorella zofingiensis* (Chlorophyta) *Appl. Microbiol. Biotechno*l. 64:848-854.

Del Campo J. A., Garcia-Gonzalez M., Guerrero M. G. (2007) Outdoor cultivation of microalgae for carotenoid production: current state and perspectives. *Appl Microbiol Biotechnol* 74:1163–1174.

Demming-Adams B., Gilmore A. M., Adams W. W. (1996). *In vivo* functions of carotenoids in higher plants.*FASEB J.* 10: 2149-2153.

Deniaud E., Fleurence J., Lahaye M. (2003) Preparation and chemical characterization of cell wall fractions enriched in structural proteins from *Palmaria palmate* (Rhodophyta). *Bot Mar.*46:366–77.

Derelle E., Ferraz C., Lagoda P., Eychenié S., Cooke R., Regad F., Sabau X., Courties C., Delseny M., Demaille J., Picard A., Moreau H. (2002). DNA libraries for sequencing the genome of Ostreococcus tauri (Chlorophytae; Prasinophyceae): the smallest free-living eukaryotic cell. *J Phycol*38, 1150-1156.

Derelle E., Ferraz C., Rombauts S., Rouze P., Worden A.Z., Robbens S., Partensky F., Degroeve S., Echeynie S., Cooke R., Saeys Y., Wuyts J., Jabbari K., Bowler C., Panaud O., Piegu B., Ball S.G., Ral J.P., Bouget F.Y., Piganeau G., de Baets B., Picard A., Delseny M., Demaille J., van de Peer Y., Moreau H. (2006). Genome analysis of the smallest free-living eukaryote *Ostreococcus tauri* unveils many unique features. *Proceedings of the National Academy of Sciences USA* 103, 11647-11652.

Disch A., Schwender J., Müller C., Lichtenthaler H. K. & Rohmer, M. (1998). Distribution of the mevalonate and glyceraldehyde phosphate/pyruvate pathways for iso- prenoid biosynthesis in unicellular algae and the cyano-bacterium *Synechocystis* PCC 6714. *Biochem. J.* 333:381- 388.

Droop M. R. (1954). Conditions governing haematochrome formation and loss in the alga *Haematococcus pluvialis Flotow. Arch. Microbiol.* 20:391-397.

Eisenreich W., Bacher A., Arigoni D., Rohdich F. (2004). Biosynthesis of isoprenoids via the non-mevalonate pathway. *Cell. Mol. Life Sci.* 61, 1401–1426.

Esteban R., Balaguer L., Manrique E., Rubio de Casas R., Ochoa R., Fleck I., Pintó-Marijuan M., Casals I., Morales D., Jiménez M.S., Lorenzo R., Artetxe U., Becerril J.M., García-Plazaola J.I. (2009). Alternative methods for sampling and preservation of photosynthetic pigments and tocopherols in plant material from remote locations. *Photosynth Res.*101:77–88.

Fabregas J., Aran J., Morales E. D., Lamella T., Otero A. (1997) Modification of sterol concentration in marine microalgae. *Phytochemistry.* 46:1189–1191.

Fan L., Vonshak A., Rachel G., Hirshberg J., Cohen Z., Boussiba S. (1995). The biosynthetic pathway of astaxanthin in a green alga *Haematococcus pluvialis* as indicated by inhibition with diphenylamine. *Plant Cell Physiol.* 36: 1519–1524.

Fernandez-Sevilla J.M., Acien-Fernandez F. G., Molina-Grima E. (2010). Biotechnological production of lutein and its applications. *Appl. Microbiol Biotechnol.* 86:27–40.

Ferreira A. F., Ribeiro L.A., Batista A.P., Marques P. A. S. S., Nobre B. P. Palavra A. M. F., Pereira P. S., Gouveida L., Silva C. (2013). A biorefinery from *Nannochloropsis* sp. Energy and CO_2 emissions and economics analyses. *Boiresour Technol.* 130 235-244.

Foyer C.H., Descourvieres P. Kunert K. J. (1994). Protection oxygen radicals: an important defense mechanism studied in transgenic plants. *Plant, Cell Environ.*17: 507-523.

Frommolt R., Werner S., Paulsen H., Goss R., Wilhelm C., Zauner S., Maier U.G., Grossman A.R., Bhattacharya D., Lohr M. (2008). Ancient recruitment by chromists of green algal genes encoding enzymes for carotenoid biosynthesis. *Mol. Biol. Evol.* 25, 2653–2667.

Garcia-Gonzalez M., Moreno J., Manzano J. C., Florencio F. J., Guerrero M.G. (2005). Production of *Dunaliella salina* biomass rich in 9-cis-b-carotene and lutein in a closed tubular photobioreactor.*J Biotechnol.*115:81–90.

Gerber G. B., Leonard A., Hantson P. (2002) Carcinogenicity, mutagenicity and teratogenicity of manganese compounds. *Crit RevOncol Hematol.*42:25–34.

Granado F., Olmedilla B., Blanco I. (2003) Nutritional and clinical relevance of lutein in human health. *Br J Nutr* 90:487–502.

Grossman A.R., Harris E.E., Hauser C., Lefebvre P.A., Martinez D., Rokhsar D., Shrager J., Silflow C.D., Stern D., Vallon O., Zhang Z. (2003). *Chlamydomonas reinhardtii* at the at the Crossroads of GenomicsEukaryot Cell. 2(6): 1137–1150.

Grossman, A. R. (2005). Paths toward Algal Genomics. *Plant Physiol.* 137:410-427.

Grunewald K., Eckert M., Hirschberg J., Hagen C. (2000). Phytoene desaturase is localized exclusively in the chloroplast and up-regulated at the mRNA level during accumulation of secondary carotenoids in *Haematococcus pluvialis* (Volvocales, Chlorophyceae). *Plant Physiol.* 122:1261-1268.

Grung M., Metzger P., Liaaen-Jensen S. (1994). Algal carotenoids: secondary carotenoids of algae. Secondary carotenoids in the green alga *Botryococcus braunii*, race, new strain. *Biochem. Syst. Ecol.* 22:25-29.

Guedes A. M., Amaro H. M., Malcata F. X. (2011). Microalgae as Sources of Carotenoids. *Mar. Drugs*.9, 625-644.

Guerin M., Huntley M.E., Olaizola M. (2003). *Haematococcus* astaxanthin: Applications for human health and nutrition. *Trends Biotech*. 21, 210–216.

Hallmann A. (2007). Algal Transgenics and Biotechnology. Transgenic Plant Journal 1(1): 81-98.

Hammond B. R.; Wooten B.; Snodderly D.M. (1998). Preservation of visual sensitivity of older individuals: Association with macularpigment density. *Invest. Ophthalmol. Vis. Sci*. 39, 397–406.

Han D., Li Y., Hu Q. (2013). Astaxanthin in microalgae: pathways, functions and biotechnological implications. *Algae*. 28(2): 131-147.

Hejazi M. A., De Lamarliere C., Rocha J. M. S., Vermuë M., Tramper J., Wijffels R. H. (2002). Selective extraction of carotenoids from the microalga *Dunaliella salina* with retention of viability. *Biotechnol Bioeng*.79:29–36.

Hejazi M. A., Wijffels R. H. (2004). Milking of microalgae. *Trends Biotechnol.*, 22(4), 189-94.

Heller F. R., Descamps O., Hondekijn J. C (1998). LDL oxidation: Therapeutic perspectives. *Atherosclerosis*, 137 (Suppl.), S25–S31.

Hideg E. (2006). Detecting stress-induced reactive oxygen species in plants under UV stress. In: Ghetti F., Checcucci G. Bornman J. F. (Ed) Enviromental UV radiation: impact on ecosystems and human health and preductive models. Netherlands: Spronger. pp 147-157.

Ho S., Chan M., Liu C., Chen C. Y., Lee W. L., Lee D. J., Chang J. S. (2014). Enhancing lutein productivity of an indigenous microalga *Scenedesmus obliquus* FSP-3 using light-related strategies. *Bioresour. Technol.* 152: 275–282.

Hornero-Méndez D., Britton G. (2002). Involvement of NADPH in the cyclization reaction of carotenoid biosynthesis. *FEBD Lett*.515:133–6.

Huang J. C., Chen F., Sandmann, G. (2006). Stress-related differential expression of multiple β-carotene ketolase genes in the unicellular green alga *Haematococcus pluvialis. J. Biotechnol.*, 122: 176–185.

Huang, J. C., Liu J., Li Y., Chen F. (2008). Isolation and characterization of the phytoene desaturase genes a potential selective marker for genetic engineering of the astaxanthin-producing green alga *Chlorella zofingiensis* (Chlorophyta).*J. Phycol.* 44: 684–690.

Ip P. F., Wong K–H, Chen F. (2004). Enhanced production of astaxanthin by the green microalga *Chlorella zofingiensis* in mixotrophic culture. *Process Biochem.* 39:1761-1766.

Ip P. F., Chen F. (2005). Employment of reactive oxygen species to enhance astaxanthin formation in *Chlorella zofingiensis* in heterotrophic culture. *Process Biochem*.40:3491-3496.

Jeffrey S. W., Mantoura R. F. C., Wright S. W. (1997). Phytoplankton pigments in oceanography. In: Guidelines to modern methods. UNESCO.

Jaren-Galan M., Nienaber U., Schwartz S. J.(1999). Paprika (Capsicum annuum) oleoresin extraction with supercritical carbon dioxide.*J Agric Food Chem*.47:3558–64.

Johnson E. A., An G.H. (1991). Astaxanthin from microbial sources. *Crit. Rev. Biotechnol.* 11:297-326.

Johnson E. A., Schroeder, W. A. (1996). Microbial carotenoids. *Adv. Biochem. Eng. Biotechnol.* 53:119-178.

Kajiwara S., Kakizono T., Saito T., Kondo K., Ohtani T., Nishio N., Nagai, S., Misawa N. (1995). Isolation and functional identification of a novel cDNA for astaxanthin biosynthesis from *Haematococcus pluvialis*, and astaxanthin synthesis in *Escherichia coli. Plant Mol. Biol.*29:343-352.

Kajiwara S., Fraser P. D., Kondo K., Misawa N. (1997). Expression of an exogenous isopentenyl diphosphate isomerase gene enhances isoprenoid biosynthesis in *Escherichia coli. Biochem. J.* 324:421-426.

Kang, K. S., Lim C.J., Han T.J., Kim J.C., Jin C.D. (1999). Changes in the isozyme composition of antioxidant enzymes in response to aminotriazole in leaves of *Arabidopsis thaliana.J Plant Biology* 42:187-193.

Karboune S., Neufeld R., Kermasha S. (2005). Immobilization and biocatalysis of chlase in selected organic solvent systems. *J Biotechnol.*120:273–83.

Khachik F., De Moura F. F., Zhao D. Y., Aebischer C. P., Bernstein P. S. (2002). Transformations of selected carotenoids in plasma, liver, and ocular tissues of humans and in nonprimate animalmodels. Invest. *Ophthalmol. Vis. Sci.* 43, 3383–3392.

Kim D. H., Kim J. H., Bae S. E., Seo J. H., Oh T. K., Lee C. H. (2005). Enhancement of natural pigment extraction using *Bacillus* species xylanase.*J Agric Food Chem.* 53:2541–5.

Kim W. J., Kim J. D., Kim J., Oh S. G., Lee Y. W. (2008). Selective caffeine removal from green tea using supercritical carbon dioxide extraction.*J Food Eng.*89:303–309.

Kim J., Smith J.J., Tian L.; DellaPenna, D. (2009). The evolution and function of carotenoid hydroxylases in *Arabidopsis. Plant Cell Physiol.* 50, 463–479.

Klippel K. F., Hiltl D. M., Schipp B. (1997). A multicentric, placebo-controlled, double-blind clinical trial of b-sitosterol (phytosterol) for the treatment of benign prostatic hyperplasia. *Brit J Urol.*80:427–432.

Klipstein-Grobush K., Launer L., Geleijnse J. M., Boeing H., Hofman A., Witteman J. C. (2000). Serum antioxidant and atherosclerosis. *The rotterdam study. Atherosclerosis.* 148, 49–56.

Koh W. P., Yuan J., Wang R., Lee Y. P., Lee B. L., Yu M.C., Ong C.N. (2011) Plasma carotenoids and risk of acute myocardial infarction in the Singapore Chinese Health Study. *Nutr. Metab. Cardiovasc. Dis.* 21, 1–6.

Kohlmeier L., Hastings S. B. (1995). Epidemiologic evidence of a role of carotenoids in cardiovascular disease prevention. *Amer. J. Clin. Nutr.* 62 (Suppl.), 1370S–1377S.

Kohlmeier L., Kark J. D.; Gomez-Gracia E., Martin B. C., Steck S. E., Kardinaal A., Ringstad J., Thamm M., Masaev V., Riemersma R., Martin-Moreno, J. M., Huttunen J. K., Kok F.J. (1997). Lycopene and myocardial infarction risk in the EURAMIC study. *Amer. J. Epidemiol.*146, 618–626.

Kristenson M., Zieden B., Kucinskiene Z., Elinder L. S., Bergdahl B., Elwing B., Abaravicius A., Razinkoviene L., Calkauskas H., Olsson, A. G. (1997). Antioxidant state and mortality from coronary heart diases in Lithuanian and Swedish men: Concomitant cross sectional study on men aged 50. BMJ 1997, 314, 629–633.

Kritchewsky S. B. (1991) β-carotene, carotenoids and the prevention of coronary heart disease.*J. Nutr.* 129, 5–8.

Kuk Y. I., Shin J.Z., Burgosc N.R., Hwang T.E., Hand O, Cho B.H., Junga S., Guh J.H. (2003). Antioxidative enzymes offer protection from chilling damage in rice plants. *Crop Sci.* 43:2109-2117.

Lamers P. P., Van de Laak C. C., Kaasenbrood P. S., Lorier J., Janssen M., DeVos R. C., Bino R. J., Wijffels R. H. (2010). Carotenoid and fatty acid metabolism in light-stressed *Dunaliella salina*. *Biotechnol Bioeng* 106(4):48–638.

Landrum J. T., Bohne R. (2001). Lutein, zeaxanthin and the macular pigment. *Arch. Biochem. Biophys.* 385, 28–40.

Lee M. Y. (2002). Effect of Na_2SO_3 on the activities of antioxidant enzymes in geranium seedlings. *Phytochemistry* 59:493-499.

Lee P. C, Schmidt-Dannert C. (2002) Metabolic engineering towards biotechnological produc- tion of carotenoids in microorganisms. *Appl Microbiol Biotechnol* 60:1–11.

Lee Y. K., Ding. S. Y. (1994). Cell cycle and accumulation of astaxanthin in *Haematococcus lacustris* (Chlorophyta). *J. Phycol.* 30:445-449.

Lemoine Y., Schoefs B. (2010). Secondary ketocaroteno astaxanthin biosynthesis in algae: a multifunctional response to stress. *Photosynth. Res.* 106:155-177.

Leya T., Rahn A., Lütz C., Remias D. (2009). Response of arctic snow and permafrost algae to high light and nitrogen stress by changes in pigment composition and applied aspects for biotechnology. FEMS *Microbiol. Ecol.* 67:432-443.

Li Y., Huang J., Sandmann G., Chen, F. (2008). Glucose sensing and the mitochondrial alternative pathway are involved in the regulation of astaxanthin biosynthesis in the dark-grown *Chlorella zofingiensis* (Chlorophyceae). *Planta* 228:735-743.

Li Y., Huang J., Sandmann G., Chen F. (2009). High-light and sodium chloride stress differentially regulate the biosynthesis of astaxanthin in *Chlorella zofingiensis* (Chlorophyceae). *J. Phycol.* 45:635-641.

Li Y., Sommerfeld M., Chen F., Hu Q. (2010). Effect of photon flux densities on regulation of carotenogenesis and cell viability of *Haematococcus pluvialis* (Chlorophyceae).*J. Appl. Phycol.* 22:253-263.

Lichtenthaler H.K. (1999). The 1-deoxy-D-xylulose-5-phos- phate pathway of isoprenoid biosynthesis in plants. Annu. *Rev. Plant Physiol. Plant Mol. Biol.* 50:47-65.

Lichtenthaler H.K. (2004). Evolution of carotenoid and isoprenoi biosynthesis in photosynthetic and non-photosynthetic organism. In *Proceedings of the 16th International Plant Lipid Symposium.* Budapest Hungary

Lichtenthaler H.K., Rohmer M., Schwender, J. (1997). Two independent biochemical pathways for isopentenyl di- phosphate and isoprenoid biosynthesis in higher plants. *Physiol. Plant.* 101:643-652.

Liu B.H., Lee Y.K. (2000). Secondary carotenoids formation by the green alga *Chlorococcum* sp. *J. Appl. Phycol.*12:301-307.

Lidebjer C., Leanderson P., Ernerudh J., Jonasson L. (2007). Low plasma levels of oxygenated carotenoids in patients with coronary artery disease. *Nutr. Metab. Cardiovasc.* Dis. 17, 448–456.

Lohr M., Schewender J., Polle J. E. W. (2012) Isoprenoid biosynthesis in eucariotic phototrophs: A spotlight on algae. *Plant Science*. 185-186, 9-22.

Lloyd-Jones D., Adams R., Carnethon M., De Simone G., Ferguson T. B., Flegal K., Ford E., Furie K., Go A., Greenlund K., et al. (2009) Heart disease and stroke statistics update: A report from the American Heart Association Statistics Committee and Stroke Statistics Subcommittee. Circulation 2009, 119, 480–486.

Lotan T., Hirschberg J. (1995). Cloning and expression in *Escherichia coli* of the gene encoding β-C-4-oxygenase that converts β-carotene to the ketocarotenoid canthaxanthin in *Haematococcus pluvialis. FEBS Lett.* 364:125-128.

Ma L., Lin X.M. (2010). Effects of lutein and zeaxanthin on aspects of eye health.*J. Sci. Food Agr.* 90, 2–12.

Ma R.Y.N., Chen F. (2001). Enhanced production of free trans-astaxanthin by oxidative stress in the cultures the green microalga *Chlorococcum* sp. *Process Biochem.* 36:1175-1179.

Macias-Sanchez M. D., Mantell C., Rodríguez M., Martínez de la Ossa E., Lubi L. M., Montero O. (2005). Supercritical fluid extraction of carotenoids and chl a from *Nannochloropsis gaditana.J Food Eng.* 66:245–51.

Marques A., Miranda J., Batista A. P., Gouveida L. (2011) Microlagae biotechnological applications: nutrition, health and environment. *In* Johansen, M.N. (Ed). Microalgae: Biotechnology, Microbiology and Energy. Nova Science Publishers Inc. pp 1-60.

Martínez-Férez, I.M., Vioque, A. (1992) Nucleotide sequence of the phytoene desaturase gene from *Synechocystis* sp. PCC 6803 and characterization of a new mutation which confers resistance to the herbicide norflurazon. *Plant Mol. Biol.* 18: 981–983.

Martinson T. A., Plumley F. G. (1995). One-step extraction and concentration of pigments and acyl lipids by sec-butanol from in vitro and in vivo samples. *Anal biochem.* 228:123–30.

Masamoto K., Wada H., Kaneko T., Takaichi S. (2001). Identification of a gene required for cis-to-trans carotene isomerization in carotenogenesis of the cyanobacterium *Synechocystis sp.* PCC 6803. *Plant Cell Physiol.* 42: 1398–1402.

Manzan A. C., Toniolo F. S., Bredow E., Povh N. P. (2003). Extraction of essential oil and pigments from *Curcuma longa* (L.) by steam distillation and extraction with volatile solvents. *J Agric Food Chem.* 51:6802–6807.

Matsuzaki M., Misumi O., Shin I.T., Maruyama S., Takahara M., MiyagishimaS.Y., Mori T., Nishida K., Yagisawa F., Nishida K., Yoshida Y., Nishimura Y., Nakao S., Kobayashi T., Momoyama Y., Higashiyama T., Minoda A., SanoM., Nomoto H., Oishi K., Hayashi H., Ohta F., Nishizaka S., Haga S., Miura S., Morishita T., Kabeya Y., Terasawa K., Suzuki Y., Ishii Y., Asakawa S., TakanoH., Ohta N., Kuroiwa H., Tanaka K., Shimizu N., Sugano S., Sato N., NozakiH., Ogasawara N., Kohara Y., KuroiwaY. (2004). Genome sequence ofthe ultrasmall unicellular red alga Cyanidioschyzon merolae 10D. *Nature* 428,653-657.

Mercer P., Armenta R. E. (2011). Developments in oil extraction from microalgae. *Eur. J. Lipid Sci. Technol.*, 113(5), 539-547.

Miki, W. (1991). Biological functions and activities of animal carotenoids. *Pure Appl. Chem.* 63:141-146.

Minoda A, Sakagami R, Yagisawa F, Kuroiwa T, Tanaka K. (2004). Improvement of culture conditions and evidence for nuclear transformation by homologous recombination in a red alga, Cyanidioschyzon merolae 10D. *Plant Cell Physiol.* 45:667-71.

Mittler R., Zilinskas B. A. (1993). Detection of ascorbate peroxidase activity in native gels by inhibition of the ascorbate-dependent reduction of nitroblue tetrazolium. *Anal Biochem* 212:540-546.

Miziorko H. M. (2011). Enzymes of the mevalonate pathway of isoprenoid biosynthesis. Arch. Biochem. Biophys. 505, 131–143.

Muller-Feuga A., le Guedes R., Pruvost J. (2003). Benefits and limitations of modeling for optimization of *Porphyridium cruentum* cultures in an annular photobioreactor. *J Biotechnol.* 103 153–163.

Murthy K., Vanitha A., Rajesha J., Swamy M., Sowmya P., Ravishankar G. A. (2005). In vivo antioxidant activity of carotenoids from *Dunaliella salina*, a green microalga. *Life Sci.*76, 1381–1390.

Mussgnug J. H., Thomas-Hall S. Rupprecht J., Foo A., Klassen V., McDowall A., Schenk P.M., Kruse O., Hankamer B. (2007). *Engineering photosynthetic light capture: impacts on improved solar energy to biomass conversion.* 5(6): 802-814.

Naik P. S., Chanemougasoundharam A., Paul Khurana S.M., Kalloo G. (2003). Genetic manipulation of carotenoid pathway in higher plants. *Current Sci.* 85:1423–30.

Ninet L., Renault J. (1979). *In*: Peppler H. J., Perlman D. (Eds) *Microbiol technology*, 2nd Ed. Academic, New York, pp 529–530.

Nokookar K., Moradshadi A., Hosseini L. (2005). Physiological responses of *Dunaliella salina* and *Dunaliella tertiolecta* to copper toxicity. *Biomol. Eng*. 22: 141-146.

Ojima K., Breitenbach J., Visser H., Setoguchi Y., Tabata K., Hoshino T., van den Berg, J., Sandmann, G. (2006). Cloning of the astaxanthin synthase gene from *Xanthophyllomyces dendrorhous* (*Phaffia rhodozyma*) and its assignment as a β-carotene 3-hydroxylase/4-ketolase. *Mol. Gen. Genomics* 275: 148–158.

Oren A. (2005). A hundred years of *Dunaliella* research: 1905–2005. *Biomed. Central.* 1(2):14

Orosa M., Torres E., Fidalgo P., Abalde J. (2000). Production and analysis of secondary carotenoids in green algae. *J. Appl. Phycol.* 12:553-556.

Paniagua-Michel J., Capa-Robles W., Olmos-Soto J., Gutierrez-Millan L. E. (2009). The carotenogenesis pathway via the isoprenoid-β-carotene interference approach in a new strain of *Dunaliella salina* isolated from Baja California Mexico. *Mar Drugs* 7:45–56.

Pasquet V., Chérouvriera J., Farhata F., Thiérya V., Piota J M. et al., (2010). Study on the microalgal pigments extraction process: Performance of microwave assisted extraction. *Process Biochemistry.* 46: 59–67.

Patrick L. (2000). Beta-carotene: the controversy continues. *Altern Med Rev*.5:530–45.

Piccaglia R., Marotti M., Grand S. (1998). Lutein and lutein ester content in different types of *Tagetes patula* and *T. erecta. Industrial Crops and Products* 8*:* 45–51

Plaza M., Herrero M, Cifuentes A, Ibañez E. (2009). Innovative natural functional ingredients from microalgae. *J Agric Food Chem.*57:7159–7170.

Polle JEW, Tran D, Ben-Amotz A. (2009). Chapter 1: history, distribution, and habitats of algae of the genus *Dunaliella* Teodoresco (Chlorophyceae). The book: the alga Dunaliella: biodiversity, physiology, genomics and biotechnology. *Science Publishers*, pp 1–14.

Preisig H. R., Andersen R. A. (2005). Historical review of algal culturing techniques *In*: Algal Culturing Techniques. Ed by Andersen R. A. London: Elsevier Academic *Press*. pp 1–12.

Pulz O., Gross W. (2004). Valuable products from biotechnology of microalgae. *ApplMicrobiol Biotechnol* 65:635–648.

Qin S., Liu G. X., Hu Z. Y. (2008). The accumulation and metabolism of astaxanthin in *Scenedesmus obliquus* (Chlorophyceae). *Process Biochem*. 43:795-802.

Rabbani S., Beyer P., Linting J, Hugueney P., Kleinig H. (1998) Induced b-carotene synthesis driven by triacylglycerol deposition in the unicellular algae *Dunaliella bardawil. Plant. Physiol.* 3:1239-1248.

Rabinowitch H. D., Privalle C. T., Fridovich I. (1987). Effects of paraquat on the graan algae *Dunaliella salina:* protection by the mimic of superoxide dismutase, desferal-Mn (IV). *Free Rad Biol.* 3: 125-131.

Ramos-Valdivia A. C., Van der Heijden R., Vapoorte R. (1997). Isopentenyl diphosphate isomerase a core enzyme in isoprenoid biosynthesis. A *review* of its biochemistry and function. *Nat. Prod Rep.* 14 (6) 591-603.

Remias D., Lütz-Meindl U., Lütz C. (2005). Photosynthesis, pigments and ultrastructure of the alpine snow alga *Chlamydomonas nivalis. Eur. J. Phycol.* 40:259-268.

Riccioni G. (2009). Carotenoids and cardiovascular disease. *Curr. Atheroscl. Rep.*11, 434–439.

Roberts R.L., Green J., Lewis B. (2009). Lutein and zeaxanthin in eye and skin health. *Clin Dermatol.* 27(2):195–201

Rohmer M. (1999). The discovery of a mevalonate-independent pathway for isoprenoid biosynthesis in bacteria, algae and higher plants. *Nat. Prod. Rep.* 16:565–574.

Rodriguez-Bernaldo de Quirós A., Costa H. S. (2006). Analysis of carotenoids in vegetable and plasma samples: a review. *J Food Compos Anal.*19:97–111.

Rostagno M.A., Palma M., Barroso C.G. (2007). Ultrasound-assisted extraction of isoflavones from soy beverages blended with fruit juices. *Anal Chim Acta.*597:265–72.

Saad Md., Yusof Y.A., Ngah W.Z. (2006). Comparison between locally produced Chlorella vulgaris and *Chlorella vulgaris* from Japan on proliferation and apoptosis of liver cancer cell line, HepG2. Malaysian J Biochem & Molec Biol. 13(1):32–36.

Sachindra N.M., Bhaskar N., Mahendrakar N. S. (2005). Carotenoids in crabs from marine and fresh waters of India. LWT *Food Sci Technol.*38:221–5.

Sánchez F., Fernández J. M., Acien F. G., Rueda A., Perez-Parra J., Molina E. (2008a) Influence of culture conditions on the productivity and lutein content of the new strain *Scenedesmus almeriensis. Process Biochem.* 43(4):398–405.

Sánchez J. F., Fernández-Sevilla J. M., Acién F. G., Cerón M. C., Pérez-Parra J., Molina-Grima E. (2008b). Biomass and lutein productivity of *Scenedesmus almeriensis*: influence of irradiance, dilution rate and temperature. *Appl Microbiol Biotechnol.* 79(5):719–729

Sánchez-Estudillo L., Freile-Pelegrin Y., Rivera-Madrid R., Robledo D., Narvàez-Zapata J.A. (2006). Regulation of two photosynthetic pigment-related genes during stress-induced pigment formation in the green algae, *Dunaliella salina Biotech. Lett.*28:787-791.

Sandmann G., Romer S., Fraser P.D. (2006). Understanding carotenoid metabolism as a necessity for geneticengineering of crop plants. *Metab. Eng.*, 8: 291–302.

Sansawa H., Endo H. (2004) Production of intracellular phytochemicals in *Chlorella* under heterotrophic conditions. *J Biosci Bioeng.* 98:437–444.

Schwender J, Seemann M, Lichtenthaler HK, Rohmer M. (1996). Biosynthesis of isoprenoids (carotenoids, sterols, prenyl side-chains of chlorophylls and plastoquinone) via a novel pyruvate/glyceraldehyde 3-phosphate non-mevalonate pathway in the greenalga *Scenedesmus obliquus. Biochem J.* 316:73–80.

Schwender J, Gemunden C, Lichtenthaler H (2001) Chlorophyta exclusively use the 1-deoxyxylulose 5-phosphate/2-C-methyleryth- ritol 4-phosphate pathway for the biosynthesis of isoprenoids. *Planta.*212:416–423.

Sesso H. D., Buring J. E., Norkus E. P., Gaziano J. M. (2005). Plasma lycopene, other carotenoids, retinol and the risk of cardiovascular disease in men. *Amer. J. Clin. Nutr.*81, 990–997.

Shi X., Zhang X., Chen F. (2000). Heterotrophic production of biomass and lutein by *Chlorella protothecoides* on various nitrogen sources. *Enzyme Microb Technol* 27:312–318.

Shin H.W., Lee, M.Y. (2003). Cadmium-induced changes in antioxidant enzymes from the marine alga *Nannochloropsis oculata*. *J Appl Phycol* 15:13-19.

Shrager J., Hauser C., Chang C.W., Harris E.H., Davies J., McDermott J., Tamse R., Zhang Z., Grossman A.R. (2003). *Chlamydomonas reinhardtii* Genome Project. A guide to the generation and use of the cDNA information. Plant Physiology 131, 401-408

Skórzynska-Polit, E., Krupa, Z. (2006). Lipid peroxidation in cadmium-treated *Phaseolus coccineus* plants. *Arch Environ ContamToxicol* 50:482-487.

Steinbrenner J., Linden H. (2001). Regulation of two carotenoid biosynthesis genes coding for phytoene synthase and carotenoid hydroxylase during stress-induced astaxanthin formation in the green alga *Haematococcus pluvialis. Plant Physiol.* 125:810-817.

Steinbrenner J., Linden H. (2003). Light induction of carotenoid biosynthesis genes in the green alga Haematococcus *pluvialis*: regulation by photosynthetic redox control. *Plant Mol. Biol.* 52:343-356.

Steinbrenner J., Sandmann G. (2006). Transformation of the Green Alga *Haematococcus pluvialis* with a Phytoene Desaturase for Accelerated Astaxanthin Biosynthesis. *Appl. Environ. Microbiol.* 72(12): 7477-7484.

Street D. A., Comstock G. W., Salkeld R. M., Schuep, W., Klag M. (1994). Serum antioxidants and myocardial infarction: Are low levels of carotenoids and α-tocopherol risk factors for myocardial infarction? *Circulation*. 90, 1154–1161.

Sun N., Wang Y., Li Y. T., Huang J. C., Chen F. (2008). Sugar-based growth, astaxanthin accumulation and carotenogenic transcription of heterotrophic *Chlorella zofingiensis* (Chlorophyta). *Process Biochem.*43:1288-1292.

Takaichi S., Mochimaru M. (2007). Carotenoids and carotenogenesis in cyanobacteria: Unique ketocarotenoids and carotenoid glycosides. *Cell. Mol. Life Sci.*, 64, 2607–2619.

Travieso L., Benítez, F., Sánchez E., Borja R., Colmenarejo M. F. (2006). Production of biomass (algae–bacteria) by using a mixture of settled swine and sewage as substrate. *J. Environ. Sci. Health* 41, 415–429.

Tran D., Haven J.H., Qiu W.G. (2009). An update on carotenoid biosynthesis in algae: phylogenetic synthase evidence for the existence of two classes of phytoene synthase. Planta 229:723–729.

Vershinin A. (1999). Biological functions of carotenoids: diversity and evolution. *Biofactors* 10:99-104.

Vidhyavathi R., Venkatachalam L., Sarada R., Ravishankar G.A. (2009). Regulation of carotenoid biosynthetic gene expression and caroenoid accumulation in the green alga *Haematococcus pluvialis* under nutrient stress conditions. *J Exp Bot* 59:1409–1418

Vila M., Couso I., León, R. (2008). Carotenoid content in mutants of the chlorophyte *Chlamydomonas reinhardtii* with low expression levels of phytoene desaturase. *Process Biochem*. 43: 1147–1152.

Wang B., Li Y., Wu N., Lan Q. C. (2008). CO_2 bio-mitigation using microalgae. *Appl Microbiol Biotechnol.* 79:707–718.

Wang C. W., Oh M. K., Liao J. C. (1999). Engineered isoprenoid pathway enhances astaxanthin production in *Escherichia coli. Biotechnol. Bioeng*. 62:235-241.

Wang J., Sommerfeld M., Hu Q. (2009). Occurrence and environmental stress responses of two plastid terminaoxidases in *Haematococcus pluvialis* (Chlorophyceae) *Planta* 230:191-203.

Wang L., Weller C.L. (2006). Recent advances in extraction of nutraceuticals fromplants. *Trends Food SciTechnol*. 17:300–12.

Wei D., Chen F., Chen G., Zhang X., Liu L., Zhang H. (2008). Enhanced production of lutein in heterotrophic *Chlorella protothecoides* by oxidative stress. Sci China C Life Sci.51(12):1088-1093.

Wijffels R.H., Barbosa M.J., Eppink M.H. (2010). Microalgae for the production of bulk chemical and biofuels. *Biofuels Bioprod Bioref.* 4: 287-295.

Wijffels R.H., Eggink G., Springer J. (2012). Algae and genetic modification. Research, production and risks. Technopolis proup. The Netherlands Commission on Genetic Modification (COGEM) report CGM 2012-05 ONDERZOEKSRAPPORT.

Woo SO., Yum S.S, Park H.S, Jung L.H., Chan S., So-Jung Kim., Taek-Kyun L. (2007). Molecular parameters for assessing marine biotoxicity: gene expressions of Rockfish (*Sebastes schlegeli*) exposed to polycyclic aromatic hydrocarbons. *Mol Cell Toxicol* 3:267-272.

Wu D. Y., Wright D. A., Wetzel C., Voytas D. F., Rodermel S. (1999). The immutans variegation locus of Arabidopsis defines a mitochondrial alternative oxidase homolog that functions during early chloroplast biogenesis. *Plant Cell* 11:43-55.

Ye Z. W., Jiang J. G., Wu H. (2008). Biosynthesis and regulation of carotenoids in Dunaliella: progresses and prospects. *Biotechnol Adv* 25:352–360.

Yla-Herttuala S. (1991). Macrophages and oxidized low density lipoproteins in the pathogenesis of atherosclerosis. *Ann. Med*. 23, 561–567.

Zorn H., Breithaupt D. E., Takenberg M., Schwack W., Berger R. G. (2003). Enzymatic hydrolysis of carotenoid esters of marigold flowers (*Tagetes erecta* L.) and red paprika (*Capsicum annuum* L.) by commercial lipases and *Pleorotus sapidus* extracellular lipase. *Enzyme Microb Technol*. 32:623–628.

In: Microalgae and Other Phototrophic Bacteria
Editor: Luis G. Torres Bustillos
ISBN: 978-1-63482-078-3

Chapter 13

UTILIZATION, ISOLATION AND CHARACTERIZATION OF *CHLORELLA VULGARIS* FOR CARBON AND WASTE WATER SEQUESTRATION TREATMENT

P. S. Syed Shabudeen*
Chemistry and Environmental Science, Kumaraguru College of Technology, Coimbatore, Tamilnadu

ABSTRACT

A simple unicellular green microalgae species are collected and isolated from the brackish and fresh water bodies of Coimbatore, Erode regions of Tamilnadu state of India.This species has been evaluated for its potential use in the production of biodiesel, carbon sequestration studies. This micro algae organism has been isolated and identified by genomic DNA, 16S rRNA gene amplifications and sequencing techniques. The growth rate has been estimated with spectro photometric measures. The Light Microscope and Scanning Electron Microscope are used to examine the morphology and the Fourier Transform Infrared Spectroscopic studies infer the presence of various functional groups in these algae.The rate of carbon sequestration and the improvement to the quality of the waste water is obvious by growth rate determination of algae bio-mass. The quality of waste waters are improved with respect to pH, conductivity, total hardness, calcium and magnesium hardness, total dissolved solids (chlorides, sulphates, phosphates, nitrates), Iron, Total Alkalinity (carbonate, bicarbonate and hydroxide alkalinities), Fluoride levels and with respect to Chlorella vulgaris growth. The improved quality of waste water has revealed that the chlorella vulgaris species are effective in treating the waste waters which are generated by human consumptions. From this research work it has been proved that this chlorella vulgaris species proves its efficiency of 100% removal of nitrate, phosphate loads at a time interval of 6 days with remarkable carbon sequestration abilities. The extractive mechanism of lipids and production of bio-diesel by transesterification from these algae species signifies its importance with respect to environmental and socio-economic concerns of any nation.

* Corresponding author: E-mail: shabu_cbe@yahoo.com.

INTRODUCTION

Green microalgae chlorella vulgaris species are collected from the Lakes at Erode and Coimbatore at Tamil Nadu State of India. The Morphological studies are examined with Light Microscope and Scanning Electron Microscope. The PCR studies revealed that the collected specimen is chlorella vulgaris. The Fourier Transform Infrared Spectroscope studies infer the presence of various functional groups in these chlorella vulgaris a bio-mass and this species indicates a positive growth rate in the waste water medium with very good absorption capacity of 100% with respect nitrate and phosphate condent ofthe waste water mediums at an interval of 6 days period.The microalgae *Chlorella* sp. are known to have various other uses for the treatment of wastewater, production of biodiesel, production of electricity using microbial fuel cells, animal food supplements and providing valuable extracts for chemical products [Becker, 2004]. Although the traditional taxonomic characteristics of Chlorella vulgaris indicated that morphological biochemical properties are used in its identification, the cell size and shape are variable and largely depend upon varying age, nutrition and environmental factors. The traditional identification of micro algae is achieved by microscopic studies identification results, which are questionable to certain degree.

Isolation of Algae

After the growth of algae species for 2-4 weeks in the streak plates, the cultures are heterogeneous at this stage, the colonies are homogenized and is achieved by a serial dilution process. These samples are plated onto BG-11 and are observed under a light microscope and these isolated colonies are selected for the studies based upon the rapid growth rates. The morphological homogeneity and with color diversities they are tentatively identified as Chlorella vulgaris. The individual cells of the colonies were in the range of 10μm.Cells are green colour, unicellular, spherical in shape is show in Figure 1.

Characterization of Micro Green

The collected algae sample from various regions are cultured by strike plate procedure and these cultures are primarily identified with light microscope and these sample are isolated and identified by DNA isolation and PCR studies [Syed Shabudeen P. S. et al., 2013].

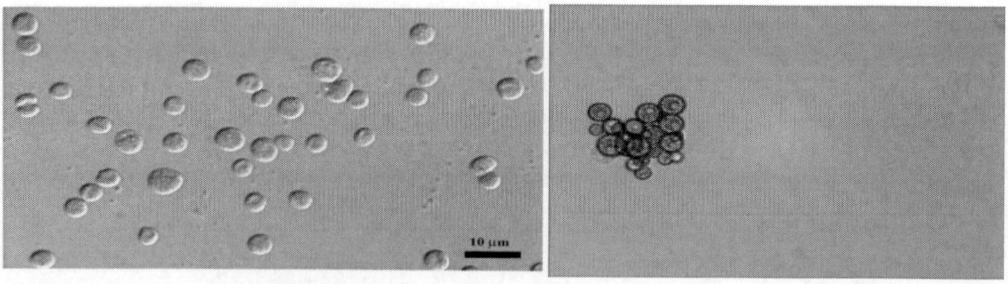

Figure 1. Light microscopic images of Cholorella vulgaris.

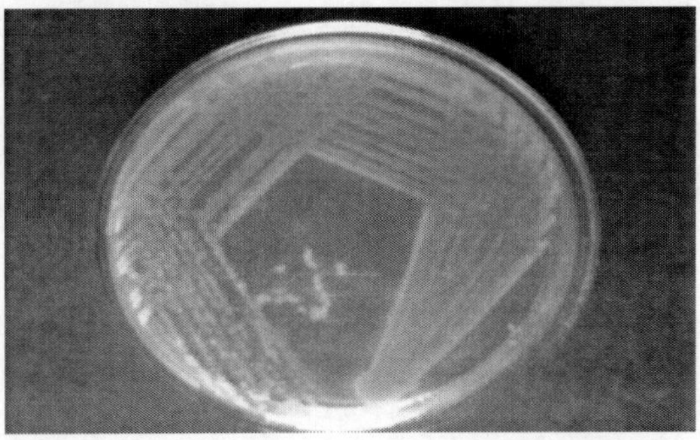

Figure 2. Pure culture obtained from streak plate method DNA Isolation.

Serial Dilution and Streak Plate Method

The sample was taken and serially diluted and these cultures are taken in the streak plates with which MRS agar is already placed with these plates and pure colonies are developed.

The10 μl of Sodium acetate (3M) solution at 5.2 pH is taken with 275 μl of 95% v/v ethanol.This solution of sodium acetate and ethanol is added to the samples in the streak plates (Figure 2). The DNA present in these samples of streak plates are mixed by flipping and placed in dry ice period and centrifuged at 15,700 x g at 4°C (Hermle Z233 MK-2 High Speed Refrigerated Centrifuge) both of these operations are carried out separately for 15 minutes duration. The supernatant is removed and the precipitates are made into pellets and washed with 1ml of 95% v/v cold ethanol and again centrifuged for 5-6 minutes and finally the supernatants are discarded and these pellets of DNA are formed and are kept in an Iso temp vacuum oven at 37°C for 10 minutes duration at a vacuum of -25 inches of Hg to obtain the dried DNA pellets. To this dried pellets 80 μl of TE buffer solution is added separately and the DNA concentration in these pellets are determined with a ND-1000 Spectrophotometer (Nanodrop, Wilmington, DE). These studies are carried out with a minimum of two trials with DNA samples of 2 μl volume.

PCR Amplication for 16s rRNA These DNA are amplified by using universal 16S rRNA primers A 8F and A 1492R (Ocimum Bio solutions Ltd) and the PCR reaction is performed in a 50 μL containing 0.1 ng of template DNA, PCR Master Mixer, 10 pmol concentration of each primer and 0.025U of Taq DNA polymerase enzymes. The final volume is adjusted with sterilized Mili-Q water. A PCRthermo cycler (Bio-rad) is used to amplify the reactions through an initial denaturation step at 94°C for 2 minutes and followed by 25 cycles at 94°C for 1 an another min. Then the temp is adjusted at 52.3°C for 1 min and with an extension of 72°C for 1 min, and then finally extended at the 72°C for 2 minutes. Finally, the amplified PCR products are stored at -20°C for further purification and are to be used in the downstream application.

Primers	Target for amplification	Oligonucleotide 5'-3'
A 8F	16SrRNA	AGAGTTTGATCCTGGCTCAG
A 1492R	16SrRNA	TACGGCTACCTTGTTACGACTT

AgaroseGel Electrophoresis for PCR Product

About 5 μl aliquot of PCR amplified product is loaded on 1.4% agarose in 1X TAE buffer at 50 V for 45 min and the PCR products were visualized in a UV trans illuminator. Lambda DNA double digested with EcoR I and Hind III is used as the marker.

PCR Studies

In molecular biology for several decades PCR is one of the most widely used methods in which attempts are made to optimizing is to be continued and efforts are made to meet its specific objectives. Templates known for being is difficult to sequence which includes with high guanine-cytosine (G/C) content, high adenine-thymine (A/T) content, as well as sequences with marked secondary structure or large regions of homo polymers [Stirling, D., 2003]. With regards to sequences with high G/C content, several approaches are attempted to solve this problem. Perhaps the most successful method for improving results is the inclusion of certain organic additives in the reaction mixture, such as DMSO, betaine, polyethylene glycol, glycerol and formamide [Chakrbarti, R. et al., 2003].The effect DMSO exerts in the PCR amplification of some GC-rich sequences is a largely studied [Sidhu, M. K. et al., 1996]. In the PCR studies of cholorella vulgaris by using the designed primer pair, the 16S rDNA sequence of the cultivated chlorella is successfully amplified. This PCR product (Figure 3) is cloned and sequenced, and it revealed that the algal strain determined is Chlorella vulgaris and confirms when it is blasted on the NCBI database. According to the electrophoresis profiles, nine viral ds RNA bands were present in Chlorella cells. The sizes of these nine ds RNA segments ranged from 250 to 2500 bp, and the 1500 segment 5 was much brighter than the others thus it is selected for further analysis.

These informations are then put into the practice for the positive identiy of environmental strain the Chlorella genus present infreshwater unicellular green algae. When compared with both an internal standardand GenBank submissions, it gained a 97% match with the strain Chlorella vulgaris and have been identified.It is the solvent degrading and solvent tolerant strain and the genetic sequences are gained by using these primers only at W 600 bp in length, it can be successfully used for positive identiy of the organism. It also infer the phylogenetic associations with similar accuracy with phylogenetic reconstructions involving the use of complete 16S rRNA gene sequences [Rainey. F., 2000]. Ithas shown that by using various bioinformatics tools of a phototrophic eucaryotes and these primers are originally designed by Nubel et al. in the cyanobacterial studies. These are ideal for the studies of cyanobacterial communities and also used as a tool in the identification of phototrophic eucaryotes. This can be useful for the positive identification of new species and also for phylogenetic studies.

Fourier Transforms Infrared Spectroscopy

The IR spectrum of dried algal biomass was recorded on Nicolet IR spectrometer at room temperature. The dried algal powder was blended with potassium bromide (KBr) powder, and pressed into tablets before measurement. A region of 4000–400 cm^{-1} was used for scanning.

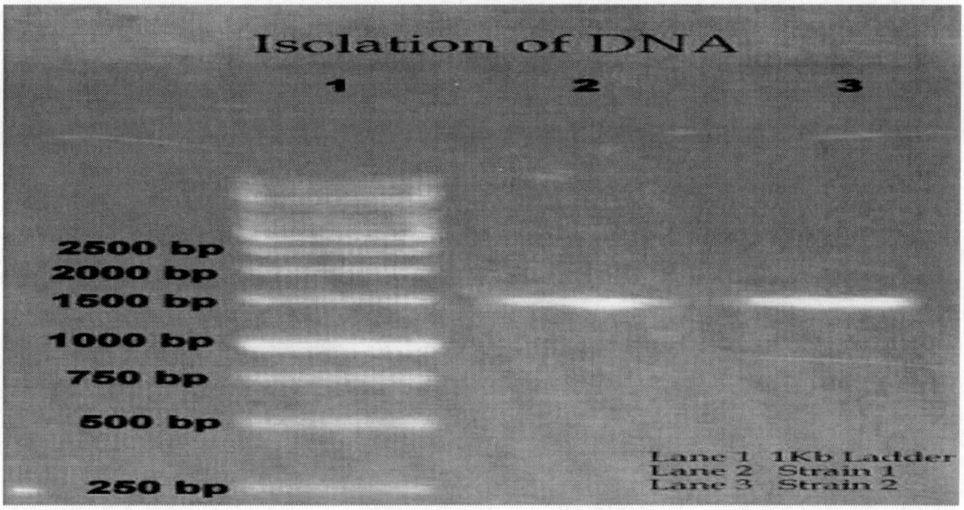

Figure 3.The PCR result.

FTIR Spectroscopic Studies

Most methods used in measuring algal nutritional and physiological changes are limited for detecting whole community responses because of the relatively large quantity of material needed for analysis. The spectroscopy studies has been developed and becomes a very powerful and flexible technique for the identification of micro organisms [Indhumathi, P et al., 2013]. The FTIR Spectroscopy has been widely useful and provide the information on range of vibrationally active functional groups (including O-H, N-H, C=O, =C-H, -CH2, -CH3, C-O-C, and >P=O) in all biological specimens. Although the technique has been largely used with isolated macromolecules and molecular complexes such as nucleic acid [Liquir and Taillandier, 1996], proteins [Stuart, 1996], lipids [Lewis, 1996], polysaccharides [Brandenburg and Seydel, 1996]. The FTIR spectroscopy has been successfully established as a reliable with user friendlyfast tool in the identification of the microalgae [Bastert et al., 1991].

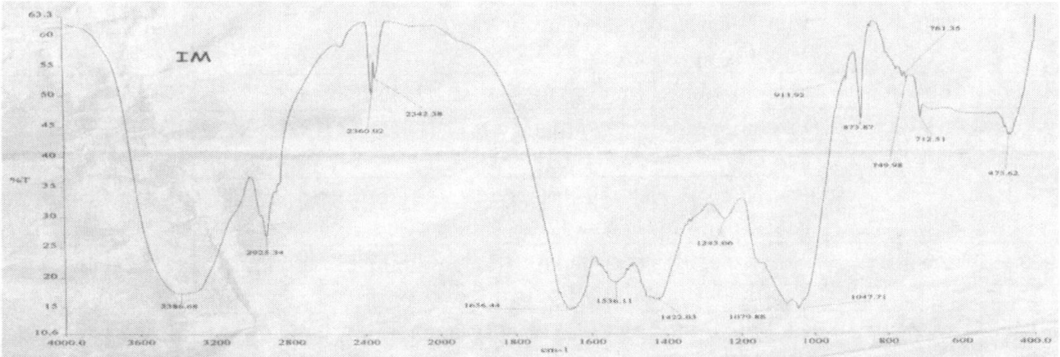

Figure 4. Fourier Transforms Infrared spectra for chlorella vulgaris.

FTIR spectra (Figure 4) in relation to specific groups (Table 1). Each peak assigned a functional group. The molecular assignments of bands are based on published data phytoplankton, bacteria and other biological materials. In this study chlorella vulgaris protein spectra characterized by strong peaks 1656 cm^{-1} (amide I) and 1536 cm^{-1} (amide II). These bands were due primarily to C=O stretching vibration and a combination of N-H and C-H Stretching vibrations in amide complexes. Lipid and carbohydrates were characterized by strong vibrations the C-H 2925cm^{-1}, C-O-C of polysaccharides at 1079cm^{-1}, 1047cm^{-1} respectively. The carbohydrates are the strongest absorbers between 1200 cm^{-1} and 1000 cm^{-1}. Several other classes of compounds, such as nucleic acids have functional groups with absorption bands in the same region of the spectrum.The strongest peaks 1536 cm^{-1} and 1422 cm^{-1} shows that bending modes of methyl groups of protein. The peak at 1243 cm^{-1} shows the presence of carboxylic acid in the algae. This study reveals that existing peaks suggested that lipid content is very high and also indicates the presence of carbohydrate, nucleic acid in chlorella vulgaris.

Scanning Electron Microscopy

The algal cells are first observed under light microscope for their morphological features and other cellular details, the cells were further studied using scanning electron microscope (SEM-Figure 5) according to the method of Fowke et al. The dried sample was screened by SEM (Scanning Electron Microscopy) for their absolute morphological studies. The basic steps for SEM sample preparations are fixing it with buffered aldehyde, post fixing it in Osmium tetra oxide, dehydrating it in ethanol, drying it with air dryer, mounting it on a specimen stub, coating it with Carbon and examining under the HRSEM (Quanta FEG 200).

Table 1. Tentative assignment of bands found in FTIR spectra of chlorella vulgaris

Band	Main peak cm^{-1}	Typical band	Wave number range cm^{-1}
1	3386	WaterV(O-H) stretching ProteinV(N-H) stretching	3029-3639
2	2925	Lipid –carbohydrate mainly V_{as} (CH_2) and V_s(CH_2) stretching	2809-3012
3	1656	Protein amideI band mainly V(C=O) stretching	1583-1709
4	1536	Protein amideII band mainly σ(N-H)bending V(C-N) stretching	1481-1585
5	1422	Proteinσ_{as} (CH_2) and σ_s(CH_3) bending of methyl lipid$_{as}$ (CH_2) bending of methyl	1425-1477
6	1243	Nucleic acid (other phosphate containing compounds) V_{as}> P=0 stretching of phosphodiesters	1191-1356
7	1079	Carbohydrate V (-O-C) of polysaccharides. Nucleic acid (other phosphate containing compounds) V_{as}> P=0 stretching of phosphodiesters	1072-1099
8	1047	Carbohydrate V(C-O-C) of polysaccharides	980-1072

Figure 5. Scanning Electron Microscope images of chlorella vulgaris.

The tertiary treatment of waste waters are achieved by C. vulgaris, this is not been demonstrated that the observed growth promotion may be due to improve capabilities of microalgae to remove nutrients from natural wastewater. The microbial carrier chosen in this study were alginate beads. Immobilization of microalgae in polysaccharide gels is an experimental way to use these microorganisms for wastewater treatment [Chevalier P et al., 1985]. The major difficulty is collecting enormous populations of cells developed during the treatment with, hamper the regular microalgae treatments.

CULTIVATINGCHLORELLAVULGARISAND SEQUESTRATING THE CARBON, WASTE WATER

Materials and Methods

The collected cultured algae biomass is taken in the preliminary treated college hostel waste water. The quality parameter are analyzed and tabled (Table 1) [Syed Shabudeen PS., et al., 2013]. The variation in quality of this waste bath water after the growth period of 2, 4, 6, 8, 10&12 days are determined by using the spectroscopic measurements and validated by measuring the dry mass. After a growth period of 12 days, a portion of this algae biomass is filtered and dried and preserved for analysis and to obtain the derivatives. The parameters such as percentage transmission, concentration, and optical density are the parameters to determine growth rate of algae species by spectro photometric methods.

Table 2. Water Quality Parameters

S. No	parameter	Sample water	After 2 days	After 4 days	After 6 days	After 8 days	After 10 days	After 12 days
1	PH	8.5	8.4	8.7	8.4	8.0	7.9	7.8
2	Conductivity	3.38	3.03	3.53	3.67	4.21	5.35	6.64
3	TDS	2.31	2.06	2.45	2.54	2.94	3.67	4.42
4	Phosphate	0.069	0.005	0.003	Nil	nil	Nil	Nil
5	Nitrate	0.857	0.801	0.387	nil	Nil	Nil	Nil
6	Iron	Nil	Nil	Nil	Nil	Nil	Nil	Nil
7	Fluoride	Nil	Nil	Nil	Nil	Nil	Nil	Nil
8	Chloride	280	223	280	309	436	684	649
9	Total hardness	1423	1150	1600	1350	2150	2970	3660
10	Calcium hardness	620	300	210	310	300	250	390
11	Magnesium hardness	803	850	1390	1040	1850	2720	3270
12	Sulphate	385	142	280	633	1059	1896	2422
13	Total alkalinity	705	600	310	270	420	235	330
14	Phenothalin alkalinity	67	75	35	20	15	10	Nil
15	Methyl orange alkalinity	638	525	275	250	405	225	330

The *Chlorella* vulgaris is grown in the pH range of 4-7. Highest lipid content was found in case of 0.02M salinity, 60 mg/L of bicarbonate salt concentration and 40% of CO2. [Indhumathi. P. et al., 2014]. Thus these strains are useful for large-scale, dense cultivation is possible with industrial discharged gases to be fixed CO_2 directly and to reduce the global warming. Thus the created cell biomass is useful for producing valuable bio-compounds.

The pre treated bath waste water from the hostel is used as a medium to grow the biomass assay. As per the linear growth kinetics studies, the waste water medium supports the growth of the biomass algae. The quality parameters of the waste water are determined and the values are tabled in Table 2. It reveals that the quality of waste water is considerably improved after the growth of the algae. The p^H, conductivity, Total hardness, Calcium hardness, Magnesium hardness, Total dissolve solids, Chloride, Sulphate, Phosphate, Nitrate, Iron,Total Alkalinity, Carbonate Alkalinity, Bicarbonate Alkalinity, Hydroxide Alkalinity, Fluoride are determined. This reveals with CO_2 sequestration studies, the decrease in TDS, Total hardness, Chloride, Sulphate, Phosphate, Nitrate, Iron etc, indicates the uptake of nutrients by algae for its growth. It is evident that, waste water qualities are improved. This indicates that, micro algae can be utilized to treat bath waste waters from hostels. The linear growth of algae is the effective tool to sequester carbon. This waste water is completely screened to remove bio-mass assay and it is subjected to pass through activated carbon columns to obtain odor less, transparent clear water. This is recycled water can be used for agricultural purpose, cleaning purpose, Industrial purpose.

Growth of Chlorella Vulgaris

The growth of biomass is continuously monitored by conducting analysis of spectro-photometer which includes the percentage of transmission, concentration and optical density and are considered as growth parameters. Simultaneously, weight of the dried biomass is recorded. This is shown in Table 3 are is based on the spectroscopic analysis of the biomass and its dry weight measurements. The Evolution 201 spectrophotometer is used for spectrophotometer analysis of a solution of any concentration. The output is available on the digital display in the forms of optical density (Absorbance), percentage transmission (%T) and concentration (C). The instrument operates at wavelength of 340 nm to 960nm. Figure 6 shows the percentage of transmission as function of growth period and Figure 7 shows the optical density of algae biomass as function of growth period obtained from the Spectroscopic Analysis of Biomass for Kinetic Study. Figure 8 shows variation in algae density as a function of time exhibits a linear growth kinetics.

Table 3. Spectroscopic Analysis of Algae Biomass for Kinetic Study and Dry Weight Measurements

Days	% Transmission	Concentration	Optical Density	Dry Weight Measurements(gm)
0	96	15	0.0	0.9
2	95	21	0.025	1.7
4	82	80	0.08	2.7
6	67	86	0.121	3.5
8	57	98	0.134	4.4
10	45	109	0.154	5.6
12	38	117	0.165	6.4

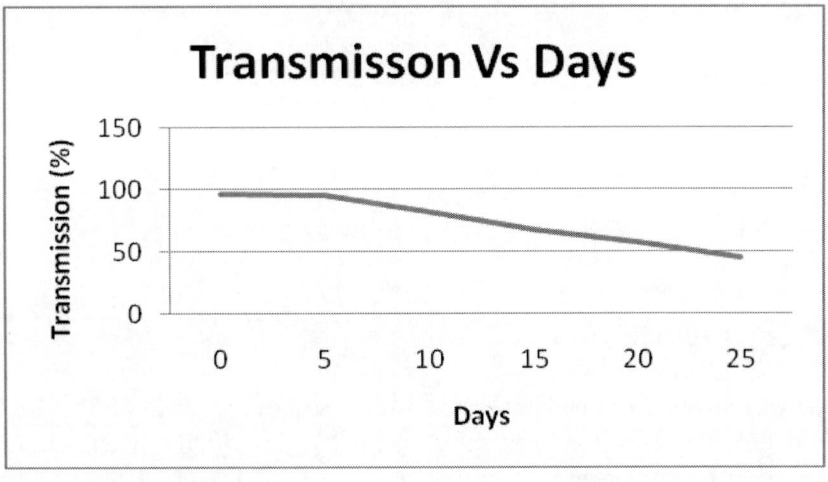

Figure 6. Transmission Vs Days.

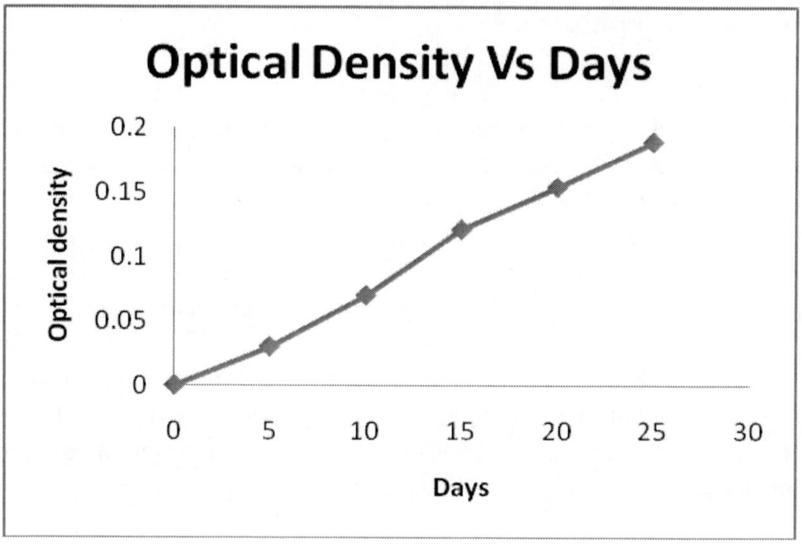

Figure 7.Optical Density Vs Days.

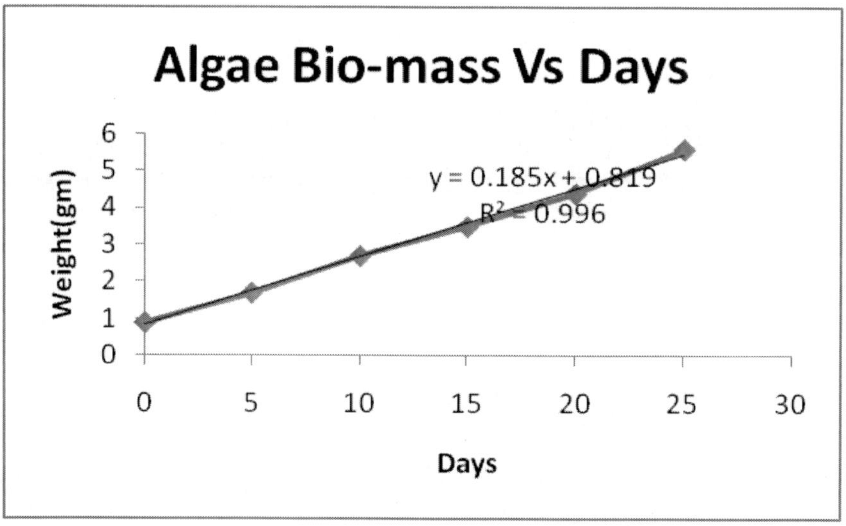

Figure 8. Algae Bio-mass Vs Days.

Analysis of Chlorophyll

The algae was mixed with methanol and distilled water(1:1), heat 60°C 30mins in water bath.after cool at room temperature then centrifuge it 10 mins, 5000 rpm extract was analyzed by UV-Visible spectrophotometer.

In microalgae, the ratio of chlorophyll to biomass has been reported to range from 0.1% to 5% of dry weight. The concentration of chlorophyll varies with the cell concentration. Variation of chlorophyll concentration follows the same pattern as the growth of cells (Young et al., 1996). Hence, the highest concentration of chlorophyll will be obtained at the highest cell concentration or at the end of the exponential phase of growth (Figure 9).

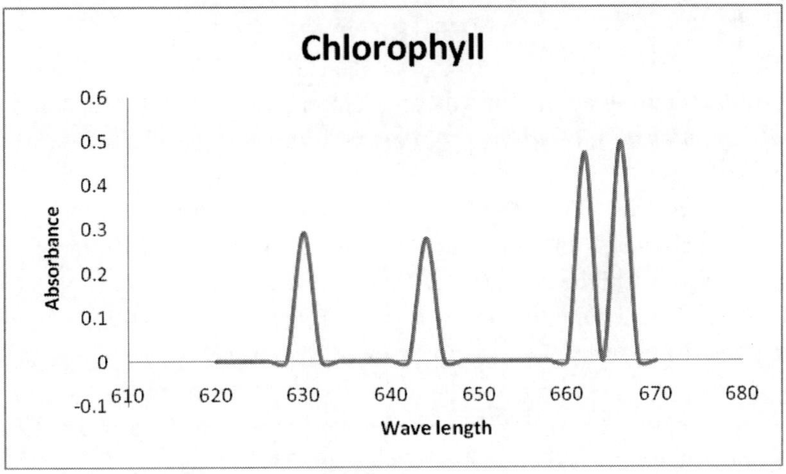

Figure 9.chlorophyll content of chlorella vulgaris.

Nutrient Removal from Waste Water

The main commercial processes for removing phosphorus from wastewater effluents are chemical precipitation with iron, alum, or lime [Donnert D et al., 1999] achieving over 95% removal, and to a lesser extent biological treatment [Stratful I et al., 1999]. Practical biological methods of removal are far less efficient, ranging between 20% and 30% of P with various microorganisms, while up to 90% removal with some bacterial species has been recorded in laboratory tests.In this study, as shown in table-2, it is revealed with the fact that 100% absorption of nitrate and phosphate from wastewater becomes possible.

CONCLUSION

Bio-mass growth rate, Isolation and characterization of Microalgae by utilizing hostel waste water medium has been investigated and the results validate its ability in sequestrating the carbon and waste waters. Lipid-producing microalgae species are isolated which increases its ability to produce bio-diesel. The FTIR spectral results indicates the rich content of protein, carbohydrate, nucleic acids which favour its role to produce value added bio-products.

ACKNOWLEDGMENT

The author Prof. Dr. Syed Shabudeen. P. S, is personally thankful to the Department of Science and Technology, Government of India, New Delhi for their support to this research project with funding, and to The Kumaraguru College of Technology, Coimbatore, India for the supports given to this research work. I personally acknowledge the research team Junior Research fellows Ms. Indhumathi Ponnusamy, Mr. Soundarrajan. M. Mr. A. Ashok and my fellow colleague co-project investigator Dr. Shoba US, Senior Associate Professor.

REFERENCES

Becker, W. (2004).Microalgae in human and animal nutrition. In: Richmond, A editor. *Handbook of microalgal culture: biotechnology and applied phycology*. Blackwell Publishing (312-346).

Brandenburg, K. & Seydel, U. (1996). Fourier Transform Infrared Spectroscopy of cell surface polysaccharides. In Manstsch HH and Chapman D. Infrared Spectroscopy of biomolicules, Wiley:chichester. 203-278.

BasterjKorting, H. C., Traenkle, P. &Schmalreck, A. F. (1999). Identification of dermatophytes BY Fourier Transform Infrared Spectroscopy. Mycoses ISSN 09333.*42*, 525-528.

Chakrabarti, R. & Schutt, C. E. (2001).The enhancement of PCR amplification by low molecular weight amides. *Nucleic Acids Res*., *29*, 2377-2381.

Chevalier, P. & De la No.ue, J. (1985). Wastewater nutrient removal with microalgae immobilized in carrageenan. *Enzyme MicrobiolTechnol*, *7*, 621–4.

Donnert, D. & Salecker, M. (1990).Elimination of phosphorus from municipal and industrial waste water.*Water Sci Techol*, *40*,195–202.

Indhumathi, Ponnuswamy.,Soundararajan, Madhavan., Syed Shabudeen, P. S. & Shoba, U. S.(2014). Resolution of Lipid Content from Algal Growth in Carbon Sequestration Studies. *International Journal of Advanced Science and Technology*, *67*, 23-32.

Indhumathi, P., Soundararajan, M., Syed Shabudeen P. S., Shoba. U. S. & Suresh E. (2013) Utilization, isolation and characterization of Chlorella vulgaris for carbon sequestration and Waste water treatment by performing FTIR spectral Studies, *asianjr. of microbiol. Biotech. Env. Sc*., *15*, (4) 661-666.

Indhumathi, Ponnuswamy, Soundararajan, Madhavan, Syed Shabudeen, P. S. & Shoba, U. S. (2013)., Isolation and Characterization of Green Microalgae for Carbon Sequestration, Waste Water Treatment and Bio-fuel Production, *International Journal of Bio-Science and Bio-Technology*, *5*, 2, 17-23.

Lewis, R. N. & McElhancy, R. N. (1996). Fourier Transform Infrared Spectroscopy in the study of hydrated lipids and lipid bilayer membranes in: Manstsch HH and Chapman D. Infrared Spectroscopy of biomolicules, Wiley: chichester. 159-202.

Liquir, J. &Tailander, E. (1996).Infrared Spectroscopy of nucleic acids.in:Manstsch HH and Chapman D. Infrared Spectroscopy of biomolicules,Wiley:chichester. 131-158.

Rainey, F. (2000) Reducing the competitors,isolating the losers and slow-growers. Presented at the Society for Industrial Microbiology Annual Meeting, The Town and Country Hotel, San Diego, CA.

Sidhu, M. K., Liao, M. J. & Rashidbaigi, A. (1996). Dimethyl sulfoxide improves RNA amplification. *Biotecniques*.*21*, 44-47.

Stratful, I., Brett, S., Scrimshaw, M. B. & Lester, J. N. Biological phosphorus removal, its role in phosphorus recycling. *Environ Technol*, 1999, *20*, 681–95.

Stirling, D. (2003) Technical notes for sequencing difficult templates. In: PCR Protocols (Ed. Bartlett J. M. S., Stirling D.), pg. 401. Humana Press, Totowa, New Jersey.

Stuart, B. (1997). Biological applications of Infrared Spectroscopy.willey: chichester.25-180.

Syed Shabudeen P. S., Soundrarajan M. & Indumathi P. (2013) algae biomass growth kinetic study in waste water medium using spectroscopic analysis, *Journal of Environmental Research And Development,7*, 4A, 1496-1500.

Young, A. J., Tsavalos, A. J. & Harker, M. (1996).Autotrophic growth and carotenoid production of *Haematococcuspluvalis*in a 30 liter air-lift photobioreactor. *Journal of fermentation and bioengineering,* 113-118.

In: Microalgae and Other Phototrophic Bacteria
Editor: Luis G. Torres Bustillos ISBN: 978-1-63482-078-3

Chapter 14

BIOGAS PRODUCTION AND UPGRADING USING MICROALGAE

***Alma Lilia Toledo Cervantes*[1] *and Marcia Guadalupe Morales Ibarría*[2]**
[1]Universidad Autónoma Metropolitana UAM- Iztapalapa, México
[2]Departamento de Procesos y Tecnología,
Universidad Autónoma Metropolitana UAM-Cuajimalpa, México

ABSTRACT

In this chapter the potential of biogas production and upgrading using microalgae cultures is presented. During the last decade, there has been a growing interest in investigating the energy potential of biofuels obtained from microalgae cultures and microalgal biomass has been forecasted as a new generation of feedstock for biofuel production (biodiesel, bioethanol, biogas, biohydrogen, etc.). However, economically feasible and sustainable energy production from microalgae requires enhancement of biomass conversion into energy. Anaerobic digestion is a mature biomass conversion technology for biofuels production and microalgal biomass conversion into biogas is gaining increased interest because it has the potential to improve the energy balance of the biofuel production process as well as reduce the production costs and the environmental impact. On the other hand, calorific values of biogas depend on the amount of methane in the gas and a high CO_2 content reduces its calorific value, therefore, microalgae culture leads to a double-purpose processing system, where the biogas is upgraded through CO_2 removal and the biomass is obtained for its production, resulting in a comprehensive technology.

1. INTRODUCTION

Biogas is produced when certain bacteria decompose biological matter such as animal manure, organic wastes and biomass in an anaerobic environment. This gas is rich in methane

(CH_4) and suitable to be used for heat and/or electricity generation and wastes such as: biosolids and liquor process that might be used as fertilizer.

The anaerobic digestion and fermentation processes use mixtures of bacteria to hydrolyze and breakdown the organic biopolymers (i.e., carbohydrates, lipids and proteins) into monomers, which are further converted into a methane-rich gas via fermentation. Carbon dioxide (CO_2) is the second major component of biogas and, like other interfering impurities has to be removed.

Microalgae emerge as a source of biomass with an enormous variety of components that can be anaerobically digested to produce biogas and they can also be used in biogas purification process to remove the CO_2 in biogas streams.

Anaerobic digestion of the microalgal biomass is also a key process that can contribute on the economic and energy balance in a biorefinery approach for biodiesel production from microalgal oils. Residues of microalgae, remaining from lipid extraction for biodiesel production are rich in fermentable-nutrients. The conversion of the biomass into methane is a process that can recover more energy than the obtained from other biofuels. Nevertheless, there are still some aspects that require attention; 1) the biodegradability of microalgae which can be low depending on species, 2) the high cellular protein content that results in ammonia release, and 3) the presence of sodium when marine microalgae are involved. Different strategies have been proposed to increase the methane conversion yield including physicochemical pretreatments and co-digestion. In addition, the biogas production through anaerobic digestion could be very effective for processing microalgae biomass produced from a wastewater treatment plant where the microalgae are grown under uncontrolled conditions and to obtain energy from wastes.

2. Biogas Production

The anaerobic digestion is a series of biological processes performed by bacteria and archaea that degrade the organic matter into biogas in the absence of oxygen. The treatment of biological wastes through anaerobic digestion is a cleaner and environmentally safe process to obtain a renewable energy source, methane. Biogas production is a well-established technology for generating bioenergy and it is considered as one of the most environmentally beneficial processes for replacing fossil fuels (Jeihanipour et al., 2013; Weiland, 2010).

Biogas has two main components: CH_4 and CO_2, but also have smaller amounts of hydrogen sulphide (H_2S), ammonia (NH_3) and traces of hydrogen, molecular nitrogen, carbon monoxide, saturated or halogenated carbohydrates and oxygen may also be present. The CH_4 content in the biogas mixture depends on the oxidation state of the organic carbon present in the initial feedstock, the more reduced the initial substrate is, the more CH_4 will be produced (Gujer and Zehnder, 1983). The feedstock should be balanced with respect to the carbon and nitrogen content, usually the C:N ratio is 20-30. Biogas is composed of CH_4 (40–75%) and CO_2 (25–60%), depending on the source. Also, the water content is 5–10%, and may also contain trace amounts of other components such as H_2S (0.005–2%), siloxanes 0–0.02%, halogenated hydrocarbons <0.6%, NH_3<1%, hydrogen, oxygen (O_2) 0–1%, carbon monoxide (CO) <0.6%, nitrogen (N_2) 0–2% and particles (Ryckebosch et al., 2011).

The energy content in the biogas is determined by the methane concentration (1 kWh per m^3 of biogas with 10% of CH_4). H_2S and its oxidation products are the major contaminants with a maximum permitted concentration of 5 ppm. CO_2 is a non-burnable compound, and its presence reduces the biogas heating value. In addition, the CO_2 removal is an expensive process. There are different specifications for biogas depending on the application, especially as natural gas in stationary appliances or to be fed to a pipeline network.

The possible applications of biogas include:

- Household fuel for cooking like natural gas.
- Heat or steam production by burning.
- Electricity generation combined with heat and power production.
- Vehicle fuel (upgrading to biomethane is necessary).
- Fuel cell to generate electricity.
- Production of chemicals.

The biogas composition depends on the type of organic compound in the substrate. The theoretical production of methane in anaerobic digestion can be calculated according to the following equation (Deublein and Steinhauser, 2008). Nevertheless, it is important to keep in mind that this theoretical approach does not take into account requirements for cell maintenance and anabolism, only considers the conversion of the organic matter into methane, carbon dioxide and ammonia. It will be discussed later, the difficulties in degradation of intracellular components due to the cell wall structure of some microalgae.

$$C_cH_hO_oN_n + yH_2O \rightarrow xCH_4 + nNH_3 + (c - x)CO_2$$

where

$$x = \frac{4c + h - 2o - 3n}{8} \ and \ y = \frac{4c - h - 2o + 3n}{4}$$

The specific methane yield expressed in liters of CH_4 per gram of volatile solids (VS) can thus be calculated as (Sialve et al., 2009):

$$CH_4 = \frac{4c + h - 2o - 3n}{12c + h + 16o + 14n} * V_m$$

where V_m is the normal molar volume of methane or 22.14 L at 0°C and 1 atm.

The ratio of CH_4 to CO_2 can therefore be calculated as follow:

$$r = \frac{4 - G}{4 + G} \ where \ G = \frac{-h + 2o + 3n}{c}$$

The anaerobic digestion technology involves several steps (Figure 1) carried out by different microorganisms, which stand in syntrophic interrelation and have different requirements on the process (Angelidaki et al., 1993). The metabolic reactions in anaerobic digestion involve hydrolysis, acidification, acetogenesis and methanogenesis.

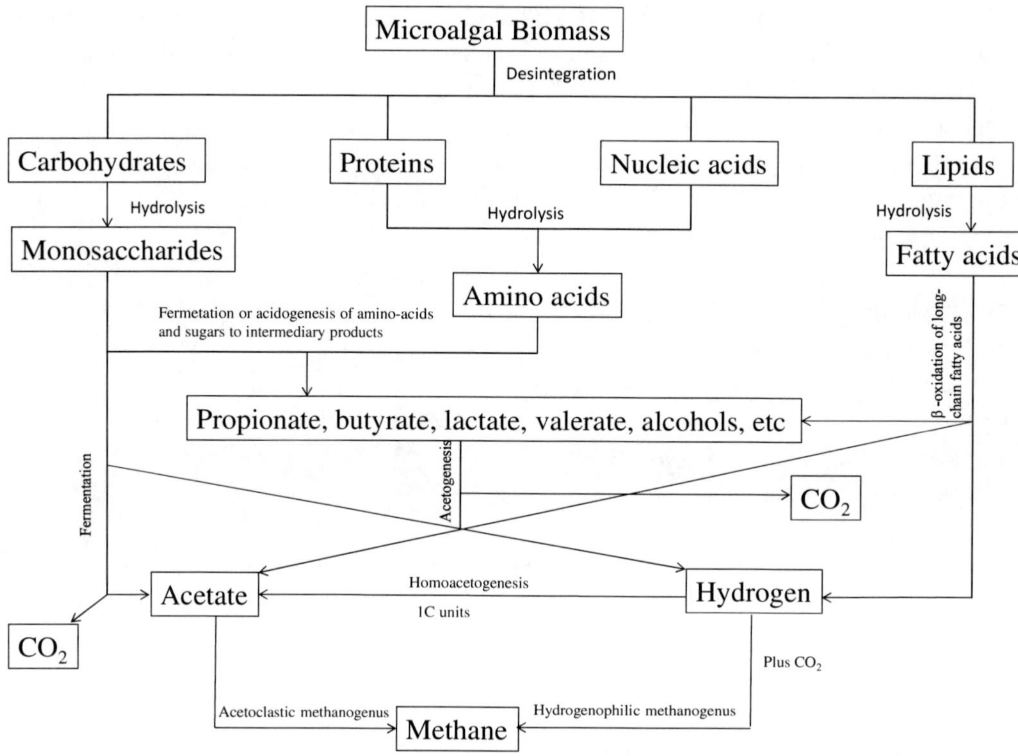

Figure 1. Anaerobic digestion from microalgal biomass (Modified from Bohutskyi and Bouwer, 2013).

The initial step is the disintegration of the biomass involving steps such as lysis, non-enzymatic decay, phase separation and physical breakdown (Batstone et al., 2002). In the case of complex organic matter like lignocellulosic material, pretreatment steps are necessary to enhance hydrolysis. Undissolved compounds like cellulose, proteins, or fats are cracked slowly into monomers within several days.

Hydrolytic microorganisms excrete enzymes, e.g., cellulase, cellobiase, xylanase, amylase, lipase, and protease, and are responsible for the initial decomposition of the polymers and monomers and produce mainly acetate, hydrogen and volatile fatty acids (VFAs) such as propionate, valerate and butyrate. Most of the bacteria involved in hydrolysis are strict anaerobes (such as *Bacteriocides, Clostridia, and Bifidobacteria*) and facultative anaerobes (such as *Streptococci* and *Enterobacteriaceae*) (Weiland, 2010).

Later, in fermentation or acidogenic stage, a group of microorganisms convert the simple monomers to intermediary products such as acetate, hydrogen, formate, and a mixture of VFAs, alcohols and other simpler organic compounds. During acidogenesis, large amounts of CO_2 and hydrogen are produced. Following, the produced VFAs are transformed to acetic acid, CO_2 and hydrogen by the acetogenic bacteria. They are slow growing microorganisms that can be inhibited by the accumulation of hydrogen that must be depleted by the hydrogen-consuming bacteria, such homoacetogenic bacteria such as *Acetobacterium woodii* and *Clostridium aceticum* in other further steps.

At the end of the degradation chain, two groups of methanogenic bacteria produce CH_4: the acetoclastic methanogens that consume acetic acid and the hydrogen-utilizing methanogens that use hydrogen and CO_2. The acetoclastic methanogens produce approximately 70% of CH_4 in the biogas. These bacteria are strict anaerobes and require a lower redox potential for growth than most of the other anaerobic bacteria. Acetoclastic methanogens are slow growing microorganisms and are particularly sensitive to pH. Only few species are able to degrade acetate into CH_4 and CO_2 e.g., *Methanosarcina barkeri, Metanonococcusmazei*, and *Methanotrix soehngenii.*

Among the most important factors affecting the anaerobic digestion process are: pH, temperature, feedstock composition, nutrients, the presence of toxic or inhibitory substances and the organic loading rate. The pH influences the hydrolysis rate (extracellular enzymes activity) and the methane formation. Different groups of methanogens have different ranges of optimum pH. Methane formation takes place within a pH interval, from about 6.5 to 8.5 with an optimum interval between 7.0 and 8.2 (Khanal, 2008), so anaerobic digesters are usually maintained in the range of 7–8. The process is severely inhibited if the pH decreases below 5 or rises above 8.5. The pH variations in anaerobic digestion process are caused by the ammonia accumulation during degradation of protein (pH value increases) and the accumulation of VFAs (pH value decreases). The methanogenic bacteria are more affected by the low pH and the concomitant accumulation of intermediate acids (acetic, propionic and butyric acids) than acidogens that exhibit its maximum activity at pH 5.5–6.5 (Wang et al., 1999; Mösche and Jördening, 1999).

In anaerobic digestion, the acetoclastic methanogens are the most sensitive microorganisms to temperature. Temperature in combination with other factors influences the selection of the microorganisms prevailing in an anaerobic digestion system. A quick temperature change of 2–3°C causes an accumulation of VFAs (El-Mashad et al., 2004). A significant temperature drop could affect the activity of all anaerobic microorganisms, but the activity is recovered after temperature stabilization (El-Mashad et al., 2004; M-c Wu et al., 2006).

Strict anaerobic microorganisms (include *Clostridia*, methanogens, sulphate reducers and homoacetogens) require redox potential (ORP) levels of around –400 mV. ORP reflects the availability of oxidants, such as oxygen or nitrate ions and reductants such as hydrogen. The most favorable ORP for fermentation and acid production is from −100 to −300 mV. Methanogenesis requires an ORP <−300 mV when CO_2 is used as an electron acceptor and methane is formed (Gerardi, 2003). A high ORP (>50 mV) indicates the presence of free oxygen; in anaerobic environments the traces of oxygen are rapidly consumed by the facultative anaerobes of the consortium. An ORP between 50 to −50 mV is characteristic of an anoxic environment with nitrates and nitrites. And, at ORP lower than −50 mV, the environment in the digester is strongly reducing. Redox potential could be an obvious control parameter in fermenters, but complicated dynamics and variability in reactors make the interpretation difficult (Weiland, 2010).

Ammonia is the degradation product of nitrogenous compounds such as proteins and amino acids and non-ionised ammonia inhibits methanogens. There are contradictory reports on the levels of tolerance to ammonia; some studies showed that ammonia is toxic for methanogenic microorganisms at concentration of 1.5–1.7 g N/L at pH 7.4 and above (Koster and Lettinga, 1984). Others demonstrated that ranges from 1.7 to 14 g/L of total ammonia concentration causing a decrease of 50% in the methane production (Chen et al., 2008); and, concentrations above 80 mg/L of ammonium seems to be responsible for complete inhibition (Kroiss, 1985). Generally, the toxicity of ammonia and sulfide is related to the presence of metals, increase in temperature, and changes of pH in digesters. Neutral forms of ammonia and hydrogen sulfide are more toxic, possibly because they can penetrate rapidly the cell membrane (Speece, 1983; Braun et al., 1981).

Some organic compounds are toxic for anaerobic digestion: alkyl benzenes, halogenated benzenes, nitrobenzenes, phenol and alkyl phenols, halogenated phenols, nitrophenols, alkanes, halogenated aliphatics, alcohols, halogenated alcohols, aldehydes, ethers, ketones, acrylates, carboxylic acids, amines, nitriles, amides, pyridine and its derivatives (Chen et al., 2008). High salts concentration (e.g., NaCl above 35 g/L) affects also methane generation. Nevertheless, it has been observed that, when the NaCl concentration was increased gradually, methanogens can be adapted to concentrations up to 65 g/L (de Baere et al., 1984).

The organic loading rate, hydraulic and solids retention times are other essential characteristics of anaerobic digestion process that must be adjusted to the carried out process in fermenters, which are usually operated at optimum temperatures between 38 and 42°C (Angelidaki, 2003). The VFAs can serve as an efficient indicator of process imbalances. A ratio of propionic acid: acetic acid >1 indicate the digester failure (Weiland, 2008) and the concentration of the butyrate and isobutyrate could be also a reliable tool for indication of process failure (Ahring et al., 1995).

3. METHANE POTENTIAL OF MICROALGAL BIOMASS

During the last decade, there has been a growing interest in the generation of energy from microalgae mainly in the production of biodiesel. Economically feasible and sustainable energy production from microalgae requires: 1) optimization of algal growth, 2) maximization of lipid content and 3) enhancement of biomass conversion into energy. High content of lipids, starch and proteins and the lack of recalcitrant lignin make microalgal biomass a promising substrate for anaerobic microorganisms (Schenk et al., 2008). However, microalgal energy production is still in research phase. To achieve an industrial energy production from microalgae, it requires economically viable massive biomass production technologies and biomass energy generation technologies.

Microalgae cultivation for biofuels is not viable in economic and environmental terms, since freshwater and fertilizers are needed. So, in large-scale energy generation processes, it has been suggested that microalgae must be cultured in wastewater treatment plants. In addition, as in the anaerobic conversion of algal biomass to energy carriers does not require cost-intensive drying of the biomass less energy in required in the whole energy generation processes. Microalgae as a feedstock for biogas production is not knew, it has been studied since the fifties (Golueke et al., 1956). Several studies have noticed a correlation between the structure of cells of the analyzed microalgae and biomass susceptibility to degradation under anaerobic conditions and intensity of biogas production. All easily biodegradable species of algae, which allowed achieving important technological applications, were characterized by a lack of cell wall. Thus, the cell structure and biomass composition should be examined for anaerobic digestion and methane production

It is well known that microalgae biomass composition is directly related to the growth conditions. Microalgae, under certain conditions, have the capability to accumulate important amounts of carbohydrates or lipids. One of the challenges in biogas production from microalgae is the significant variation in biochemical composition among different genera or similar species, which depends on several environmental factors, such as temperature, salinity, light intensity, and nutrient availability. For the production of biogas from microalgae, it is necessary to know the calorific value of the biomass along with the digestion potential, which is linked to the cell wall composition of the microalgae. There are two basic types of microalgal cells: prokaryotes and eukaryotes. Prokaryotic cells lack membrane-bounded organelles and occur in the cyanobacteria. Eukaryotic microalgae are often surrounded by a cell wall composed of polysaccharide. The average composition of microalgae is $CO_{0.48}H_{1.83}N_{0.11}P_{0.01}$ (Grobbelaar, 2004). In general, algal cell walls are made up of two kinds of components: the fibrillar component, which forms the skeleton of the wall, and the amorphous component, which forms a matrix within which the fibrillar component is embedded. The most common kind of fibrillar component is cellulose. The amorphous mucilaginous components are polysaccharides as galactans (polymers of galactose), residues of L-arabinose, D-xylose, D-glucuronic acid and L-rhamnose (Lee, 1999).

The amount of biogas production is correlated to the algal biomass degradation potential. Some microalgae species like *Chlorella, Nannochloropsis* and *Scenedesmus* have resistant trilaminar membrane-like structure containing non-hydrolysable sporopollenin-like biopolymer-algaenan (Gelin et al., 1999; Atkinson et al., 1972). The overall cell wall structure has complex organization with three distinct layers: rigid internal micro fibrillar, medial trilaminar, and external columnar (for green algae *Coelastrum*). *Chlorella* and *Scenedesmus* have internal rigid cell walls either glucose- mannose type or glucosamine-type. In contrast, *Chlamydomonas reinhardtii* has a cell wall composed of proteins and glycoproteins (Lee, 1999). Studies with some microalgae such as: *Chlamydomonas reinhardtii*, *Chlorella kessleri, Euglena gracilis*, *Spirulina (Arthrospira) platensis, Scenedesmus obliquus* and *Dunaliella salina*, have been demonstrated that the potential quantity of biogas is strongly dependent on the species and on the pre-treatment. *C. reinhardtii* revealed being the most efficient with a production of 587 mL of biogas per g of volatile solids (VS). In addition, the resistant cell wall of *Scenedesmus obliquus* could remain undamaged after 6 months of digestion (Mussgnug et al., 2010).

Methane yields from microalgae change due to variation in cellular protein, carbohydrate and lipid content, cell wall structure, and process parameters such as the reactor type, digestion temperature and the pH. For biogas production, the microalgae species should have a high degradation potential and low amount of indigestible compounds (Mussgnug et al., 2010). Theoretical yields are, 0.415, 0.851 and 1.014 L CH_4/g VS for carbohydrate, protein, and lipid, respectively (Sialve et al., 2009). Microalgae with high lipid content are preferred because lipids provide the highest biogas yield since they have the lowest oxidation state and largest theoretical methane yield, which is more than twice the methane yield from proteins, glycerol, and carbohydrates. Carbohydrates and proteins show much faster conversion rates but lower gas yields (Weiland, 2010). The mineral composition of microalgae biomass provides the nutrients requirements of the anaerobic digestion process. Besides carbon, nitrogen and phosphorus, which are major components in microalgae composition, nutrients such as iron, cobalt, and zinc are also found and known to stimulate methanogenesis.

Table 1. Theoretical methane yields from different microalgae species

Microalgae species	**Methane yield (mL/gVS)**	**References**
Arthrospira platensis	293	Mussgnug et al., 2010
Arthrospira maxima	330	Varel et al., 1988
Arthrospira maxima	190	Samson and Leduyt, 1983a
Arthrospira maxima	400	Samson and Leduyt, 1986
Phaedodactylum tricornutum	270-350	Zamalloa et al., 2012
Arthrospira maxima (Pretreated with ultrasonication)	170	Samson and Leduy, 1983a
Arthrospira maxima (Pretreated with Heat)	210-240	Samson and Leduy, 1983a
Arthrospira maxima (In co-digestion)	250-360	Samson and Leduy, 1983b
Scenedesmus obliquus	170-210	Zamolloa et al., 2012
Chlamydomonas reinhardtii	387	Mussgnug et al., 2010
Dunaliella salina	323	Mussgnug et al., 2010
Scenedesmus obliquus	178	Mussgnug et al., 2010
Chlorella kessleri	218	Mussgnug et al., 2010
Euglena gracilis	325	Mussgnug et al., 2010
Chlorella spp. (Lipid extracted)	268	Ehimen et al., 2009
Chlorella spp. (Drying and grinding)	>400	Ehimen et al., 2009
Chlorella vulgaris	286	Lakaniemi et al., 2011
Chlorella vulgaris	147-240	Ras et al., 2011
Isochrysis sp.	408	Frigon et al., 2013
Scenedesmus dimorphus	397	Frigon et al., 2013
Scenedesmus spp. (Lipid extracted and pretreated with alkali and heat)	323	Yang et al., 2011
Scenedesmus sp.-AMDD	410	Frigon et al., 2013
Mixture of *Chlorella*, *C. reinhardtii,* unknown specie and *Pseudokirchneriella subcapitata*	220-390	De Schamphelaire and Verstraete, 2009
Tetraselmis (Dry)	260	Marzanoet al., 1982
Tetraselmis (Fresh)	310	Marzano et al., 1982
Chlorella sorokiniana	212	Polakovicova et al., 2012
Dunaliella tertiolecta	24	Lakaniemi et al., 2011
Nannochloropsis oculata (Lipid extracted)	130	Park and Li, 2012
Nannochloropsis oculata	204	Buxy et al., 2013

Anaerobic digestion of microalgal biomass has been investigated in batch and fed-batch systems as well as in continuously stirred tank reactors (De Schamphelaire and Verstraete, 2009; Sialve et al., 2009). In biogas production from microalgae, the retention times required to obtain high methane yields are relatively long, 20–30 days (Ras et al., 2011; Zamalloa et al., 2011).

As is shown in the table 1, the methane yield during batch digestion of different microalgae and cyanobacteria species has a wide range of values. Yield values for *Chlamydomonas reinhardtii*, *Chlorella kessleri*, and *Scenedesmus obliquus* were 0.387, 0.218, and 0.178 L/g VS, respectively (Mussgnug et al., 2010). *Arthrospira platensis* and *Arthrospira maxima* species methane yield varied from 290 to 330 mL/gVS corresponding to 68–77% of the theoretical value (Chen, 1987; Varel et al., 1988). On the other hand, biogas production levels of 180.4 mL/(g day) of biogas could be achieved using a two-stage anaerobic digestion process with different algae reaching methane concentration of 65% (Vergara-Fernandez et al., 2008).

4. Strategies to Increase Methane Yields from Microalgal Biomass

Co-Digestion

An option to increase the biogas production yields is using *co-digestion* (Callaghan et al., 1999). Co-digestion refers to the simultaneous anaerobic digestion of multiple organic wastes and it is used to increase methane production from low yielding or difficult to digest materials. Among the benefits of using co-digestion are included: the dilution of potential toxic compounds, an improved balance nutrients, an increased load of biodegradable organic matter, a synergistic effect of microorganism and a higher biogas yield (Agdag and Sponza, 2007; Braun and Wellinger, 2003). Co-digestion of different substrates (as manure) has been recognized as an attractive approach that has economic and environmental benefits (Holm-Nielsen and Seadi, 2004).

The addition of biomass organic matter can increase the loading rate and methane yield up to 60–100%. Biomass with low nitrogen includes: municipal solid waste, paper, sisal pulp, straw, grasses, and wood wastes. Higher nitrogen wastes include: sewage sludge, chicken or livestock manure, slaughterhouse, meat or fish processing wastes (Stroot et al., 2001; Callaghan et al., 2002; Sosnowski et al., 2003; Alvarez and LidÈn, 2008). The increase in biogas production is mainly due to a better carbon and nutrient balance as consequence of increasing the organic loading rate (Mshandete et al., 2004).

Co-digestion of microalgae, normally improves the digestion process through the synergistic effects produced. Generally, anaerobic sludge from domestic sewage treatment plant or marine anaerobic sediment is used to startup anaerobic digestion of the microalgal biomass (Schramm and Lehnberg, 1984; Samson and Leduy, 1982; Chynoweth et al., 1981). Nevertheless, it has to be considered that heavy metals (i.e., chromium, iron, cobalt, copper, zinc, cadmium and nickel) can be present in significant concentrations in sewage.

The residues from the biodiesel production, such as glycerol could be used as co-digestion materials. Interestingly, it has been demonstrated the biogas could increase from 100 to 200% when glycerol is used as co-substrate because it is a rapidly biodegradable compound (Mata-Alvarez et al., 2000; Fountoulakis and Manios, 2009).

Recently, *Scenedesmus* residues generated after the extraction of amino acids (SRA) and lipids (SRL) were used as substrates for anaerobic digestion and, methane production was compared from the anaerobic digestion of raw *Scenedesmus* biomass and co-digestion of microalgae residues with carbon rich substrates. Methane yields of SRA and SRL in mono-digestion were 272.8 mL CH_4 /g VS and 212.3 mL CH_4 /g VS, respectively. Methane yield of raw biomass anaerobic digestion was 140.3 mL CH_4 /g VS, the low value suggests that organic compounds were not available for bacteria, remaining within the cell (Ramos-Suárez and Carreras, 2014).

An increase on the methane yield has been observed after the extraction of metabolites, because it has an effect on the C/N ratio of residual biomass compared to the raw biomass. The increase in C/N is beneficial for the development of the digestion process by favoring bacterial activity.

Microalgal Biomass Pretreatments

The most common rate-limiting step for anaerobic digestion of complex organic substrates is the hydrolysis step. Pretreatment of algal biomass is one of the strategies used for conditioning and increasing the algal digestibility, the methane yield, and the degradation rate. It has been tested different pretreatment methods applicable when solid feedstock is involved. In general, pretreatment methods can be classified into the following groups:

- Mechanical: grinding, milling, homogenization, ultrasonic treatment, liquid shear (Baier et al., 1997; Barjenbruch and Kopplow, 2003).
- Biological: enzymatic treatment (Ge et al., 2010; Lv et al., 2010).
- Chemical: acid or alkali hydrolysis, ozonation, hydrogen peroxide treatment (Bruni et al., 2010; Goel et al., 2003; Zheng et al., 2009).
- Thermal: drying, steam pretreatment, hydrothermolysis (Barjenbruch and Kopplow, 2003; Neyens and Baeyens 2003; Valo et al., 2004).
- Electrical: electro-Fenton (Erden and Filibeli, 2010; Khoufi et al., 2006).
- Irradiation: gamma ray, electron-beam, microwave (Eskicioglu et al., 2007a; 2007b; Lafitte-Trouque and Forster, 2002).
- Combination: thermochemical, wet oxidation (Kim et al., 2010; Penaud et al., 1999).

The important requirements to select pretreatment methods are: to preserve the total organic matter content and to prevent the formation of inhibitory materials.

Mechanical pretreatment disintegrates and/or grinds solid particles of the substrates, thus releasing cell compounds and increasing the specific surface area. Physicochemical pretreatment involves the use of acid, alkaline or oxidative conditions, at ambient or high temperature. Biological pretreatment methods such as addition of microorganisms or cellulolytic microorganisms (bacteria and fungi), specific enzymes such as peptidase, carbohydrolase and lipase, or cell lysate can also increases the substrate digestibility and enhance biodegradability (Ariunbaatar et al., 2014). However, in general physicochemical methods yield higher efficiencies.

Mechanical pretreatment is usually applied before any other pretreatment to increase the surface area providing better contact between substrate and anaerobic bacteria (Carrere et al., 2010; Skiadas et al., 2005; Elliot and Mahmood, 2012). Also, mechanical pretreatment of lignocellulose materials increased the hydrolysis and methane yield by 5–25% (Delgenes et al., 2003). Studies have concluded that a larger particle radius results in a lower chemical oxygen demand (COD) and a lower methane production rate (Esposito et al., 2011). Therefore, mechanical pretreatments such as sonication, lysis-centrifugation, liquid shearing, collision, high-pressure homogenization, maceration, and liquefaction are conducted in order to reduce the substrate particle size.

Chemical pretreatment is used to achieve the destruction of the organic compounds by strong acids, alkalis or oxidants (Li et al., 2012). The effect of chemical pretreatment depends on characteristics of the biomass. Oxidative methods such as ozonation have shown to improve the hydrolysis rate and thus the biogas production. The chemical pretreatment must be applied to hard biodegradable feedstock. Easily biodegradable substrates containing high amounts of carbohydrates culminate in accumulation of VFAs, which leads to failure of the methanogenesis step (Ariunbaatar et al., 2014).

The thermal pretreatment disintegrates the cell membranes resulting in solubilization of organic compounds (Bien et al., 2004; Ferrer et al., 2008; Marin et al., 2010; Protot et al., 2011). However, temperature selection and treatment times are crucial for effectiveness. Thermal pretreatment at high temperatures (>170°C) might lead to the creation of chemical bonds and result in the agglomeration of the particles (Bougrier et al., 2006) as in the Mallaird reaction between carbohydrates and amino acids (Carrere et al., 2010; Elliot and Mahmood, 2012; Hendriks and Zeeman, 2009; Penaud et al., 1999; Pinnekamp, 1989).

The combination of pretreatments may result beneficial, but in certain cases like high temperature, microwaves and hydrogen peroxide pretreatment (Shahriari et al., 2012) could cause a decrease in biogas production (Rafique et al., 2010; Carrere et al., 2009). However, alkaline pretreatment coupled with thermal methods at a lower temperature could result in a higher biogas production (Carrere et al., 2009).

The physical methods used for microalgal biomass include: high-pressure homogenizers, microwaves, bead beating, freezing, sonication, and autoclaving. The chemical lysis is a different approach to the biomass processing, where chemical agents acid or alkaline are added (e.g., hydrochloric or sulfuric acid, sodium hydroxide) in order to hydrolyze the biomass into its constituent molecules (Molina Grima et al., 2004). The chemical pretreatment methods include: chemical lysis with NaOH, HCl, H_2SO_4 or solvent addition.

The effect of biomass loading, temperature, acid concentration and the number of extraction cycles was determined by Miranda et al. (2012) for the microalgae *Scenedesmus obliquus*. They determined that the best results were obtained with acid hydrolysis by H_2SO_4, however this study was done for ethanol production from biomass carbohydrates. Other alternatives such as ultrasound (Janczyk et al., 2007) or high-pressure homogenization (Komaki et al., 1998) pretreatments enhanced the digestibility of *C. vulgaris*. Chen and Oswald (1998) increased the CH_4 yield in 33% by heat pretreating microalgal biomass at 100°C for 8 h. Higher solubilization can also be achieved with lower temperatures, but longer treatment times are needed. However, the amount of energy consumed in the pretreatment could be higher than the corresponding energy gained from CH_4 production (Yen and Brune, 2007). Mussgnug et al., 2010 have studied drying as a pretreatment for algae before anaerobic digestion. They found that drying decreased the amount of biogas production by 20%. In spite of the many cell disruption methods tested for microalgae, the most efficient method has not yet been unequivocally confirmed.

5. Biogas Upgrading

The most common contaminants in biogas are sulfur compounds such as H_2S and mercaptans (e.g., methanethiol) generated during the anaerobic fermentation of proteins. The H_2S is a non-desirable compound because its bad smell and affectation on the energy-recovery processes. If the biogas is used directly without H_2S removal, it would corrode engines, pipelines and biogas storage structures. Biogas combustion containing H_2S generates the highly corrosive, unhealthy and environmentally hazardous sulfur dioxide (SO_2) and sulfuric acid (H_2SO_4) that corrodes pipeline metal parts, storage tanks, compressors, and engines. Concentrations >300–500 ppm of H_2S can produce the above-mentioned compounds (Holm-Nielsen and Al Seadi, 2004). Biogas produced by co-digestion of manure with energy crops or harvesting residues can contain levels of H_2S between 100 and 3,000 ppm (Weiland, 2010). Another corrosive contaminant in biogas is NH_3 that represents a health risk but it is not considered as harmful as H_2S. Siloxanes are the third most important contaminant. Its presence during combustion is harmful because form glassy microcrystalline silica. All other components in biogas are considered innocuous (Abatzoglou and Boivin, 2009).

The biogas produced from anaerobic digestion is a potential fuel for power generators, but quality of the raw biogas is not high enough to be used. Biogas must accomplish the same standards for fossil natural gas in order to be use as vehicle fuel or for injection in the natural gas network (Ryckebosch et al., 2011). CO_2, H_2S, and the other non-combustible component contents are regulated. On the other hand, calorific values of biogas depend on the amount of CH_4. A high CO_2 content in biogas reduces its calorific value and increases carbon monoxide and hydrocarbon emissions after combustion, even if desulfurized biogas is used as engine fuel. Besides, a high CO_2 content makes the compression and transport of desulfurized biogas uneconomical.

In order to obtain biomethane from biogas, it has to be cleaned to remove the trace components and to adjust the calorific value. Upgrading is generally performed in order to meet those standards. After the clean-up process, the final product is referred to as biomethane, typically containing 95–97% CH_4 and 1–3% CO_2. Biogas purification methods include four basic steps for removal of: water vapor, H_2S, CO_2, and finally the siloxane and other trace gases. Humidity needs to be removed because the presence of water promotes the formation of sulphur oxidation products. Common treatments for H_2S and CO_2 removal can be divided into two categories: 1) Physicochemical processes (reactive or non-reactive absorption; reactive or non-reactive adsorption) like chemical absorption in aqueous solutions; chemical adsorption of H_2S on solid adsorbent; and scrubbing with solvents or other liquid phases or, 2) Biological processes (contaminants are consumed by microorganisms and converted into less harmful compounds).

The typical technologies for biogas cleaning include scrubbing by solvents or an aqueous alkaline solution, absorption, and oxidation on solid sorbents, chelation, precipitation in the form of poorly soluble metal sulfides, and biological removal (Tippayawong and Thanompongchart, 2010; Horikawa et al., 2004; Osorio and Torres, 2009). Other physicochemical methods include: the use of iron containing compounds (iron chloride, iron oxide), activated carbon, water scrubbing, dimethylether of polyethylene glycol (or selexol) or NaOH scrubbing. Also, some biogas applications require the CO_2 removal. These methods include: pressure swing absorption on zeolites, selective membrane separation, cryogenic separation polyethylene glycol absorption, (Tippayawong and Thanompongchart, 2010; Abatzoglou and Boivin, 2009; Yang et al., 2008). Andriani et al. (2014) published an extensive review on optimization production and upgrading biogas through CO_2 removal using various techniques.

In addition, it has being developed a wide range of technologies based on biological mechanisms to remove CO_2 (Toledo-Cervantes et al., 2013) and H_2S (González-Sánchez et al., 2014; Montebello et al., 2012), which can be applied to biogas in order to obtain high quality biomethane. There are biological methods for H_2S removal using sulphide-oxidizing microorganisms (*Thiobacillus*) grown in a micro-aerophilic environment on CO_2. In this process, the H_2S is converted to sulphate or elemental sulphur depending on the oxygen availability. This transformation could be performed in biofilter, bio-trickling filter and bio-scrubber units (Abatzoglou and Boivin, 2009).

Biogas scrubbing with water from a high rate algal pond (HRAP) was suggested in 1960 (Oswald and Golueke, 1960). In this process the CO_2, H_2S and other trace gases are scrubbed from the biogas, leaving pure methane. The CO_2 transferred into the HRAP water could contribute to overcome the carbon limitation for algal growth and increase both the algal productivity and nutrient assimilation (Goldman et al., 1972). Mandeno et al. (2005) proposed scrubbing biogas with HRAP water to avoid the use and disposal of expensive chemicals using two apparatus to improve gas–liquid contact and thus gas transfer: an "in-pond angled gutter" and a "counter-current pit". Results showed that the counter current pit allowed that CO_2 in biogas was reduced from 40% to <5%. The performance of the in-pond angled gutter was less effective due to bubble coalescence that reduced the total bubble surface area available for gas transfer.

Specific microalgae like *Chlorella vulgaris* has being used to remove CO_2 from biogas produced in anaerobic digesters. The microalga was grown in an open pond using synthetic mineral medium and a diluted piggery waste at a dilution of 1:9 showing COD value of 2 (30 mg L^{-1}), the biogas was bubbled into the ponds through a BiOLIFT device. The raw biogas was composed primarily of 55–71% of CH_4, 44–48% of CO_2 and less than 1% of H_2S, and the purified biogas contained 88–97% of CH_4, 2.5–11.5% of CO_2 and less than 0.5% of H_2S (Conde et al., 1993). Mann et al. (2009) tested the strain *Chlorella vulgaris* SAG 211-11b growth in a photobioreactor for conditioning biogas with a composition of CH_4 (58%), CO_2 (42%) and H_2S (438 ppm). As a result the removal of biogas components CO_2 and H_2S was 97.1% and 100%, respectively.

Similarly, Kao et al. (2012a, 2012b) investigated the use of outdoor microalgae culture system to upgrade biogas produced from the anaerobic digestion of swine wastewater. They isolated a mutant strain, *Chlorella* sp. MB-9, which was able to tolerate high CH_4 and CO_2 concentrations. The CH_4 concentration in the biogas increased from 70% up to 85–90%, and 70% of the CO_2 in desulfurized biogas was captured.

A different upgrading system was proposed using an anaerobic-methanogenic bacterium (Strevett et al., 1995). *Methanobacterium thermoautotrophicum* was used to consume CO_2 from biogas as a carbon source and H_2 as an energy source, this bacterium has also a specific requirement for H_2S, so both unwanted components were removed at temperatures of 65–

70°C. A synthetic biogas with CH_4 (50–60%), CO_2 (30–40%) and H_2S (1—2%) was mixed with H_2 to reach a final mole fraction of H_2:CO_2 of 0.79:0.21. The gas mixture was fed to the hollow fibers packed with the *Methanobacterium thermoautotrophicum* and it was obtained an effective CO_2 and H_2S removals and doubled the initial CH_4 content.

On the other hand, small concentrations (0–4%) of oxygen in biogas are harmless, however when oxygen is between 6 and 12% and CH_4 content is 60%, explosion could occur depending on the temperature. Considering this, it has been recently suggested the use of anoxic biotrickling filter to remove the H_2S of the biogas without any dilution with air and the methane concentration was not reduced (Fernández et al., 2014).

6. PROCESS INTEGRATION

The high cost of biofuel production and low efficiency of captured energy are the major factors that limit the large-scale use of algae for biofuels. The integration of the biofuel production to other high-value products (e.g., food supplements, pharmaceuticals) and waste minimization or nutrient recycling (e.g., wastewater treatment) in a biorefinery has been suggested to generate an energy and economically attractive technology for biofuel generation (Sánchez-Tuirán et al., 2013).

The biorefinery approach is a combination of technologies looking for zero waste process by conversion of biomass into several final products (Toledo-Cervantes and Morales, 2014). A strategy to establish a microalgae-based biorefinery should consider: 1) To isolate microalgae strains rich in the target products, 2) To establish the cultivation conditions and operation strategies, 3) To define the conversion processes of whole microalgae/defatted microalgal biomass to biofuels, 4) To define the biomass harvesting and post-harvesting technologies, 5) To define the sequence extraction of co-products and processing the whole biomass or the pre-extracted product to final products, 6) To maximize the process integration of stream and recycling of materials, reducing wastes, 7) To perform a Life cycle analysis (LCA) (Toledo-Cervantes and Morales, 2014).

Razon et al. (2011) presented a net energy analysis for a system to produce biodiesel and biogas for two microalgae (*Haematococcus pluvialis* and *Nannochloropsis* sp). They reported large energy deficits even with highly optimistic assumptions and concluded that for a financially viable process, the energy products must be viewed as by-products and the biodiesel production has to be taking place within a multifunctional biorefinery system.

The integration of microalgae growth with anaerobic digestion can significantly improve the economic and energy balance of such a promising platform technology. Utilization of the digestate reduces the need to produce inorganic fertilizers and, therefore, reduces further the fossil fuel consumption required for their production. Also, coupling the treatment of nutrient-rich wastewater with algal growth followed by conversion to methane represents a potentially attractive biofuel production process with reduced impact on the environment (Douskov et al., 2010; Clarens et al., 2010). So, combining anaerobic digestion with microalgae culture, the process can become sustainable using the digestate as nutritive medium for microalgae culture and more energy-efficient by using the biogas as biofuel for electricity production that can be sold as a by-product or used on site to reduce the fuel requirements. A potential source of biomass for anaerobic digestion is the residual biomass after protein extraction for human and animal nutrition. When proteins are extracted from microalgae, sugars and lipids remains in the residual biomass and its potential for energy production is very high (Ramos-Suárez and Carreras, 2014). Residual microalgal biomass after lipid extraction for biodiesel production is also a potential substrate for production of CH_4 (Yang et al., 2011; Ehimen et al., 2009). This was also concluded by Collet et al. (2011), who performed a life-cycle assessment (LCA) of biogas production from *Chlorella vulgaris,* and the results were compared to microalgal biodiesel and to first generation biodiesel.

Sialve et al. (2009) stated that the anaerobic digestion of algae might be the optimal strategy for the energy recovery. However, they detailed that lipid extraction when biomass contains less than 40% of lipids combined with anaerobic digestion of the residues is not effective in terms of energy or costs. Under this situation, anaerobic digestion of the whole biomass appears to be a good strategy on an energy balance basis, for the energetic recovery of biomass. Harun et al. (2011) suggested that more energy could be generated from the production of methane from microalgae (14.04 MJ/kg of dry weight of microalgae), rather than biodiesel from microalgal lipids (6.6 MJ/kg) or ethanol from fermentable microalgal carbohydrates (1.79 MJ/kg). Biogas also can provide higher energy output when the production is combined with the biodiesel (16.14 MJ/kg) and it is slightly lower with ethanol (13.49 MJ/kg). Nevertheless, these results are far from the theoretical energy content of biofuels from microalgae that are: for ethanol 30 MJ/kg, biodiesel 41 MJ/kg, methane 55 MJ/kg, hydrogen 138 MJ/kg, oil 36 MJ/kg, and the whole biomass depending on lipid content 20-27 MJ/kg (Suali and Sarbatly, 2012).

Among requirements for an energetically viable microalgal biogas production (Milledge and Heaven, 2014):

1. Favourable climatic conditions.
2. Achievement of yields equivalent to ~3% photosynthetic efficiency (25 g m^{-2} day^{-1}).
3. Incorporate energy sources of CO_2 and nutrients from flue gas and wastewater at low or no cost.
4. Mesophilic rather thermophilic digestion.
5. Adequate conversion of the organic carbon to biogas $\geq$60%.
6. Include an organic flocculant that is readily digested or microalgal communities that settle easily.
7. Additional concentration after flocculation or sedimentation.
8. Minimization of pumping of diluted microalgal suspension.

The production of other high-energy fuels and compounds such as acetone, butanol, and ethanol (ABE), could also be coupled with methane production from microalgae. Ellis et al. (2012) reported ABE fermentation by *Clostridium saccharoperbutylacetonicum* from a mixture of pretreated microalgal biomass. The reached yield was 0.244 g ABE/g microalgal biomass, of which the bulk 0.201 g/g was butanol.

On the other hand, Zamalloa et al. (2011) evaluated the potential of microalgae as feedstock for methane production from a technical and economic point of view. They considered a raceway pond, a pre-concentration unit and a high rate anaerobic digester. The costs of biomass available for bio-methanation were calculated to be in the range of 86–124 € ton^{-1}, this value is high compared to the value reported by Acién et al. (2012) (69 €/kg of biomass) but they did not consider the biogas production stage. Zamalloa et al. (2011) also concluded that from the technical point of view, the use of high rate anaerobic digesters reaching 10–20 kg COD m^{-3} d^{-1}, productivities of minimum 90 ton DM ha^{-1} y^{-1} and a percentage VS fermented of 75% is crucial to exploit the potential of microalgae biomass for production of commodity kWh-energy and from the economic point of view.

Regarding to the integration of an upgrading biogas stage. A comprehensive process of biogas production and upgrading of biogas is shown in Figure 2. Up to date few works report this approach. Tongprawhan et al. (2014) used microalgae to capture CO_2 from biogas and improve methane content and simultaneously produce lipid with the marine microalgae *Chlorella* sp. At optimal conditions, 89.3% of CO_2 was removed from biogas and the methane content was increased up to 94.7%, it was also observed a lipid productivity of 94.7 mg/(L day). Converti et al. (2009) investigated the biogas production and purification by a two-step bench-scale biological system, consisting of fed-batch pulse-feeding anaerobic digestion of mixed sludge, followed by methane enrichment of biogas by the use of the cyanobacterium *Arthrospira platensis*.

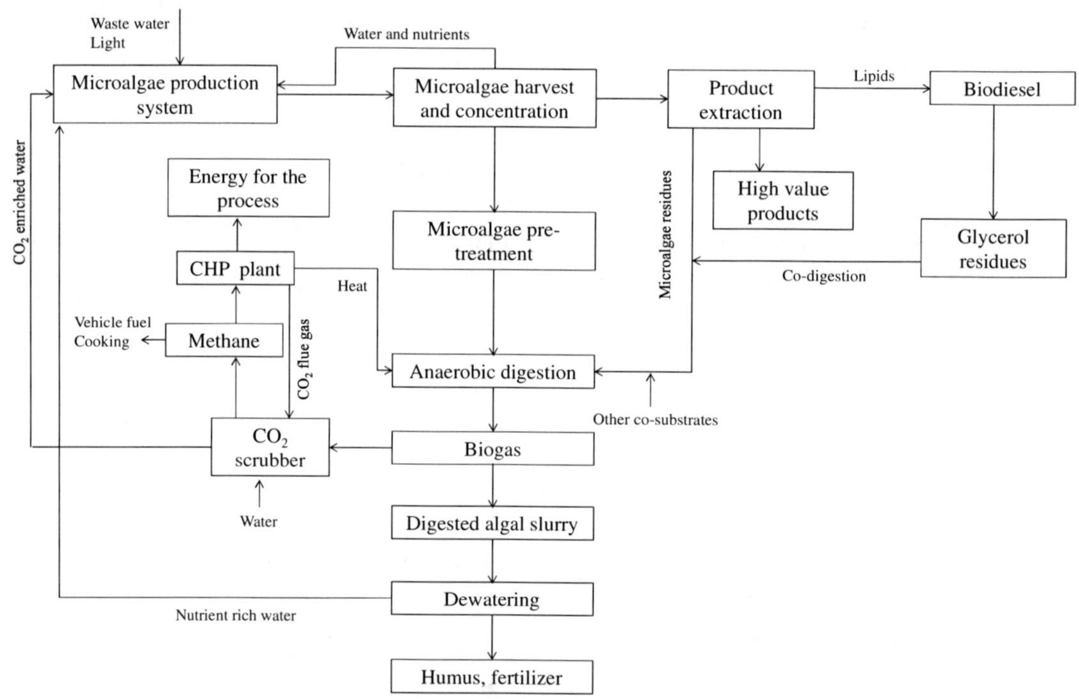

Figure 2. Processes integration in a biorefinery approach.

Recently, Bahr et al. (2014) proposed a microalgal-bacterial symbiotic process for simultaneous removal of biogas contaminants (H_2S and CO_2) at a low energy cost and low environmental impact. In this system, microalgae used solar energy to fix the CO_2 from biogas, via photosynthesis producing O_2. Coupled to this process the sulfur oxidizing bacteria present in the reactor oxidized H_2S to sulfate (Muñoz and Guieysse, 2006). Bahr et al. (2014) evaluated the potential use of a pilot HRAP interconnected via liquid recirculation with an external absorption column using an alkaliphilic microalgal-bacterial consortium. The combined HRAP-bubble column system removed 100% of the H_2S (up to 5000 ppmv) and 90% of the CO_2 supplied, with O_2 concentrations in the upgraded biogas below 0.2%. In addition, methane was produced, 0.21–0.27 L/gVS, by anaerobic digestion of the algal-bacterial biomass produced during biogas upgrading. The authors claimed that the formed methane could satisfy up to 60% of the overall energy demand for biogas upgrading (Bahr et al., 2014).

Nevertheless, the biogas purification process, using cyanobacteria or microalgae, has been little investigated and to the very best of our knowledge its application is not feasible yet.

Conclusion

The potential of microalgal biomass as a source of biogas is high, however technical improvements are still necessary and further research is still required to produce microalgae biofuels. In a microalgal biorefinery configuration, the extraction of high added value metabolites or lipids for biodiesel would produce great amount of organic residues. The integration of anaerobic digestion process is possible to exploit the entire organic biomass for energy production through biogas generation. The applications of this gaseous fuel in engines, boilers, fuel cells, vehicles, etc. will require biogas upgrading to fulfill the energy requirements. Cultivation of microalgae feeding biogas into the reactors represents a promising option to remove both CO_2 and H_2S. Nevertheless, the methane production from microalgae has several limitations to overcome such as: the high cost of microalgae production, incomplete digestibility of algal cells and unbalanced C:N ratio.

The biogas production after upgrading to biomethane can be used for further economic gains. These conditions contribute positively to the economic feasibility of microalgal biorefineries when used together with energy intensive industries.

References

Abatzoglou N., Boivin S. (2009). A review of biogas purification processes. *Biofuels, Bioprod. Bioref.* 3(1):42-71.

Acién F.G., Fernández J.M., Magán J.J., Molina E. (2012). Production cost of a real microalgae production plant and strategies to reduce it. *Biotechnol. Adv*. 30(6): 1344-1353.

Agdag O.N., Sponza D.T. (2007). Co-digestion of mixed industrial sludge with municipal solid wastes in anaerobic simulated landfilling bioreactors. *J. Hazard. Mater*. 140:75-85.

Ahring B.K., Sandberg M., Angelidaki I. (1995). Volatile fatty acids as indicators of process imbalance in anaerobic digesters. *Appl. Microbiol. Biotechnol*. 34:559-565.

Alvarez R., LidÈn G. (2008). Semi-continuous co-digestion of solid slaughterhouse waste, manure, and fruit and vegetable waste. *Renew. Energy*. 33(4):726-734.

Andriani D., Wresta A., Atmaja T.D., Saepudin A. (2014). A review on optimization production and upgrading biogas through CO_2 removal using various techniques. *Appl. Biochem. Biotechnol*. 172:1909-1928

Angelidaki I., Ellegaard L., Ahring B.K. (1993). A mathematical model for dynamic simulation of anaerobic digestion of complex substrates: focusing on ammonia inhibition. *Biotechnol. Bioeng.* 42:159-166.

Angelidaki I., Ellegaard L., Ahring B.K. (2003). Application of the anaerobic digestion process. In: Biomethanation II, Adv. *Biochem Eng/Biotechnol*, Springer, 2-33.

Ariunbaatar J., Panico A., Esposito G., Pirozzi F., Lens P. N. (2014). Pretreatment methods to enhance anaerobic digestion of organic solid waste. *Appl. Energ.*123: 143-156.

Atkinson A.W., Gunning B.E.S., John P.C.L. (1972). Sporopollenin in the cell wall of Chlorella and other algae: ultrastructure, chemistry, and incorporation of 14C-acetate, studied in synchronous cultures. *Planta.* 107(1):1-32.

Bahr M., Díaz I., Dominguez A., González-Sánchez A., Muñoz R. (2014). Microalgal-Biotechnology as a platform for an integral biogas upgrading and nutrient removal from anaerobic effluents. *Environ. Sci. Technol.* 48: 573-581.

Baier U., Schmidheiny P. (1997). Enhanced anaerobic degradation of mechanically disintegrated sludge. *Water Sci. Technol.* 36(11):137-143.

Barjenbruch M., Kopplow O. (2003). Enzymatic, mechanical and thermal pre-treatment of surplus sludge. *Adv. Environ. Res.* 7(3): 715-720.

Batstone D.J., Keller J., Angelidaki I., Kalyuzhnyi S.V., Pavlostathis S.G., Rozzi A., Sanders W.T.M., Siegrist H., Vavilin V.A. (2002). *Anaerobic Digestion Model* No. 1 (ADM1), Task Group for Mathematical Modelling of Anaerobic Digestion Processes, IWA Publishing, London.

Bien J.B., Malina G., Bien J.D., Wolny L. (2004). Enhancing anaerobic fermentation of sewage sludge for increasing biogas generation. *J. Environ. Sci. Health.* 39: 939-949.

Bohutskyi E., Bouwer E. (2013). Biogas production from algae and cyanobacteria through anaerobic digestion: A review, analysis, and research needs. In *Advanced Biofuels and Bioproducts.* 873-975. Springer New York.

Bougrier C., Albasi C., Delgenes J.P., Carrere H. (2006). Effect of ultrasonic, thermal and ozone pretreatments on waste activated sludge solubilization and anaerobic biodegradability. *Chem. Eng. Process.* 45:711-718.

Braun R., Huber P., Meyrath J. (1981). Ammonia toxicity in liquid piggery manure digestion. *Biotechnol. Lett.* 3(4): 159-164.

Braun R., Wellinger A. (2003). Potential of co-digestion. IEA Bioenergy Task 37 – *Energy from Biogas and Land fill Gas.* 15.

Bruni E., Jensen A.P., Angelidaki I. (2010). Comparative study of mechanical, hydrothermal, chemical and enzymatic treatments of digested bio fibers to improve biogas production. *Bioresour. Technol.* 101(22): 8713-8717.

Buxy S., Diltz R., Pullammanappallil P. (2013). Biogasification of marine algae Nannochloropsis oculata in materials challenges in alternative and renewable energy II: *Ceramic transactions* (eds G. Wicks J., Simon R., Zidan R., Brigmon G., Fischman S., Arepalli A., Norris, M. McCluer), John Wiley & Sons, Inc., Hoboken, NJ, USA

Callaghan F.J., Wase D.A.J., Thayanithy K., Foster C.F. (1999). Co-digestion of waste organic solids: batch studies. *Bioresour. Technol.* 67: 117-122.

Callaghan F.J., Wase D.A.J., Thayanithy K., Forster C.F. (2002). Continuous co-digestion of cattle slurry with fruit and vegetable wastes and chicken manure. *Biomass Bioenerg.* 22(1): 71-77.

Carrere H., Dumas C., Battimelli A., Batsone D.J., Delgenes J.P., Steyer J.P., Ferrer I. (2010) Pretreatment methods to improve sludge anaerobic degradability: a review. *J. Hazard. Mater.* 183:1-15.

Carrere H., Sialve B., Bernet N. (2009). Improving pig manure conversion into biogas by thermal and thermo-chemical pretreatments. *Bioresour. Technol.* 100: 3690-3694.

Chen P.H. (1987). *Factors influencing methane fermentation of micro-algae.* California University, Berkeley (USA).

Chen Y., Cheng J.J., Creamer K.S. (2008). Inhibition of anaerobic digestion process: a review. *Bioresour. Technol.* 99: 4044-64.

Chen P.H., Oswald, W.J. (1998). Thermochemical treatment for algal fermentation. *Environ. Int.* 24: 889-897.

Chynoweth D.P., Ghosh S., Klass D.L. (1981). Anaerobic digestion of kelp. In: Sorer SS, Zaborsky OR (eds) *Biomass conversion processes for energy and fuels.* Plenum Press, New York, 315-338.

Clarens A.F., Resurreccion E.P., White M.A., Colosi L.M. (2010). Environmental life cycle comparison of algae to other bioenergy feedstocks. *Environ. Sci. Technol.* 44(5):1813-1819.

Collet P., Hèlias A., Lardon L., Ras M., Goy R-A., Steyer J.-P. (2011). Life-cycle assessment of microalgae culture coupled to biogas production. *Bioresour. Technol.* 102(1): 207-214.

Conde J.L., Moro L.E., Travieso L., Sanchez E.P., Leiva A., Dupeirón R., Escobedo R. (1993). Biogas purification process using intensive microalgae cultures. *Biotechnol. Lett.* 15: 317-320.

Converti A., Oliveira R.P.S., Torres B.R., Lodi A., Zilli M. (2009). Biogas production and valorization by means of a two-step biological process. *Bioresour. Technol.* 100 (23): 5771-5776.

de Baere L.A., Devocht M., Van Assche P., Verstraete W. (1984). Influence of high NaCl and NH_4Cl salt levels on methanogenic associations. *Water. Res.* 18(5): 543-548.

De Schamphelaire L., Verstraete W. (2009). Revival of the biological sunlight-to-biogas energy conversion system. *Biotechnol. Bioeng.* 103(2): 296-304.

Delgenes J.P., Penaud V., Moletta R. (2003). Pretreatments for the enhancement of anaerobic digestion of solid wastes. In: Mata-Alvarez J (ed) Biomethanization of the organic fraction of municipal solid wastes. IWA Publishing, London.

Deublein D., Steinhauser A. (2008). *Biogas from waste and renewable resources. An introduction.* Wienheim: Wiley.

Douskov I., Kastnek F., Maleterov Y., Kastnek P., Doucha J., Zachleder V. (2010). Utilization of distillery stillage for energy generation and concurrent production of valuable microalgal biomass in the sequence: biogas-cogeneration-microalgae-products. *Energ. Convers. Manage.* 51(3): 606-611.

Ehimen E.A., Connaughton S., Sun Z., Carrington C., (2009). Energy recovery from lipid extracted, transesterified and glycerol co-digested microalgae biomass. *GCB Bioenergy.* 1: 371-881.

Elliot A, Mahmood T. (2012). Comparison of mechanical pretreatment methods forthe enhancement of anaerobic digestion of pulp and paper waste. *Water Sci. Technol.* 84: 497-505.

Ellis J.T., Hengge N.N., Sims R.C., Miller C.D., (2012). Acetone, butanol, and ethanol from wastewater algae. *Bioresour. Technol.* 111: 491-495.

El-Mashad H.M., Zeeman G., van Loon W.K., Bot G.P., Lettinga G. (2004). Effect of temperature and temperature fluctuation on thermophilic anaerobic digestion of cattle manure. *Bioresour. Technol.* 95(2):191-201.

Erden G., Filibeli A. (2010). Improving anaerobic biodegradability of biological sludges by Fenton pre-treatment: effects on single stage and two-stage anaerobic digestion. *Desalination.* 251(1–3): 58-63.

Eskicioglu C., Droste R.L., Kennedy K.J. (2007a). Performance of anaerobic waste activated sludge digesters after microwave pretreatment. *Water Environ. Res.* 79(11): 2265-2273.

Eskicioglu C., Terzian N., Kennedy K.J., Droste R.L., Hamoda M. (2007b). A thermal microwave effects for enhancing digestibility of waste activated sludge. *Water Res.* 41(11): 2457-2466.

Esposito G., Frunzo L., Panico A., Pirozzi F. (2011). Modeling the effect of the OLR and OFMSW particle size on the performances of an anaerobic co-digestion reactor. *Process Biochem.* 46: 557-65.

Fernández M., Gómez J.M., Cantero D. (2014). Biogas biodesulfurization in an anoxic biotrickling filter packed with open-pore polyurethane foam. *J. Hazard. Mater.* 264. 529-535.

Ferrer I., Ponsa S., Vasquez F., Font X. (2008). Increasing biogas production by thermal (70°C) sludge pretreatment prior to thermophilic anaerobic digestion. *Biochem. Eng. J.* 42: 186-92.

Fountoulakis M.S., Manios T. (2009). Enhanced methane and hydrogen production from municipal solid waste and agro-industrial by-products co-digested with crude glycerol. *Bioresour. Technol.* 100: 3043-3047.

Frigon J.-C., Matteau-Lebrun F., Hamani Abdou R., McGinn P.J., O'Leary S.J.B., Guiot S.R. (2013). Screening microalgae strains for their productivity in methane following anaerobic digestion. *Appl. Energ.* 108: 100-107.

Ge H., Jensen P.D., Batstone D.J. (2010). Pre-treatment mechanisms during thermophilic-mesophilic temperature phased anaerobic digestion of primary sludge. *Water Res.* 44(1): 123-130.

Gelin F., Volkman J.K., Largeau C., Derenne S., Sinninghe DamstÈ JS, De Leeuw J.W. (1999). Distribution of aliphatic, nonhydrolyzable biopolymers in marine microalgae. *Org. Geochem.* 30(2–3): 147–159.

Gerardi M.H. (2003). The microbiology of anaerobic digesters. *Wastewater microbiology series.* Wiley Interscience, Hoboken.

Goel R., Tokutomi T., Yasui H. (2003). *Anaerobic digestion of excess activated sludge with ozone pretreatment,* vol 47. IWA, London

Goldman J.C., Porcella D.B., Middlebrooks E.J. Toerien D.F. (1972). The effect of carbon on algal growth – its relationship to eutrophication. *Water Res.* 6: 637-679.

Golueke C., Oswald W., Gotass H., (1956). Anaerobic digestion of algae. *Appl. Microbiol.* 5 (1): 47-55.

González-Sánchez A., Flores Márquez T.E., Revah S., Morgan Sagastume, J.M. (2014). Enrichment and cultivation of a sulfide-oxidizing bacteria consortium for its deploying in full-scale biogas desulfurization. *Biomass Bioenerg.* 66(0): 460-464.

Grobbelaar J.U. (2004). Algal nutrition. In Richmond A, editor. *Handbook of microalgal culture: biotechnology and applied phycology.* Wiley-Blackwell.

Gujer W., Zehnder A.J.B. (1983). Conversion processes in anaerobic digestion. *Water Sci. Technol.* 12: 127-167.

Harun R., Davidson M., Doyle M., Gopiraj R., Danquah M., Forde G. (2011). Techno-economic analysis of an integrated microalgae photobioreactor, biodiesel and biogas production facility. *Biomass Bioenerg*. 35: 741-747.

Hendriks A.T.W.M., Zeeman G. (2009). Pretreatments to enhance the digestibility of lignocellulosic biomass: a review. *Bioresour. Technol.* 100:10-18.

Holm-Nielsen J.B., Al Seadi T. (2004). Manure-based biogas systems—Danish experience. In: Piet L, Hamelers B, Hoitink H, Bidlingmaier W (eds) *Resource recovery and reuse in organic solid waste management.* IWA Publishing, London, 377-394.

Horikawa M.S., Rossi F., Gimenes M.L., Costa C.M.M., da Silva M.G.C. (2004). Chemical absorption of H_2S for biogas purification. Braz. J. Chem. Eng. 21: 415-422.

Janczyk P., Franke H., Souffrant W.B. (2007). Nutritional value of Chlorella vulgaris: effects of ultrasonication and electroporation on digestibility in rats. *Anim. Feed Sci. Technol.* 132: 163-169.

Jeihanipour A., Aslanzadeh S., Rajendran K., Balasubramanian G., Taherzadeh M.J. (2013). High-rate biogas production from waste textiles using two-stage process. *Renew. Energ.* 52: 128-135.

Kao C.-Y., Chiu S.-Y., Huang T.-T., Dai L., Hsu L.-K., Lin C.-S. (2012a). Ability of mutant strain of microalgae *Chlorella* sp. to capture carbon dioxide for biogas upgrading. *Appl. Energ*. 93:176-183.

Kao C.-Y., Chiu S.-Y., Huang T.-T., Dai L., Wang G.-H., Tseng C.-P., Chen C.-H., Lin C.-S. (2012b). A mutant strain of microalga *Chlorella* sp. for the carbon dioxide capture from biogas. *Biomass Bioenerg*.36: 132-140.

Khanal S.K. (2008) Anaerobic biotechnology for bioenergy production: principles and applications. Wiley-Blackwell, Ames.

Khoufi S., Aloui F., Sayadi S. (2006). Treatment of olive oil mill wastewater by combined process electro-Fenton reaction and anaerobic digestion. *Water Res*. 40(10): 2007-2016.

Kim D.-H., Jeong E., Oh S.-E., Shin H.-S. (2010). Combined (alkaline + ultrasonic) pretreatment effect on sewage sludge disintegration. *Water Res*. 44(10): 3093-3100.

Komaki H., Yamashita M., Niwa Y., Tanaka Y., Kamiya N., Ando Y., Furuse M. (1998). The effect of processing of *Chlorella vulgaris*: K-5 on in vitro and in vivo digestibility in rats. *Anim. Feed Sci. Technol.* 70(4):363-366.

Koster I.W., Lettinga G. (1984). The influence of ammonium-nitrogen on the specific activity of pelletized methanogenic sludge. *Agric. Wastes*. 9(3): 205-216.

Kroiss H. (1985). *Anaerobe Abwasserreinigung*. Wiener Mitteilungen. 62:65–68.

Lafitte-Trouque S., Forster C.F. (2002). The use of ultrasound and gamma-irradiation as pretreatments for the anaerobic digestion of waste activated sludge at mesophilic and thermophilic temperatures. *Bioresour. Technol*. 84(2): 113-118.

Lakaniemi A.-M., Hulatt C.J., Thomas D.N., Tuovinen O.H., Puhakka J.A., (2011). Biogenic hydrogen and methane production from *Chlorella vulgaris* and *Dunaliella tertiolecta* biomass. *Biotechnol. Biofuels*. 4(34).

Lee R.E. (1999). *Phycology.* 3 ed. Cambridge University Press. United Kingdom.

Li H., Chenchen Li., Liu W., Zou S. (2012). Optimized alkaline pretreatment of sludge before anaerobic digestion. *Bioresour. Technol*. 123: 189-194.

Lv W., Schanbacher F.L., Yu Z. (2010). Putting microbes to work in sequence: recent advances in temperature-phased anaerobic digestion processes. *Bioresour. Technol.* 101(24): 9409–9414.

Mandeno G., Craggs R., Tanner C., Sukias J., Webster-Brown J. (2005). Potential biogas scrubbing using a high rate pond. *Water Sci. Technol.* 51(12): 253-256.

Mann G., Schelgel M., Schumann R., Sakalauskas A. (2009). Biogas conditioning with microalgae, *Agronomy Res.* 7: 33-38.

Marin J., Kennedy K.J., Eskicioglu C. (2010). Effect of microwave irradiation on anaerobic degradability of model kitchen waste. *Waste Manage.* 30:1772-1779.

Marzano C.M.A.D.S., Legros A., Naveau H., Nyns E. (1982). Biomethanation of the marine algae Tetraselmis. *Int. J. Sustain. Energy.* 1(4): 263-272.

Mata-Alvarez J., Macé S., Llabrés P. (2000). Anaerobic digestion of organic solid wastes: an overview of research achievements and perspectives. *Bioresour. Technol.* 74(1): 3-16.

M-c Wu., K-w S., Zhang Y. (2006). Influence of temperature fluctuation on thermophilic anaerobic digestion of municipal organic solid waste. *J. Zhejiang Univ. Sci. B.* 7(3):180-185.

Milledge J., Heaven S. (2014). Methods of energy extraction from microalgal biomass: a review. *Rev. Environ. Sci. Biotechnol.* 13(3): 301-320.

Miranda J., Passarinho P., Gouveia I. (2012) Pre-treatment optimization of *Scenedesmus obliquus* microalga for bioethanol production. *Bioresour. Technol.*104: 342–348.

Molina Grima, E., Acién Fernández, F.G., Robles Medina, A. (2004). Downstream processing of cell-mass and products. In: Richmond, A. (Ed.), *Handbook of Microalgal Culture: Biotechnology and Applied Phycology.* Blackwell, pp. 215– 252.

Montebello A.M., Fernández M., Almenglo F., Ramírez M., Cantero D., Baeza M., Gabriel D. (2012). Simultaneous methylmercaptan and hydrogen sulfide removal in the desulfurization of biogas in aerobic and anoxic biotrickling filters. *Chem. Eng. J.* 200: 237-246.

Mösche M., Jördening H.J. (1999). Comparison of different models of substrate and product inhibition in anaerobic digestion. *Water Res.* 33: 2545-2554.

Mshandete A., Kivaisi A., Rubindamayugi M., Mattiasson B. (2004). Anaerobic batch co-digestion of sisal pulp and fish wastes. *Bioresour. Technol.* 95(1):19-24.

Muñoz R., Guieysse B. (2006). Algal-bacterial processes for the treatment of hazardous contaminants: A review. *Water Res.* 40 (15): 2799-2815.

Mussgnug J.H., Klassen V., Schlüter A., Kruse O. (2010). Microalgae as substrates for fermentative biogas production in a combined biorefinery concept. *J. Biotechnol.* 150: 51-56.

Neyens E., Baeyens J. (2003). A review of thermal sludge pre-treatment processes to improve dewaterability. *J. Hazard. Mater.* 98(1–3): 51-67.

Osorio F., Torres J.C. (2009). Biogas purification from anaerobic digestion in a wastewater treatment plant for biofuel production. *Renew. Energy.* 34(10): 2164-2171.

Oswald W.J. Golueke C.G. (1960). Biological transformation of solar energy. *Adv. Appl. Microbiol.* 2: 223-262.

Park S., Li Y. (2012). Evaluation of methane production and macronutrient degradation in the anaerobic co-digestion of algae residue and lipid waste. *Bioresour. Technol.* 111: 42-48.

Penaud V., Delgenes J.P., Moletta R. (1999). Thermo-chemical pretreatment of a microbial biomass: influence of sodium hydroxide addition on solubilization and anaerobic biodegradability. *Enzyme Microb. Technol.* 25(3): 258-263.

Pinnekamp J. (1989). Effects of thermal pretreatment of sewage sludge on anaerobic digestion. *Water Sci. Technol.* 21: 97-108.

Polakovicova G., Kušnír P., Nagyová S., Mikulec J. (2012). Process integration of algae production and anaerobic digestion, *Chem. Eng. Trans.* 29: 1129-1134.

Protot A., Julien L., Christophe D., Partick L. (2011). Sludge disintegration during heat treatment at low temperature: a better understanding of involved mechanisms with a multi-parametric approach. *Biochem. Eng. J.* 54: 178-184.

Rafique R., Poulse T.G., Nizami A.-S., Asam Z.Z., Murphy J.D., Kiely G. (2010). Effect of thermal, chemical and thermo-chemical pretreatments to enhance methane production. *Energy*. 35: 4556-4561.

Ramos-Suárez J.L., Carreras N. (2014). Use of microalgae residues for biogas production. *Chem. Eng. J.* 24: 86-95.

Ras M., Lardon L., Bruno S., Bernet N., Steyer J.P. (2011). Experimental study on a coupled process of production and anaerobic digestion of *Chlorella vulgaris. Bioresour. Technol.* 102: 200-206.

Razon L.F., Tan R.R. (2011). Net energy analysis of the production of biodiesel and biogas from the microalgae: *Haematococcus pluvialis* and *Nannochloropsis. Appl. Energ.* 88: 3507-3514.

Ryckebosch E., Drouillon M., Vervaeren H. (2011). Techniques for transformation of biogas to biomethane. *Biomass Bioenerg.* 35: 1633-1645.

Samson R., LeDuy A. (1982). Biogas production from anaerobic digestion of *Spirulina maxima* algal biomass. *Biotechnol. Bioeng.* 24(8): 1919-1924.

Samson R., LeDuy A. (1983a). Influence of mechanical and thermochemical pretreatments on anaerobic digestion of *Spirulina maxima* algal biomass. *Biotechnol. Lett.* 5: 671-676.

Samson R., LeDuy A. (1983b). Improved performance of anaerobic digestion of *Spirulina maxima* algal biomass by addition of carbon-rich wastes. *Biotechnol. Lett.* 5: 677-682.

Samson R., Leduyt A. (1986). Detailed study of anaerobic digestion of *Spirulina maxima* algal biomass. *Biotechnol. Bioeng.* 28(7): 1014-1023.

Sánchez-Tuirán E., El-Halwagi M.M., Kafarov V. (2013). Integrated utilization of algae biomass in a biorefinery based on a biochemical processing platform. In *Integrated Biorefineries: Design, Analysis and Optimization* Edited by Stuart PR and El-Halwagi, M. CRC Press Boca Raton Florida. 707-726.

Schenk P.M., Thomas-Hall S.R., Stephens E., Marx U.C., Mussgnug J.H., Posten C., Kruse O., Hankamer B. (2008). Second generation biofuels: high-efficiency microalgae for biodiesel production. *Bioenerg. Res.* 1: 20-43.

Schramm W., Lehnberg W. (1984). Mass culture of brackish-water-adapted seaweeds in sewage- enriched seawater. II: fermentation for biogas production. In *Eleventh International Seaweed Symposium*. 282-287. Springer Netherlands.

Shahriari H., Warith M., Hamoda M., Kennedy K.J. (2012). Anaerobic digestion of organic fraction of municipal solid waste combining two pretreatment modalities, high temperature microwave and hydrogen peroxide. *Waste Manage*. 32: 41-52.

Sialve B., Bernet N., Bernard O. (2009). Anaerobic digestion of microalgae as a necessary step to make microalgal biodiesel sustainable. *Biotechnol. Adv*. 27: 409-416.

Skiadas I.V., Gavala H.N., Lu J., Ahring B.K. (2005). Thermal pre-treatment of primary and secondary sludge at 70°C prior to anaerobic digestion. *Water Sci. Technol.* 52: 161-166.

Sosnowski P., Wieczorek A., Ledakowicz S. (2003). Anaerobic co-digestion of sewage sludge and organic fraction of municipal solid wastes. *Adv. Environ. Res.* 7(3): 609-616.

Speece R.E. (1983). Anaerobic biotechnology for industrial wastewater treatment. *Environ. Sci. Technol.* 17(9): 416A-427A.

Strevett K.A., Vieth R.F., Grasso D. (1995). Chemo-autotrophic biogas purification for methane enrichment: mechanism and kinetics. *Chem. Eng. J. Biochem. Eng. J.* 58:71-79.

Stroot P.G., McMahon K.D., Mackie R.I., Raskin L. (2001). Anaerobic codigestion of municipal solid waste and biosolids under various mixing conditions–I. Digester performance. *Water Res.* 35(7): 1804-1816.

Suali E., Sarbatly R. (2012). Conversion of microalgae to biofuel. *Renew. Sust.Energ. Rev.* 16(6): 4316-4342.

Tippayawong N., Thanompongchart P. (2010). Biogas quality upgrade by simultaneous removal of CO_2 and H_2S in a packed column reactor. *Energy.* 35(12): 4531-4535.

Toledo-Cervantes A., Morales M., (2014). Biorefinery: Using microalgal biomass for producing energy and chemicals. In *Energy and Environment* Nowadays ISBN 978-63117-399-8. Ed. Nova publisher, Nova Science Publishers, Inc. NY, USA.

Toledo-Cervantes A., Morales M., Novelo E., Revah S. (2013). Carbon dioxide fixation and lipid storage by *Scenedesmus obtusiusculus*. *Bioresour. Technol.* 130: 652-658.

Tongprawhan W., Srinuanpan S., Cheirsilp B. (2014). Biocapture of CO_2 from biogas by oleaginous microalgae for improving methane content and simultaneously producing lipid. *Bioresour. Technol.* 170: 90-99.

Valo A., Carrère H., Delgenès J.P. (2004). Thermal, chemical and thermo-chemical pre-treatment of waste activated sludge for anaerobic digestion. *J. Chem. Technol. Biotechnol.* 79(11): 1197-1203.

Varel V.H., Chen T.H., Hashimoto A.G. (1988). Thermophilic and mesophilic methane production from anaerobic degradation of the cyanobacterium *Spirulina maxima*. *Resour. Conserv. Recycling*. 1(1): 19-26.

Vergara-Fernandez A., Vargas G., Alarcon N., Velasco A. (2008). Evaluation of marine algae as a source of biogas in a two-stage anaerobic reactor system. *Biomass Bioenerg.* 32: 338-344.

Wang Q.H., Kuninobu M., Ogawa H., Kato Y. (1999). Degradation of volatile fatty in highly efficient anaerobic digestion. *Biomass Bioenerg.* 16: 407-416.

Weiland P. (2008). Wichtige Messdaten für den Prozessablauf und Stand der Technik in der Praxis. Gülzower Fachgespräche. 27: 17-31.

Weiland P. (2010). Biogas production: current state and perspectives. *Appl. Microbiol. Biotechnol.* 85: 849-860.

Yang H., Xu Z., Fan M., Gupta R., Slimane R.B., Bland A.E., Wright I. (2008). Progress in carbon dioxide separation and capture: a review. *J. Environ. Sci.* (China) 20(1): 14-27.

Yang Z., Guo R., Xu X., Fan X., Luo S. (2011). Hydrogen and methane production from lipid-extracted microalgal biomass residues. *Int. J. Hydrogen Energy*. 36: 3465-3470.

Yen H.-W., Brune D.E. (2007). Anaerobic co-digestion of algal sludge and waste paper to produce methane. *Bioresour. Technol.* 98: 130-134.

Zamalloa C., De Vrieze J., Boon N., Verstraete W. (2012). Anaerobic digestibility of marine microalgae Phaeodactylum tricornutum in a lab-scale anaerobic membrane bioreactor. *Appl. Microbiol. Biotechnol.* 93: 859-869.

Zamalloa C., Vulsteke E., Albrecht J., Verstraete W. (2011). The techno-economic potential of renewable energy through the anaerobic digestion of microalgae. *Bioresour. Technol.* 102: 1149-1158.

Zheng M., Li X., Li L., Yang X., He Y. (2009). Enhancing anaerobic biogasification of corn stover through wet state NaOH pretreatment. *Bioresour. Technol.* 100(21): 5140-5145.

In: Microalgae and Other Phototrophic Bacteria
Editor: Luis G. Torres Bustillos
ISBN: 978-1-63482-078-3

Chapter 15

RECOVERY OF MICROALGAE BY COAGULATION-FLOCCULATION-SEDIMENTATION AND CHARACTERIZATION OF THE PRODUCED PASTES

Luis G. Torres Bustillos*
Bioprocesses Department, UPIBI- Instituto Politecnico Nacional,
Col. Barrio La Laguna-Ticoman, Del Gustavo Madero, Mexico

ABSTRACT

In recent years there has been a microalgae culture boom, with a wide range of applications, including their use as feed for crustaceans and fishes, production of pigments, special lipids, biodiesel, wastewater treatment, biomass to be converted in methane or hydrogen, biomass for energy production by pyrolysis, and many others.

In this chapter, a brief introduction to the different microalgae cell recovery processes, as well as a more specific description of the triad coagulation-flocculation-sedimentation, will be presented and discussed. Finally, the importance of characterizing the produced microalgal pastes through rheology studies will be discussed also.

Recovery of biomass can become a bottleneck of the entire process, due to the small size of algal cells, besides the fact that microalga culture broths are, in general, very diluted and for many processes this concentration must be increased by 200 fold. This increment in microalgae concentration can contribute with 20 to 30% of the total cost of producing biomass.

Among the most suitable methods for microalgae harvesting, are centrifugation, filtration, flotation, and coagulation-flocculation. These methods present advantages and disadvantages, but it is a priority to analyze the energetic cost involved in each of these methods.

Coagulation-flocculation (CF) of microalgae can be achieved in several ways and a wide range of approaches have been explored in recent years. These approaches range from traditional CF to novel ideas based on the biology of microalgae, the use of wall-deficient cells and the use of emerging technologies. After CF it is very common to add a sedimentation step.

* Corresponding author: email: ltorresbustillos@gmail.com.

The simplest equipment to carry out the sedimentation process is the sedimenter or sedimentation device. It is generally a round or square tank designed in such way that the microalgae suspension goes through the device at a speed lower than the particle`s sedimentation rate, so the residence time allows sedimentation of most of the particles. Sludge is accumulated in the base of the sedimenter.

Once the microalgae recovery process has been applied, the resulting product can be a concentrated suspension of cells or rather a humid cell paste. Using some specific procedures, a more concentrated cell dry paste can be obtained. It is necessary to characterize the concentrated pastes through rheological measurements. This step can be carried out by determining viscosity vs. shear rate profiles, as well as the storage and loss modulus and complex viscosities, which, in turn, define the paste's viscoelastic behavior.

Keywords: centrifugation, coagulation-flocculation, filtration, flotation, harvesting, microalgae, rheology

Microalgae Production Nowadays

Many products can be obtained through the application of biotechnology processes. Moreover, basic human necessities such as health, food, energy, and environment can be fulfilled using these technologies (Torres and Garcia- Peña et al., 2013). Nowadays, the necessity of new energy sources, especially renewable or at least potentially renewable sources has open new alternatives for energy production. One of them is the use of microalgae and phototropic bacteria for the direct or indirect production of biodiesel, hydrogen, methane, bioethanol, etc.

Microalgae have been exploited since ancestral ages, by Chinese people and ancient Mexicans (Aztecs) for providing food and other products to their people. Many civilizations have cultured microalgae with different purposes. In the last years, the main application for algae were as food complement (i.e., *Spirulina platensis*, among others) and as feed for the aquaculture industry (growth of shrimps and other valuable fisheries) (Posten, 2012).

In recent years, there has been a microalgae culture boom, with a wide range of applications including those already mentioned, plus production of pigments, special lipids (omegas), biodiesel, wastewater treatment (tertiary treatment), biomass to be converted in methane or hydrogen, biomass for energy production by pyrolysis, and a list of etceteras.

Microalgae can be produced massively in open or closed systems (see figure 1). These systems have both advantages and disadvantages. Some of them require more energy inputs (airlifts), others are prone to contamination (lagoons). Some of them are easily scaled (raceways) while others are more difficult to scale up (tubular bioreactors).

Though many systems can be employed for microalgae and other phototropic bacteria culture, the final cell concentrations are in average around 0.5 g/L and in extreme cases around 2-4 g/L. Microalgae can grow in heterotrophic or mixotrophic regimens, yielding higher cell concentrations.

The next step in the production process is the recovery of microalgae. It is necessary to eliminate as much water as possible, producing a humid paste. After that, the consecutive step is paste drying. It does not matter if microalgae will be used as animal feed, or for the extraction of lipids, proteins, or sugars. It is necessary to dry as much as possible the produced biomass paste. In this chapter, a brief introduction to the different microalgae cell recovery processes, as well as a more specific description of the triad coagulation-flocculation-sedimentation, will be presented and discussed. Finally, the importance of the characterization of the produced microalgal pastes through rheology studies will be discussed also.

OPTIONS FOR MICROALGAE HARVESTING

Any suitable cells harvesting method must be able to process large volumes of algal biomass. For extremely low value products, gravity and sedimentation, possibly enhanced by flocculation, may be the method of choice (Molina Grima et al., 2003). If any other valuable products, i.e., specific lipids, proteins, pigments, special molecules are to be recovered also, the horizon for harvesting processes will be much wider.

Recovery of biomass can be a bottleneck of the entire process (for biodiesel production as an example), due to the small size of algal cells (about 2-25 μm diameter), besides the fact that microalga culture broths are, in general, very diluted (c.a., 0.5 kg dry biomass/m^3) for many processes this concentration must be increased by 200 fold. This increase in the concentration of the microalgae cell can contribute with 20 to 30% of the total cost of the biomass production (Gudin and Terpentier, 1986).

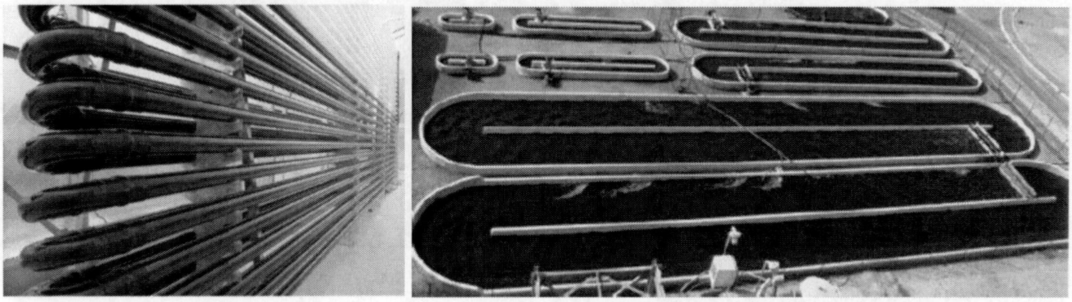

Figure 1. Closed and open systems for microalgae massive culture (left, tubular reactor; right, raceways,).

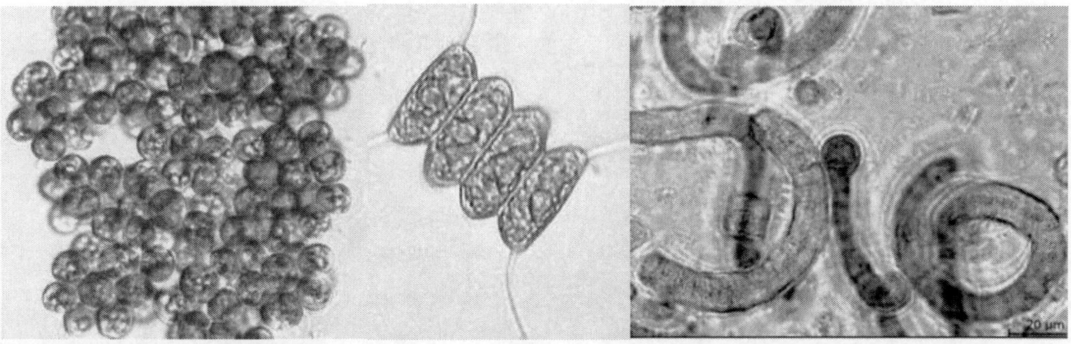

Figure 2. Recovery of microalgae depends on many factors, including the size and shape of microalga cells.

Centrifugation

Most microalgae can be harvested from suspension by centrifugation. This operation can be quite rapid, but it is definitively energy intensive. Many kinds of centrifuges in continuous or discontinuous modes can be employed, such as self-cleaning disc-stack centrifuges, nozzle discharge centrifuges, decanter bowl centrifuges, and hydrocyclones (Molina Grima et al., 2003).

Sim et al. (1988) compared the options of centrifugation (C), dissolved air flotation (DAF), and drum filtration (DF) for harvesting microalgae, employing algal suspensions of 125-1200 m^2 ponds. They concluded that C gave good recoveries, and thickened slurry, but required high capital investment and energy inputs. DAF was more economical, but if the recovered algae were to be incorporated into animal feed, the use of flocculants, such as alum, could have undesirable effects on the growth rate of animals. Finally, the continuous DF process had significant advantages in energy efficiency, economics, and chemical free operation.

Figure 3. Common centrifuge for biomass recovery.

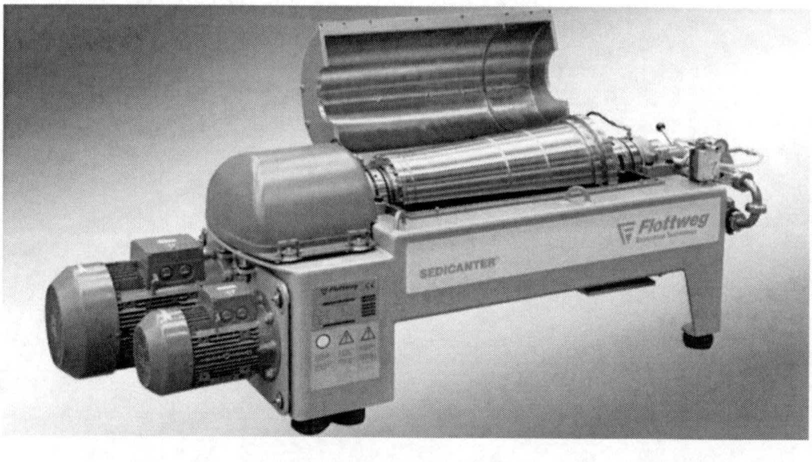

Figure 4. Specialized decanter-centrifuge for microalgae concentration.

Filtration

Filtering processes operating under pressure or vacuum are satisfactory for recovering relatively large microalgae such as *Coelastrum proboscideum* and *Spirullina platensis,* but could fail to recover organisms with bacterial dimensions (*Scenedesmus, Dunandiella, Chlorella*, among others) (Mohn, 1980).

What kind of filters can be employed? Basically pressure filters, such as Netzch chamber filters, Netzch belt press, Bellmer devices, and suction filters. Vacuum filters. such as cylindrical sieve rotators, filter baskets, non-precoat vacuum drum filters, Nivova, Walther suction filters, belt filters, Dinglinger filter thickener, and Shenck filters, could also be used.

These apparatuses can be operated in continuous or discontinuous modes in one or two steps, with concentration factors of 50-245 and energy consumptions between 0.1 and 5.9 kwh/m^3 (Mohn, 1980).

Figure 5. Belt filter for microalgae concentration.

Figure 6. Industrial filter press.

Sim et al. (1988) described the filtration process for harvesting microalgae using a weave belt filter device that can drive throughoutputs between 0.4 and 5 m^3/h. For a reverse Dutch wave filter, with a 12 μm pore size, the final algal slurry solid concentrations (2-3%) are basically a function of the belt travelled (5-20 m/min). Power requirements averaged between 0.3 and 0.5 kWh/m^3, when using suspension cells with an initial concentration of 0.5% and a final cell concentration of 3%.

Flotation

Foam flotation, also known as foam fractionation, is a process that consists in removing surface active chemicals from water and dewatering diluted solid-liquid mixtures (Coward et al., 2013). This process has been widely employed in mineral processing operations, such as the separation of specific ores from rock particles. For this process, different surface active compounds have been employed, but very seldom molecules such as fatty acids, dithiophosphates, amines, palmitic acids. A family of compounds known as xanthates has been selected for this purpose.

Coward et al. (2013) reviewed the use of dissolved air flotation for recovering species such as *Scenedesmus quadriculata, Chlorella vulgaris,* and *Chaetoceros sp.,* at a scale of 4 × 30 to 45 × 157 cm columns. They reported removal efficiencies between 76.6 and 90%.

The independent variables for the process are air flow rate, batch run times, columns height, and surfactant type and concentrations. Coward et al. (2013) proposed a model including these parameters and concluded that the highest concentration factors were gained using the following variables and their interactions: cationic cetyltrimetylammonioum bromide (CTAB), lower surfactant concentrations and CTAB combined with high column heights. Variables that increased foam residence time produced the greatest concentration factors. Energy analysis revealed that DAF consumed only 0.015 kWh/m^3, providing an advantageous cost-benefit relationship.

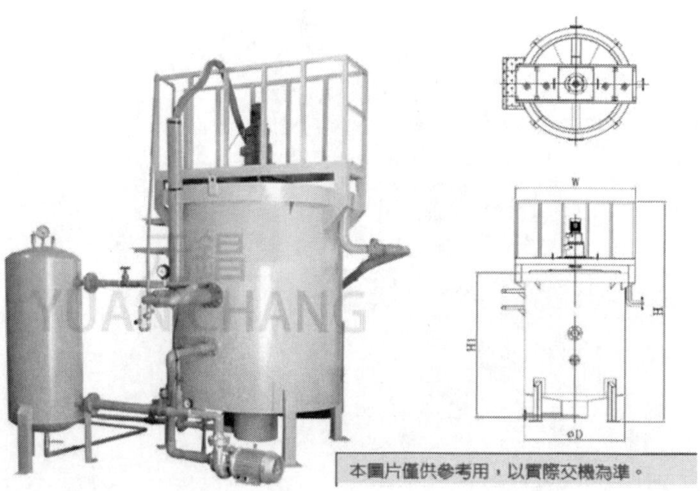

Figure 7. DAF system for microalgae recovery.

Figure 8. CF process applied to microalgal suspension.

Coagulation-Flocculation

By definition, coagulation is considered the step where changes in colloidal particles are neutralized using electrolytes such as Fe, Al, polychlorides of these metals, Ca, Mg, etc. The consequent step is called flocculation, i.e., the addition of large and branched synthetic molecules to collect those neutralized colloid particles, forming stiff and big flocs, which have much more adequate sedimentation rates than those found for discrete particles.

The coagulation-flocculation (CF) process has been extensively employed in wastewater treatment plants, but also in applications such as those required in the pulp paper industry and others. The target for wastewater treatment processes is not the produced sludge, but the clarified water. Much of the research developed for coagulation-flocculation of colloids is valid for CF in the case of algae.

Main products employed for wastewater coagulation process are Fe and Al salts, including polychlorides. For flocculation, large and branched molecules such as polyacrylamides have been employed at low doses. Torres (2013), among other authors, has proposed the substitution of these products by natural polymers that act as CF aids. Guar and locust bean gum, mesquite seed gum, *Opuntia ficus* and flax mucilages, as well as other galactomannans arising from *Annanas* seeds (i.e., *A. cherimola, diversifolia, and muricata*) have been assessed in CF of municipal (Torres et al., 2011a; Torres et al, 2011b; Carpinteyro-Urban et al., 2013; Torres et al., 2014) and industrial wastewaters (Carpinteyro-Urban et al., 2012; Torres et al., 2013; Bandala et al., 2013).

CF of suspensions of particles (including wastewater colloids or microalga cells) can often be attributed to four common mechanisms that can act alone or in combination (see figure 9) (Vandamme et al., 2013):

a) Charge neutralization
b) Electrostatic patch mechanism
c) Bridging
d) Sweeping flocculation.

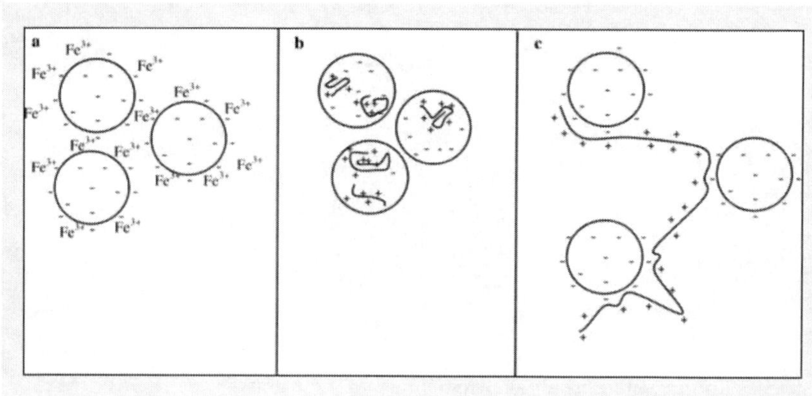

Figure 9. Charge neutralization, electrostatic patch, and bridging patterns.

In particular, CF for microalgae can be achieved in several ways and a wide array of approaches have been explored in recent years. These approaches range from traditional CF (i.e., chemical flocculation) to novel ideas based on the biology of microalgae (i.e., bioflocculation) and the use of emerging technologies (i.e., the use of magnetic particles) (Vandamme et al., 2013).

Heredia et al. (2014) studied the CF process for harvesting cells of *Neochloris oleabundans* without any CF aid and by using $FeCl_3$, $Al_2(SO_4)_3$, locust bean gum, hydroxypropyltrimetyl ammonium chloride-guar gum (HPTAC-guar) and chitosan. They reported that *N. oleoabundans* can be sedimented without the addition of any CF aid, but modifying the pH of the suspension. As shown in figure 10, volumes up to 10 mL/L of biomass can be harvested when the pH was adjusted to 8, in comparison with the non-treated sample (maximum harvesting of 5 mL/L). When the pH value was adjusted to higher values (10 and 12), the recovered volume decreased to 7.5 and 2 mL/L). This phenomenon can be easily explained if the Z potential of the suspensions is analyzed. It is well known that Z potential changes with pH and destabilizes the particles; hence, promoting their sedimentation, i.e, for Z potential values between -20 and 20 mV (Mijaylova et al. 1996).

CF efficiency can be enhanced manipulating the pH of the medium, the type of CF aid employed, and the applied dose. Heredia et al. (2014) determined that when adding chitosan at different doses and pH vales to a suspension of *N. oleoabundans*, microalgae recoveries can be as high as 89% for chitosan doses of 150 mg/L and a pH = 11, as shown at figure 11.

Molina Grima et al. (2003) reviewed the CF for different freshwater algae, employing different metal salts, synthetic polymers and chitosan. For the CF of *Anabaena* and *Asterionella*, using Al and Fe salts at doses between 0.17 and 0.37 × 10^{-2} mmol/L, total cell removals were up to 95% (for polyferric sulfate) in the case of *Anabaena,* and up to 93% (again for polyferric sulfate) in the case of *Asterionella.* Another interesting salt resulted aluminum sulfate, followed by ferric sulfate at the end (for both microalgae cultures).

These authors also reviewed the effect of the ionic strength on the removal efficiency of the CF process carried out with chitosan and two other synthetic polymers. They tested ionic strengths between 0.01 and 0.7 kmol/m^3, and they found that maximal removal efficiencies are inversely proportional to the culture ionic strength, for chitosan and the two synthetic polymers. In particular, chitosan reached removal efficiencies up to 100% at the lowest ionic strength, while for the higher ionic strength value (0.7 kmol/m^3), the maximal removal efficiency was about 22%. For the Zetag 63 polymer (cationic polyacrylamide with molecular weight of 10,000 kDa), removal efficiencies were between 10 and 85%. Finally, for the Zetag 92 polymer (cationic polyacrylamide with molecular weight of 20,000 kDa), minimum and maximal removal efficiencies were 5 and 80%.

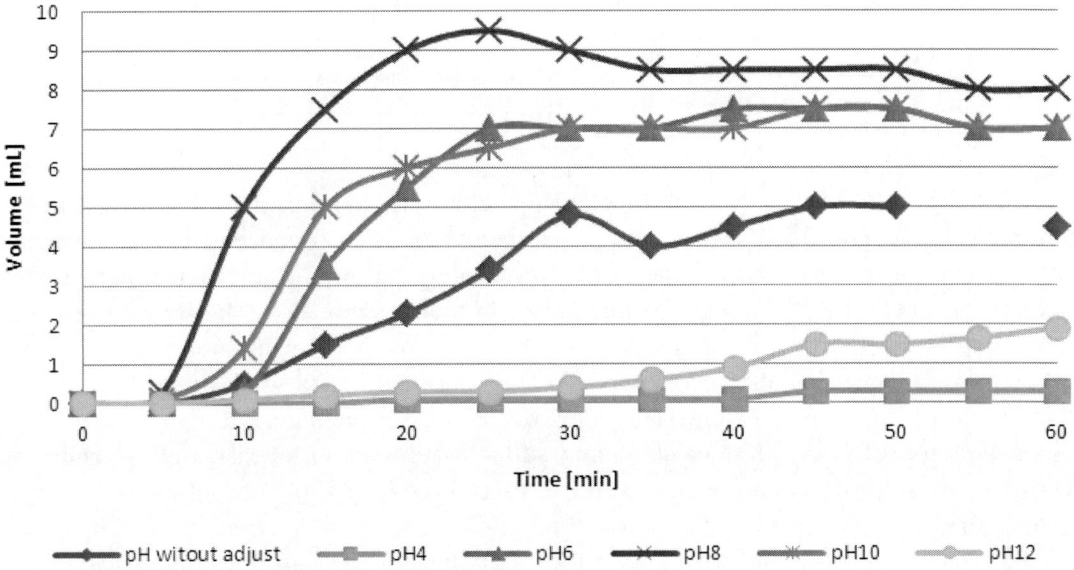

Figure 10. pH effect on *N. oleoabundans* sedimentation without addition of coagulants-flocculants (adapted from Heredia et al, 2014).

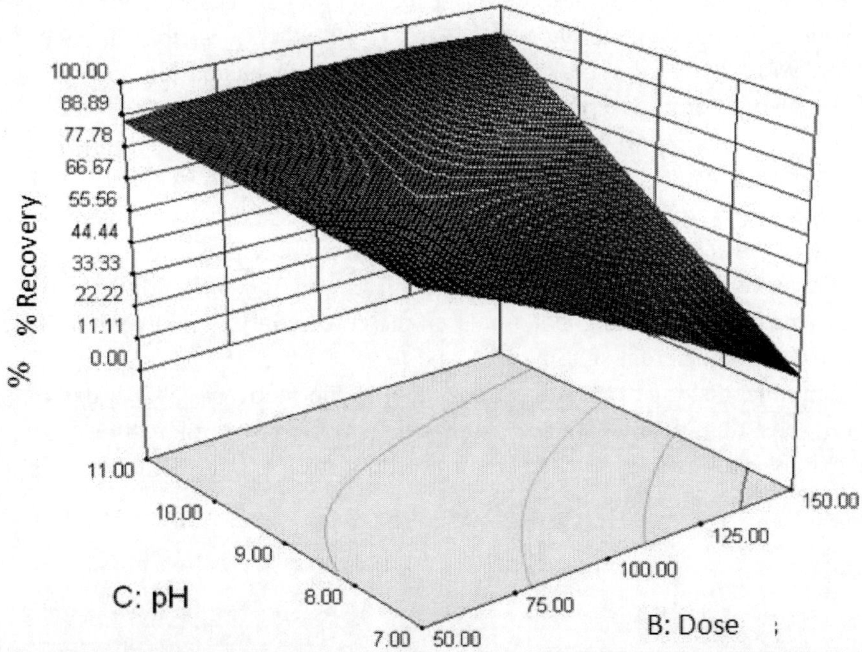

Figure 11. Response surface for the recovery percentage vs. pH of the medium and dose for *N. oleabundans* (adapted from Heredia et al., 2014).

Table 1. Minimum flocculant doses and concentration factors obtained to recover 90% of *Muriellopsis sp.* biomass for different synthetic flocculants. Adapted from Granados et al. (2012)

Polyelectrolyte code	Minimum dose (mg/L)	Concentration factor
EM1	15	37
EM16	10	26
EM22	16	37
FB1	12	36

Granados et al. (2012) evaluated metal salts, chitosan, and polyelectrolytes used in the wastewater treatment processes for microalgae recovery. They found that the use of metal salts or chitosan is not efficient, whereas polyelectrolytes allow for an efficient recovery of biomass, at doses of 2 to 25 mg/g of microalgae biomass. They also state that the required doses depend on the type of polyelectrolyte and the microalga strain, but cationic polyelectrolytes are generally recommended. The concentration factors obtained are higher than 35 in most cases, as shown in Table 1 for *Muriellopsis sp.* cultures.

In the case of iron chloride, aluminum sulfate, and iron sulfate, as well as chitosan, recoveries were lower than 30% even at doses of 100 mg/L, giving concentration factors of up to 10.

Torres et al. (2014) studied the CF process applied to five different microalga strains grown in 1-L bottles, employing chitosan as CF aid. The strains and some physical characteristics of diverse strains, such as polarity, conductivity, Z-potential, mobility, average size and shape, are shown at Table 1.

CF assessments have been carried out using microalgal suspensions of *Scenedesmus obliquus*, *Nannochloropsis sp., Chlorela vulgaris, Dunaliella tertiolecta,* and *Neochloris oleoabundans*. After the CF process, cells recovery, the produced sludge volume, and the final pH of the suspensions were registered. Figure 12 summarizes these values.

As noticed, recoveries were in the range of 30 to 80% (for *D. tertiolecta* and *N. oleoabundans*, respectively). The produced sludge volumes were in the range of 15 to 70 mL/L, while final pH values were very similar for all the assessments (around 5). The initial pH values (ca. 8) were decreased, since chitosan was added as an acetic acid in water solution. The reasons why the recoveries for the different strains are not the same can be: a) the size and shape of cells b) the Z-potentials, c) the cells polarity and even the ionic strength (culture media contain different salts).

Why the produced sludges are not proportional to the recoveries? That can be due to the shape of the cells that form different kinds and strengths of micro-flocs, i.e., different apparent densities. This is a key factor when deciding what CF aid should be employed for each microalgal strain and particular conditions

Table 2. Some characteristics of the studied microalgae (adapted from Torres et al. 2013)

Strain	***Polarity***	***Conductivity mS/cm***	***Z-potential mV***	***Mobility MS/V/cm***	***Aver. Size µm***	***Microalgae shape***
Neochloris oleoabundans	*negative*	*1.70*	*-11.94*	*-0.93*	*6-25*	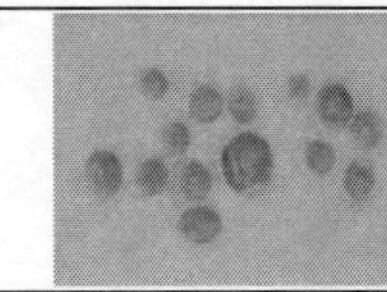
Scenedesmusobliquus,	*negative*	*3.33*	*-27.01*	*-2.11*	*10*	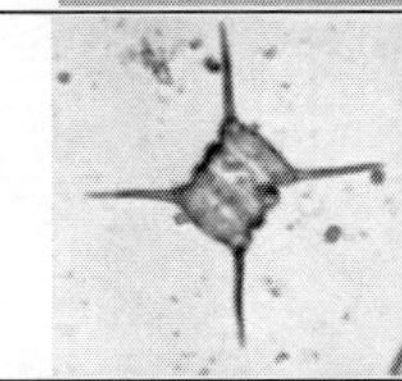
Chlorela vulgaris	*positive*	*2.66*	*53.48*	*4.18*	*3.5*	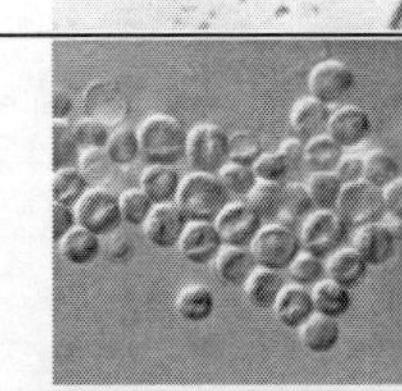
Dunaliella tertiolecta	*NM*	*NM*	*NM*	*NM*	*9-11*	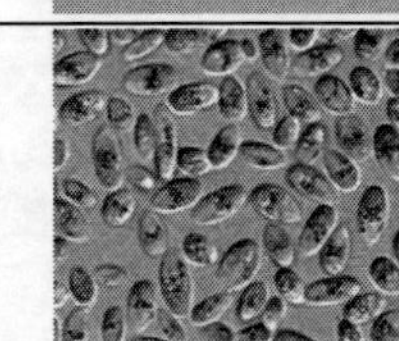
Nannochlorop-sis oculata	*NM*	*NM*	*NM*	*NM*	*2-4*	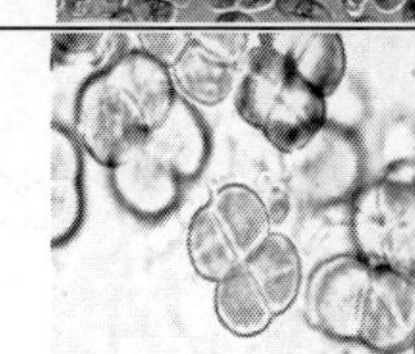

Table 2. (Continued)

Strain	*Polarity*	*Conductivity mS/cm*	*Z-potential mV*	*Mobility MS/V/cm*	*Aver. Size µm*	*Microalgae shape*
Asthospira sp. (Spirullina)	*positive*	*9.81*	*31.28*	*2.44*	*9*	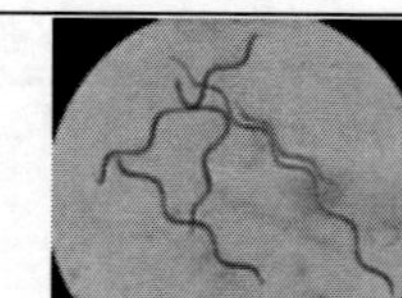
Lyngbia sp.	*negative*	*3.24*	*-28.68*	*-2.24*	*NR*	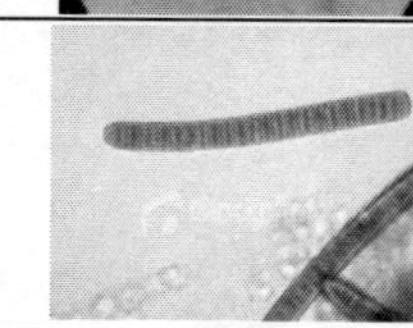

NM, not measured.

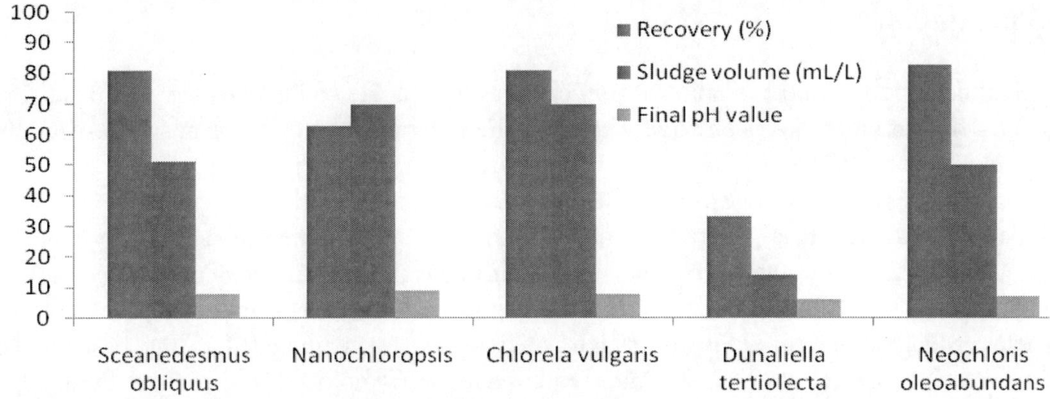

Figure 12. Efficiency of coagulation-flocculation (recovery percentage), final pH, and volume of produced sludge of each microalga (adapted from Torres et al., 2013).

Schlesinger et al. (2102) employed NaOH, CaOH, NH_4OH, $Ca(OH)_2$, and $Mg(OH)_2$ salts in order to carry out CF of microalgal suspensions, including *Chlamydomonas reynhardtii, Nannochloris, Tetraselmis, Phaeodactylum, Nannochloropsis, Pavlova, Isochrysis,* and *Synechococcus.* They observed that NH_4OH was very suitable for the CF, requiring low molar doses (4-16.2 μM) to initiate microalgae CF. pH values had an important effect on the process, but, in general, values for optimum CF process were above 8.6. Maximum pH values reported was >12 for *C . reinhardtii,* but for the rest of the strains pH values was as high as 10.2 units. Using the right products at best pH values, CF efficiencies were in the range of 40 (*Phaeodactylum with NH_4OH*) to 97% (for *Nannochloris* and *Phaeodactylum* with $Ca(OH)_2$).

Scholz et al. (2011) studied the flocculation of wall-deficient cells of *Chlamydomonas reindhartii* mutant, using calcium and methanol. They assessed different concentrations of $CaCl_2$ to get a good microalgae CF, between 0 and 62 mM, and observed that the wall-deficient microalgae settled in a maximum of 90% at a $CaCl_2$ concentration of about 15 mM, whereas the wild type microalgae reached a CF level of about 38%, even without the addition of $CaCl_2$. The addition of this salt, diminished the CF efficiency of the wild type microalgae. Regarding the use of methanol, the authors reported that best results were obtained when combining 12 mM of $CaCl_2$ and 4.6% methanol. This combination promoted a microalgae CF efficiency of about 86.4%.

Regarding bioflocculation, the process consists in the use of natural products, such as biopolymers produced by other microorganisms. The mechanism for bioflocculation is shown in figure 13. Wan et al. (2011) studied the biopolymer produced by *Solbacillus silvestris* W01, arising from an activated sludge basin, and its application to harvest *Nannochloropsis oceanica* by flocculation. They reported that the biopolymer produced by S. silvestris is a proteoglycan containing about 75% carbohydrate and 25% protein (w/w). They demonstrated that ions such as CK, KCl, $CaCl_2$, and $FeCl_3$ did not affect the CF efficiency. Best pH for the CF process was 8.7 (88.2% efficiency) when compared with a pH value of 6.7, where only 14.0% of microalgae removal was achieved. For the purified biopolymer, with the combination of the bioflocculant and some ions, higher CF efficiencies were achieved. For example, 100 mM of Al or 1 mM of Fe promoted CF efficiencies of 67 and 45%, respectively.

Sedimentation

Sedimentation has been employed for centuries to separate solids from suspensions using gravity. This operation has been widely used for the separation of microorganisms (especially those in the form of flocs, biofilms, or immobilized cells) after a biological wastewater treatment process. The simplest equipment to carry out this operation is the sedimenter or sedimentation device. It is generally a round or square tank designed in such a way that the wastewater (in this case, the microalgal suspension) goes through the device at a speed lower than the particle`s sedimentation rate, so the residence time allows sedimentation of most particles. Sludge (i.e., colloidal material or microalgae) is accumulated in the base of the sedimenter. Submerged pumps or similar equipments are employed to drain out the sludge (see figure 14).

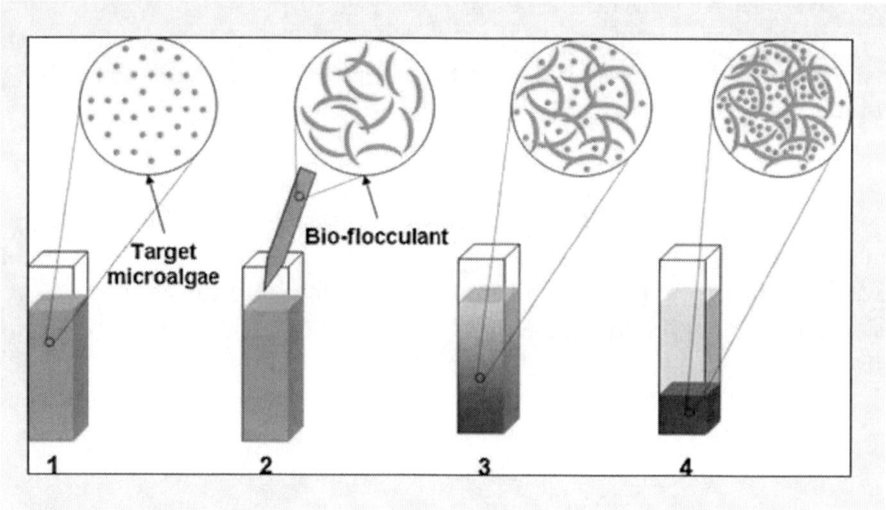

Figure 13. Principle of bio-flocculation.

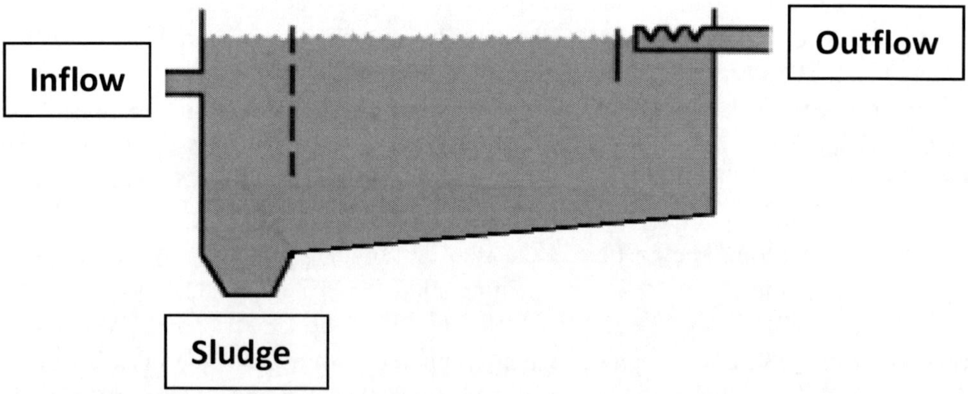

Figure 14. Simple sedimentation device.

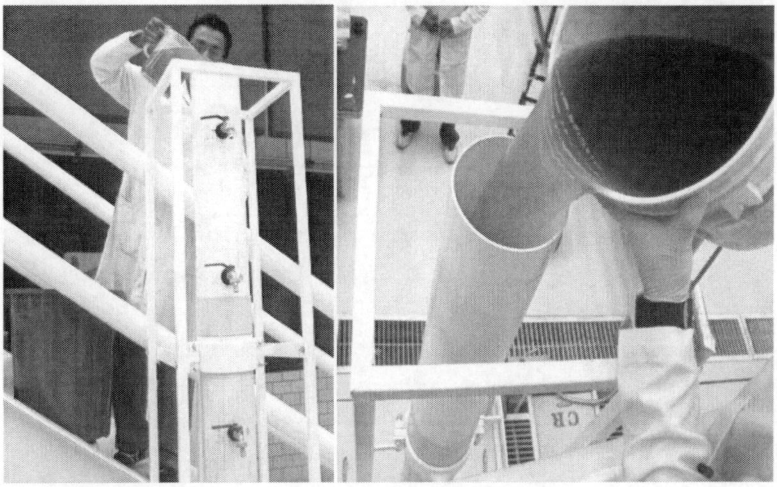

Figure 15. Thirty-eight liter column for the measurement of microalgae sedimentation rates. Photo: L. G. Torres.

Although this operation is cheap and easy to carry out, some information is necessary for the proper design of the sedimentation device. Test columns are very often employed to obtain that kind of information. As an example, Torres et al. (2013) employed a 38-L column provided with five valves or keys separated from the bottom and between themselves at certain distances, distributed along the 1.8 m height. Thirty-eight liters of a final culture of *Scenedesmus* were employed. The suspension was deposited carefully inside the column (see figure 15). Then, an appropriate amount of the CF aid, in this case a chitosan concentrated acetic acid in water solution, is added and the suspension is mixed for some minutes using a long PVC rod The rod is removed and the time is registered. A sample of the suspension (about 40 mL) of this initial time is reserved. Every 5 minutes for the first hour, and every 10-15 minutes for the subsequent time, samples were taken at each of the five valves. For each process, samples can be taken until 1 or 2 hours. Every sample is measured using a spectrophotometer, Values obtained are plotted as DO (optical density) versus time, for each valves. After that, the Do is divided by the initial DO value (DOo) and a new plot is prepared.

As an example, Figure 16 shows this plot for a *Scenedesmus* culture with no CF aid, with $FeCl_3$, and, finally, when chitosan (40 mg/L) was added. Note how different the respective DO/DOo vs time plots are.

Based on these plots, the isoconcentration curves are prepared. This means that the recovery values (80, 70, 60, 50, 40%) are plotted as a function of time, for every column height.

On figure 16, as noted, the curve for *Scenedesmus* and chitosan allows identifying the specific removal at every height of the column, On the other hand, the slope of these curves represents the sedimentation rates.

Based on these isoconcentration plots, it is possible to calculate the main dimensions of a simple sedimentation device, if a calculation base is chosen. In this case, under the assumption of treating 100,000 L of microalgae suspension, the calculation of the main sedimenter characteristics is shown on Table 3.

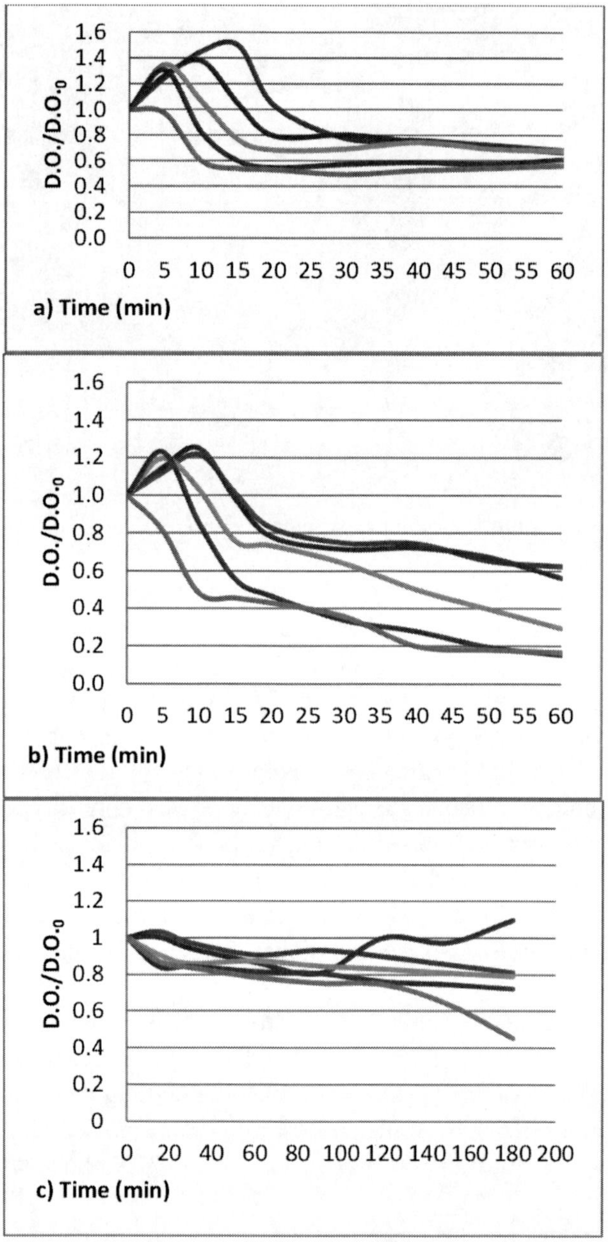

Figure 16. DO/DO$_0$ for each sedimentation process as a function of time when using *Scenedesmus sp.* culture: a) no coagulant-flocculant aid, b) ferric chloride, and c) chitosan (unpublished results).

According to the general procedure, in the calculation of a sedimentation device, for a 100,000 L flow, it is more adequate to employ two sedimenters instead of just one, allowing for preventive maintenance procedures, at given time, in one of the two equipments. From figure 16, the sedimentation rate is about 1.39 m/h. The total flow of microalgae suspension is ca. 10.8 m^3/h, giving a value for each tank of 5.4 m^3/h.

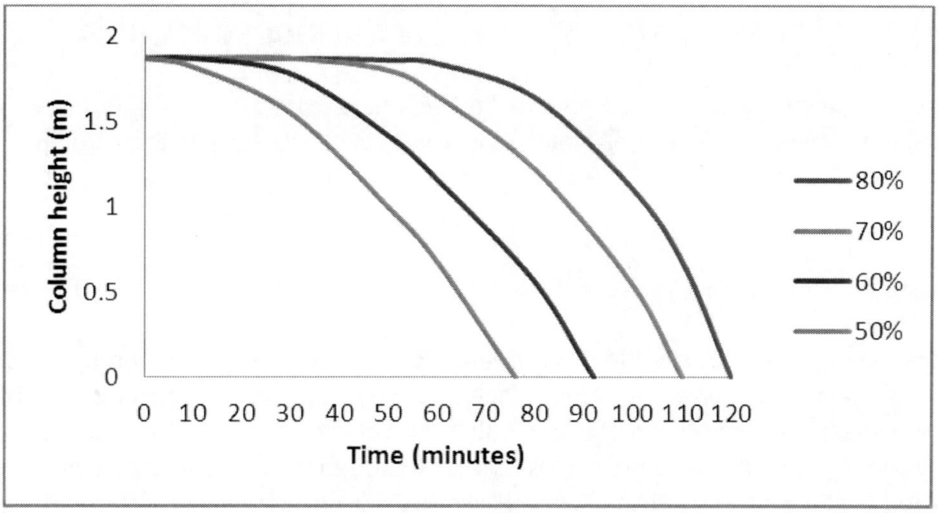

Figure 17. Iso-concentration curves for *Scenedesmus sp.* with chitosan (40 mg/L). (adapted from Torres et al., 2013).

Table 3. Summary of the simple sedimentation device calculation (based on 100,000 L of medium) (adapted from Torres et al., 2013)

Sedimentation rate	1.39 m/h
Number of tanks	2
Total flow	10.83 m^3/h
Flow in each tank	5.41 m^3/h
Retention time	2 h
Total surfacel area	3.89 m^2
Volume	2.70 m^3
Transversal area	0.68 m^2
Flow speed	7.89 m/h

For a retention time of 2 hours, the required area is about 3.9 m^2 and the total volume per tank is 2.7 m^3. Transversal area is 0.7m^2 and flow speed must be around 7.9 m/h.

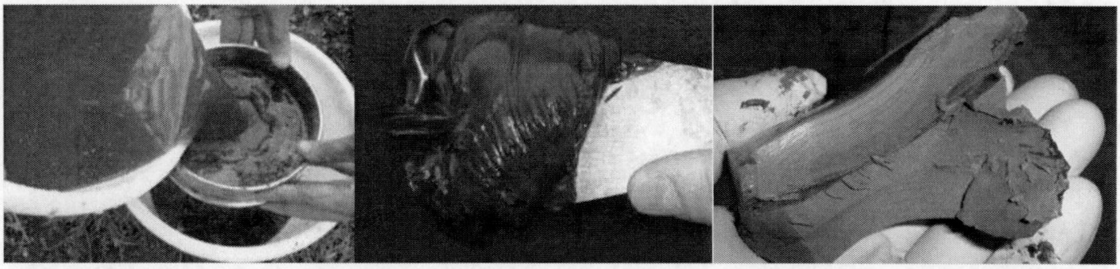

Figure 18. Concentrated suspension of microalgal biomass; a still humid paste and a more concentrated dry paste.

CHARACTERIZATION OF THE PRODUCED PASTES

Once the microalgae recovery process has been applied, the resulting product can be a concentrated suspension of cells or rather a humid cell paste. Using some specific procedures (vacuum belt filters), a more concentrated dry cell paste can be obtained (see figure 18).

Microalgae Viscometry

These concentrated suspensions or pastes need to be transported through pipes and accessories, pumped, mixed, or dried. Therefore, a rheological characterization of these suspensions is mandatory. It is well known that the culture suspensions can achieve cellular concentrations of 3 g/L (though in most cases, for different reasons, the final microalgal suspension contains no more than 1 g/L). These suspensions are merely water with salts and suspended particles, i.e., they exhibit Newtonian behavior. This means that no matter at which shear rate the microalgae suspensions are subjected, the viscosity is the same. In contrast, the decanted, filtered, precipitated, or centrifuged suspension can reach solid concentrations of up to 90% (only 10% of water). Hence, these concentrated suspensions are non-Newtonian in nature.

Among the various rheological models to represent the viscous behavior of materials, the Ostwald de Waele model (best known as the power law model) is a simple and useful model. This can be employed to model the behavior of foods, polymer solutions, emulsions, drilling fluids, sludges, and so on. Because of its simplicity (i.e., only two parameters) it has been applied to a wide range of products and conditions (concentrations, temperatures, etc.).

The power law model is as follows:

$$\mu = K\,\gamma^{n-1} \qquad \text{(Equation 1)}$$

Where μ = viscosity, Pa.s
γ = shear rate, s^{-1}
K = consistency index, $Pa.s^n$ and
n = index flow, dimensionless

Torres et al. (2013) studied the rheological characteristics of *Dunaliella tertiolecta* and *Nannochloris oculata* cultures, after dewatering by means of consecutive sedimentation of the culture, with and without addition of a CF aid (chitosan in this case). The microalgae culture was grown in a 200-L raceway for 15 days. After the cultivation process, part of each culture was submitted to CF process using 40 mg/L of chitosan. The medium was decanted and the sludge was recovered in a small recipient. This was done as much as possible, until a minimum amount of water was present in each specific culture. The final biomass concentration was calculated for both cultures using an oven. The resulting pastes were characterized using a very simple and cheap device, Brookfield-type instrument. Viscosities were measured in duplicate and data were fitted to the Power Law as shown in figure 19 for *Dunaliella tertiolecta.*

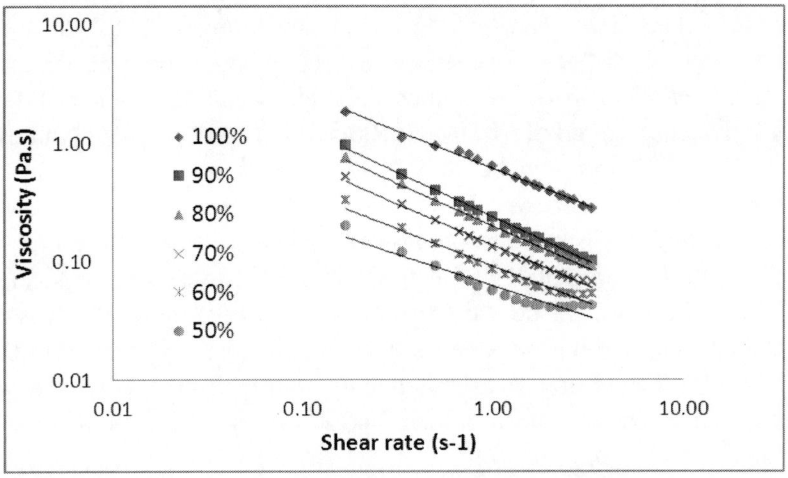

Figure 19. Viscosity *vs.* shear rate plot for *Dunaliela tertiolecta* pastes. Measurements were carried out in a Brookfield-type viscometer. Adapted from Torres et al. (2013).

As observed, the curves for 100-50% of the original biomass concentration fitted perfectly to the Power Law. The viscosities for the lower shear rate value range from hundreds to thousands of mPa.s (centipoises). The second fact is that all curves show rheofluidizing or shear thinning behavior. The values of the parameters K and n as well as the correlation factor R^2 for the *Dunaliella* and *Nannochloropsis* suspensions are presented on Table 4, in comparison to K and n values reported for other systems.

Table 4. K, n, and R^2 values from the Power Law for *D. tertiolecta* and *N. oculata* pastes in comparison with data published in the literature. Adapted from Torres et al. (2013)

Strain	Concentration g/L	K(Pa.s^n)	n	R^2	Comments	Reference
Dunaliella tertiolecta	28.22	0.895	0.42	0.997	Adding 40 mg/L of chitosan	Torres et al. (2013)
	25.39	0.268	0.52	0.915		
	22.57	0.165	0.62	0.915		
Nannochloropsis oculata	100.44	0.607	0.37	0.991	Adding 40 mg/L of chitosan	Torres et al. (2013)
	90.39	0.197	0.26	0.998		
	80.35	0.136	0.29	0.997		
	70.38	0.091	0.31	0.993		
	60.26	0.097	0.38	0.970		
	50.22	0.062	0.48	0.921		
Rhodosorus marinus	Along a 30-day culture	0.0015-0.012	0.55-0.90	NR	No polymer	Basaca-Loya et al. (2008)
Nannochloropsis sp.	0.5-80	0.0011-0.0040	0.89-1.0	0.999	No polymer	Willeman et al. (2012)
Chlorella vulgaris	0.5-80	0.0010-0.0163	0.62-1.0	0.981-0.999		
P. tricornutum	0.5-80	0.011-0.0025	1.0	0.999		

Note that K values for *Dunaliella* were in the range of 0.165 and 0.895, for concentrations between 22.5 (80% concentration) and 28.22 g/L Pa.s^n (100% concentration). In the case of *Nannochloropsis,* K values were between 0.062 and 0.607 Pa.s^n for concentrations between 50.2 g/L (50%) and 100.5 g/L (100%). These values reveal how different rheological behaviors can be shown by different microalgal strains suspensions. Regarding the n values, values between 0.42 and 0.62 (dimensionless) were obtained for the 80 and 100% concentration suspensions. At last, for *Nannochloropsis*, n values between 0.26 and 0.48 (dimensionless) were obtained for the 90 and 50% concentration suspensions. Note that R^2 values are quite good for all the adjustments, i.e., 0.91 and 0.99. It is interesting to note that the other K and n values reported in the literature are different from those reported by Torres et al. (2013). The reasons are, undoubtedly, a) the difference in the assessed strains, b) the differences in biomass concentrations, and c) the use of a biopolymer for the CF process in the mentioned work.

Microalgae Oscillatory Rheology

Using more sophisticated rheological equipments (such as a Physica rheometer 1000, a Rheomatic Expansion System RES, etc.) it is possible to measure not only the liquid-like behavior of the microalgal suspensions (μ), but also the viscoelastic behavior through parameters such as G` (the storage modulus), G`` (the loss modulus), and μ* (the complex viscosity). While G' represents the liquid behavior, G'' is related with the solid-like behavior, and μ* is the combination of both parameters.

These parameters are expressed in units of Pa.s (μ*) and Pa (G' and G''), respectively. They are measured as a function of the ω oscillation frequency (in Hz). The relationship between the parameters and the complex viscosity s given by:

$$\mu^* = \frac{\sqrt{G'^2 + G''^2}}{w} \qquad \text{(Equation 2)}$$

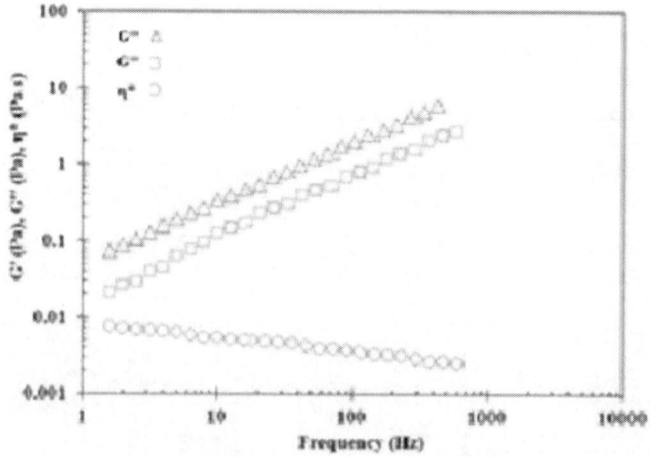

Figure 20. Oscilatory rheology for *Scenedesmus obliquus* (suspension with 9% of solids). G', G'', and μ* as a function of ω. Adapted from Adesanya et al. (2012).

Adesanya et al. (2012) characterized microalgal suspensions using two different systems. They applied oscillatory rheology on *Sceenedesmus obliquus* cultures to find out the effect of the microalgal concentrations, the effect of the ω oscillation frequency, and the effect of the physiological state of the culture (dead or alive cultures).

These data were calculated using two different rheometers.To avoid introducing confusion, only one line is extrapolated for the whole range of ω tested.

As shown in figure 20, G' and G'' increase in a parallel way as the ω frequency is increased. G' is higher than G'' for the entire w range, which indicates the existence of a viscoelastic behavior for *Scenedesmus* suspensions containing 9% of solids. Besides, μ* decreases as ω increases.

This behavior was found for *Scenedesmus* suspensions between 6.8 and 15% w/w. In another section of the paper, Adesanya et al. (2012) carried out the same kind of measurements (G', G'', and μ*) for a 3% w/w suspension, with intact microalgae (live cells) and heat-treated microalgae (dead). Results show that G', G'', and μ* for the intact microalgae were always higher than the correspondent values for the heat treated cells.

This means that motility of cells influences the values of G' and G'', which is a very interesting point. It has been reported that *Scenedesmus* cells are provided with flagella that help in the microalgal cell motility.

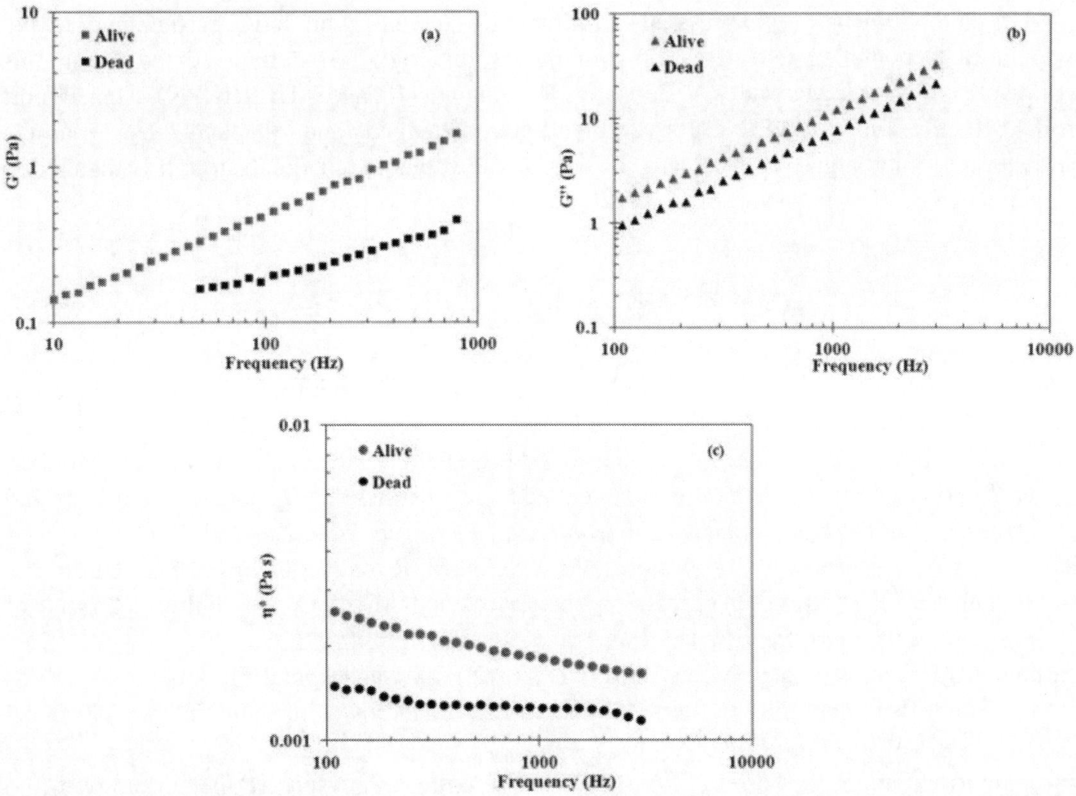

Figure 21. G' (a), G''(b), and μ* (c) as a function of ω, measured for *Scenedesmus,* live (see symbols) or dead cells. Adapted from Adesanya et al. (2012).

Conclusion

The microalgae culture process for biodiesel production may be optimized in different aspects, such as increasing the final microalgae concentration and the microalgae and lipid productivities. The energy consumed during the culture process must be minimized. There is no doubt that new microalgae strains must be assessed in order to achieve the new goals. But the harvesting step must be considered, as the cost of the downstream process can represent about the 20-30% of the biodiesel total production cost. There are relative few groups working in the optimization of harvesting operation. More multidisciplinary groups should be integrated, where engineers, chemical specialists, biologist, economists and other specialized professionals can interact to solve new challenges. The characterization of the produced microalgae pastes is in progress. Very few researchs have been yet reported because many groups are working at low scale levels. In the future, a higher scale production will provide more material for the optimization. It is a must the use of simple and low-cost operations, such as coagulation-flocculation-sedimentation to fulfill the harvesting goals.

Acknowledgments

The author wishes to express his gratitude to students who have participated in the coagulation-flocculation-sedimentation and pastes rheology research presented along this work, specially to M. Martinez, A. Pèrez, V. Heredia and J.Yañez (UPIBI-IPN). The support from SIP-IPN and SENER-CONACyT is acknowledged and thanked. The massive cultivation of microalgae (in raceways) as well as the valuable technical support is thanked to L. Fernandez (UPIBI-IPN) and his research group.

References

Adesanya, V-O., Vadillo, D. C. & Mackley M. R. (2012) .The rheological characterization of algae suspensions for the production of biofuels. *Rheology Journal. 56*, 925.

Bandala, E., Tiro, J. B., Lujan, M., Camargo, F. J., Sanchez-Salas, J. L., Reyna, S., Moeller, G. & Torres, L. G. (2013). Petrochemical effluent treatment using natural coagulants and an aerobic biofilter. *Advances in Environmental Research*, *2*(3), 229-243.

Basaca-Loya, A., Burboa, M. G., Valdez, M. A., Gamez, R., Goycoolea, F. M. & Gutierrez-Millan, L. E. (2008). Aggregation behavior and rheology of culture broths of *Rhodosorusmarinus*. *Revista Mexicana de Fisica*. S *54*(2), 119-126.

Carpinteyro-Urban, S., Vaca, M. & Torres, L. G. (2012). Can vegetal biopolymers work as coagulant-flocculant aids in the treatment of high-load cosmetic industrial wastewaters? *Water, Air & Soil Pollution*. *223*, 4925-4936.

Carpinteyro-Urban, S. & Torres, L. G. (2013). Use of response surface methodology in the optimization of coagulation-flocculation of wastewaters employing biopolymers. *International Journal of Environmental Research*. *7*(3), 717-726.

Coward, T., Lee, J. G. M. & Caldwell, G. S. (2013). Development of a foam flotation system for harvesting microalgae biomass. *Algal Research*., 2, 135-144.

Granados, M. R., Acien, F. G., Gomez C., Fernandez-Sevilla, J. M. & Molina Grima, E. (2012). Evaluation of flocculants for the recovery of freshwater microalgae. *BioresourceTechnology*. *118*,102-110.

Gudin, C. & Terpentier, C. (1986). Bioconversion of solar energy into organic chemicals by microalgae. *Advances in Biotechnology Processes*. *6*, 73-110.

Heredia, V., Fernandez, L. C. & Torres, L. G. (2014). Harvesting *Neochloris oleoabundans* by means of coagulation-flocculation with inorganic salts and bio-polymers. Submitted to Aquaculture Engineering.

Mijaylova, P., Torres, L. G., Ramirez, E., Lopez, S. & Cardoso L. (1996). Characterization and coagulation-flocculation treatability of Mexico City wastewater applying ferric chloride and polymers. *Water Science and Technology*. *34*(3-4), 235-247.

Mohn, F. H. (1980). Experiences and strategies in the recovery of biomass from mass cultures of microalgae. In, Shelef G., Soeder CI, editors. Algae biomass, Amsterdam, Elsevier, 547-571.

Molina Grima, E., Belarbi, E.-H., Acien Fernandez, F.G., Robles Medina, A. & Chisti Y. (2003). Recovery of microalgal biomass and metabolites, process options and economics. *Biotechcnology Advances*. *20*, 491-515.

Posten, C. (2012). Discovering microalgae as source for sustainable biomass. In, Microalgal Biotechnology, Potential and Production. Posten C., Walter Ch. (Editors) De Gruytere Edit. Germany.

Scholz, M., Hoshino, T., Johnson, D., Riley, M. R. & Cuello, J. (2011). Flocculation of wall-deficient cells of *Chlamydomonas reinhardtii* mutant *cw*15 by calcium and methanol. *Biomass and Bioenergy*. *35*, 4835-4840.

Shlesinger, A., Eisenstandt, D., Bae-Gil, A., Carmelly, H., Eibinder, Sh. & Gressel, J. (2012). Inexpensive non-toxic flocculation of microalgae contradicts theories; overcoming a major hurdle to bulk algal production. *Biotechnology Advances*, *30*, 1023-1030.

Sim, T. S., Goh, A. & Becker, E. W. (1988). Comparison of centrifugation, dissolved air flotation and drum filtration techniques for harvesting sewage-grown algae. *Biomass*, *16*, 51-62.

Torres, L. G. (2013). Plant polysaccharides as a new source for coagulant-flocculant aids and surfactants. In: Biotechnology: health, food, energy and environment applications. L. Torres and E.I. Garcia-Peña (editors) Nova Science Publishers. USA.

Torres, L.G., Carpinteyro-Urban, S. & Vaca, M. (2011). Use of *Prosopis laevigata* seed gum and *Opuntia ficus-indica* mucilage for the treatment of municipal wastewaters by coagulation-flocculation. Natural Resources.doi 10.4236/nr.2011

Torres, L. G., Heredia, V., Yañez, J. & Fernández, L. C. (2013a). Harvesting five different microalgae species using chitosan coagulation-flocculation. 2nd Iberoamerican Congress on Biorefineries. Jaen, España, Abril 2013.

Torres, L. G., Carpinteyro-Urban, S. & Orozco, C. (2013b). Production and characterization of *Opuntia ficus-indica* mucilage and its use as coagulant-flocculant aid for industrial wastewaters. *International Journal of Biotechnology Research*. *1*(13), 38-45.

Torres, L. G., Martínez, M., Heredia, V. & Fernández, L. C. (2014a). Harvesting microalgae by chitosan coagulation-flocculation-sedimentation and rheological characterization of the produced pastes. Submitted to International Journal of Microalgae.

Torres, L. G., Martínez, M., Carrillo, L. & Fernández, L. C. (2014b). Rheological characterization of microalgae pastes produced in an agitated reactor (in Spanish). Proceedings of the AMIDIQ Meeting. Puerto Vallarta, Mexico. May 6-10, 2014.

Torres, L. G., Cadena, G., Carpinteyro-Urban, S. & Corzo, L. J. (2014c). New galactomannans and mucilages with coagulant-flocculant activity for an environment-friendly treatment of wastewaters. *Current Advances in Environmental Science*. *2*(2), 52-58

Vandamme, D., Foubert, I. & Muylaert, K. (2013). Flocculation as a low-cost method for harvesting microalgae for bulk biomass production. *Trends in Biotechnology*. *31*(4),

Wan, Ch., Zhao, X-Q., Guo, S-L. & Alam A Md, Bai (2013). Bioflocculant production from *Solibacillus silvestris* W01 and its application in cost-effective harvest of marine microalgae *Nannochloropsis oceanica* by flocculation. *Bioresource Technology*. *135*, 207-212.

Wileman, A., Ozkan, A. & Berberoglu, H. (2012). Rheological properties of algae slurries for minimizing harvesting energy requirements in biofuel production. *Bioresource Technology*, *104*, 432-439.

EDITOR'S CONTACT INFORMATION

Dr. Luis G. Torres Bustillos

Professor

Bioprocess Department

UPIBI-Instituto Politecnico Nacional

Calle Acueducto s/n.

Col. Barrio La Laguna-Ticoman

Del Gustavo Madero

Mexico 07042 DF Mexico

Email: LTorresB@ipn.mx

INDEX

A

C

D

E

F

G

H

I

N

O

P

Q

R

S

T

U

V

W

X

Y

Z